Androgens in Gynecological Practice

Androgens in Gynecological Practice

Edited by

Leo Plouffe Jr. MD CM FACOG FRCS(C)
Vice President, Global Pharmacovigilance, and Head, Risk Management, Bayer HealthCare Pharmaceuticals, Whippany, NJ, USA

Botros R. M. B. Rizk MD MA FRCOG FRCS HCLD FACOG FACS
Professor of Obstetrics and Gynecology, and Director of the Division of Reproductive Endocrinology and Infertility, University of South Alabama, Mobile, AL, USA

CAMBRIDGE
UNIVERSITY PRESS

CAMBRIDGE
UNIVERSITY PRESS

University Printing House, Cambridge CB2 8BS, United Kingdom

Cambridge University Press is part of the University of Cambridge.

It furthers the University's mission by disseminating knowledge in the pursuit of education, learning and research at the highest international levels of excellence.

www.cambridge.org
Information on this title: www.cambridge.org/9781107041318

© Cambridge University Press 2015

Printed in the United Kingdom by TJ International Ltd. Padstow Cornwall

A catalogue record for this publication is available from the British Library

Library of Congress Cataloguing in Publication data
Androgens in gynecological practice / edited by Leo Plouffe Jr, Botros Rizk.
   p. ; cm.
Includes bibliographical references and index.
ISBN 978-1-107-04131-8 (hardback)
I. Plouffe, Leo Jr., 1957– , editor.   II. Rizk, Botros, editor.
[DNLM: 1. Adrenal Gland Diseases.   2. Androgens–adverse effects.   3. Adrenogenital Syndrome.   4. Polycystic Ovary Syndrome. WK 700]
RC659
618.1–dc23
2014045886

ISBN 978-1-107-04131-8 Hardback

....

Contents

About the editors   vii
Dedication   viii
List of contributors   ix
Preface   xi

Section 1 – Managing the basics

1. **Hirsutism: managing the basics**   1
   Héctor F. Escobar-Morreale

2. **Alopecia: managing the basics**   11
   Thamer Mubki and Jerry Shapiro

3. **Polycystic ovary syndrome: managing the basics**   25
   Keith A. Hansen

4. **Sexual dysfunction, including hypoactive sexual desire disorder: diagnosis and treatment recommendations**   38
   Ricki Pollycove and James A. Simon

Section 2 – The scientific essentials

5. **The androgen receptor: basic structure and function**   55
   Leo Plouffe Jr.

6. **Androgens in postmenopausal women: their practically exclusive intracrine formation and inactivation in peripheral tissues**   64
   Fernand Labrie

7. **Diagnostic criteria of polycystic ovary syndrome**   74
   Heather R. Burks and Robert A. Wild

8. **Androgen effects on the skin**   79
   Grace E. Kim and Alexa B. Kimball

9. **Polycystic ovary syndrome and cardiovascular risk**   89
   Heather R. Burks and Robert A. Wild

10. **Effects of androgens on female genital tract**   97
    Abdulmaged M. Traish and André T. Guay

Section 3 – Learning from the extremes

11. **Congenital adrenal hyperplasia in females**   111
    Candice P. Holliday and Botros R. M. B. Rizk

12. **Cushing's syndrome in females**   118
    Andrew S. Fischer, Christopher B. Rizk, and Sylvia Hsu

13. **Androgen-producing ovarian tumors**   136
    Patricia Carney

14. **Cognitive and behavioral impact of androgen disorders in females: learning from complete androgen insensitivity syndrome and congenital adrenal hyperplasia**   146
    Amy B. Wisniewski

15. **Testosterone replacement in the aging male: lessons learned from the Women's Health Initiative**   151
    David Muram and Craig F. Donatucci

Section 4 – The tool kit

16. **History and physical examination of polycystic ovary syndrome: detecting too much or too little**   161
    John J. Kohorst, Andrew S. Fischer, and Christopher B. Rizk

17. **Androgen assays: of mice and men**   180
    Keith A. Hansen

Contents

18. **Magnetic resonance imaging of the adrenal gland** 191
Sajal S. Pokharel and Ihab R. Kamel

19. **Androgens and DHEA in postmenopausal medicine** 200
Lila E. Nachtigall and Jeffrey A. Goldstein

20. **Lifestyle, diet, and exercise in polycystic ovary syndrome** 208
Elizabeth Burt and Rina Agrawal

21. **Polycystic ovary syndrome ovulation induction** 216
Carolyn J. Alexander

22. **Ovulation induction in women with polycystic ovary syndrome** 226
Shawky Z. A. Badawy and Botros R. M. B. Rizk

Index 233

The color plate section can be found between pages 116 and 117.

About the editors

Leo Plouffe, MD CM is currently Vice President, Head of Risk Management, Global Pharmacovigilance at Bayer Health Care Pharmaceuticals following his tenure as Head of Women's Health Care US Medical Affairs, also at Bayer. Dr Plouffe started his career as a research scientist over 25 years ago at the Medical College of Georgia, department of Obstetrics and Gynecology, Reproductive Endocrinology and Genetics Section. He achieved the rank of tenured professor before joining Eli Lilly Pharmaceuticals, where he held positions of increasing responsibility in medical affairs and clinical development.

His research interests encompass pediatric and adolescent gynecology, menopausal medicine, contraception, and patient safety. He has authored over 100 scientific papers, book chapters, and abstracts. He is past vice-president of the North American Society for Pediatric and Adolescent Gynecology, and has served on a number of committees and special interest groups at ASRM, ACOG, and SGI. He is a reviewer for several journals in women's health.

Botros R. M. B. Rizk, MD is Professor and Chief of the Division of Reproductive Endocrinology and Infertility of the Department of Obstetrics and Gynecology at the University of South Alabama, Mobile, AL, USA. He is also Program Director and Medical and Scientific Director of the USA IVF and ART Program, and Laboratory Director and Clinical Consultant at the Reproductive Endocrinology Laboratory of the University of South Alabama.

Professor Rizk has worked as a research scientist and clinician in reproductive medicine for more than 25 years. His main research interests include the modern management, prediction, and the genetics of ovarian hyperstimulation syndrome (OHSS), as well as the role of vascular endothelial growth factor and interleukins in the pathogenesis of severe OHSS. He has authored more than 300 peer-reviewed published papers, book chapters, and abstracts. Professor Rizk has edited and authored 18 medical textbooks on OHSS, infertility and assisted reproduction, endometriosis, ultrasonography in reproductive medicine, ovarian stimulation, and the future of ART.

He is past chair of the ASRM international membership committee, member of the scientific advisory board of the Mediterranean Society for Reproductive Medicine, President Elect of the Middle East Fertility Society, on the editorial board of the Middle East Fertility Society Journal, and reviewer for several national and international journals, as well as for the European Society of Human Reproduction and Embryology Task Force on perinatal outcome and congenital malformations of intracytoplasmic sperm injection.

Dedication

To my mentors on this subject, Drs Morrie M. Gelfand, Robert B. Greenblatt, Robert AH Kinch and Paul G. McDonough.

From Drs Gelfand and Greenblatt, I received the passion and enthusiasm to study this topic. From Drs Kinch and McDonough, I received the skillset and discipline to apply evidence-based medicine in the care of these patients.

On a personal level, to my wife Eve, who has supported me through this project as she has through every step of our life journey.

Leo Plouffe Jr.

To the RIZK family for their tremendous contribution to Egypt; my great grandfather Rizk the chief treasurer in charge of the Royal mint and in charge of all financial matters during the reign of Ali Bey the Great in the second half of the eighteenth century, his grandson Rizk Ghobrial Rizk appointed the Governor of East Delta by Mohamed Ali in 1814, my grandfather Botros Rizk after whom I was named, my father Mitry Botros Rizk, and all the uncles and cousins.

Botros R. M. B. Rizk

Chapter 10 is dedicated to the memory of my dear friend and colleague Andre T. Guay, MD and for his immense contributions to the advancement of clinical endocrinology, especially sexual function in men and women.

Abdulmaged M. Traish

Contributors

Rina Agrawal MD PhD
Centre for Reproductive Medicine, London, UK

Carolyn J. Alexander MD
Department of Obstetrics & Gynecology,
Cedars-Sinai Medical Center, Los Angeles, CA, USA

Shawky Z. A. Badawy MD
Department of Obstetrics & Gynecology, SUNY
Upstate Medical University, Syracuse, NY, USA

Heather R. Burks MD
Department of Obstetrics & Gynecology, University
of Oklahoma Health Sciences Center, Oklahoma City,
OK, USA

Elizabeth Burt MBBS
University College London, London, UK

Patricia Carney MD
Bayer HealthCare Pharmaceuticals, Whippany,
NJ, USA

Craig F. Donatucci MD
Division of Urology, Duke University Medical Center,
Durham, NC, USA

Héctor F. Escobar-Morreale MD PhD
Department of Endocrinology & Nutrition; Hospital
Universitario Ramón y Cajal, Universidad de Alcalá,
IRYCIS, CIBERDEM, Madrid, Spain

Andrew S. Fischer
Baylor College of Medicine, Houston, TX, USA

Jeffrey A. Goldstein DO
Bayer HealthCare Pharmaceuticals, Whippany,
NJ, USA

André T. Guay MD
Center for Sexual Function, Lahey Clinic Northshore,
Peabody, MA, USA

Keith A. Hansen MD
Department of Obstetrics & Gynecology, University
of South Dakota Sanford School of Medicine, Sioux
Falls, SD, USA

Candice P. Holliday JD
Department of Obstetrics & Gynecology, University
of South Alabama, Mobile, AL, USA

Sylvia Hsu MD
Department of Dermatology, Baylor College of
Medicine, Houston, TX, USA

Ihab R. Kamel MD PhD
Department of Radiology & Radiological Sciences,
Johns Hopkins University School of Medicine,
Baltimore, MD, USA

Grace E. Kim BS
Harvard Medical School, Boston, MA, USA

Alexa B. Kimball MD MPH
Department of Dermatology, Massachusetts General
Hospital, Boston, MA, USA

John J. Kohorst BA
Mayo Medical School, Rochester, MN, USA

Fernand Labrie MD PhD
Molecular Endocrinology & Oncology Research
Center, Laval University Hospital, Québec City,
Québec, Canada

Thamer Mubki MD
Department of Dermatology, Al Imam Mohammad
Ibn Saud Islamic University, Riyadh, Saudi Arabia

David Muram MD
Eli Lilly & Company, Lilly Corporate Center,
Indianapolis, IN, USA

Lila E. Nachtigall MD
Department of Obstetrics & Gynecology, NYU School
of Medicine, New York, NY, USA

Leo Plouffe Jr. MD CM
Bayer HealthCare Pharmaceuticals, Whippany,
NJ, USA

Sajal S. Pokharel MD PhD
Department of Radiology, University of Colorado
School of Medicine, Aurora, CO, USA

Ricki Pollycove MD
Department of Obstetrics, Gynecology and
Reproductive Sciences, CPMC, and VCF Department
of Psychiatry, University of California San Francisco,
San Francisco, CA, USA

Botros R. M. B. Rizk MD MA
Department of Obstetrics & Gynecology, University
of South Alabama, Mobile, AL, USA

Christopher B. Rizk BA
Baylor Medical College, Houston, TX

Jerry Shapiro MD
Department of Dermatology, New York University
Langone Medical Center, New York City, NY, USA

James A. Simon MD
Department of Obstetrics & Gynecology, George
Washington University School of Medicine,
Washington, DC, USA

Abdulmaged M. Traish PhD
Institute of Sexual Medicine, Boston University
School of Medicine, Boston, MA, USA

Robert A. Wild MD PhD MPH
Department of Obstetrics & Gynecology, University
of Oklahoma Health Sciences Center, Oklahoma City,
OK, USA

Amy B. Wisniewski PhD
Department of Urology, University of Oklahoma
Health Sciences Center, Oklahoma City, OK, USA

Preface

A wide range of disease states that affect women are associated with dysfunctional androgen production, either deficiency or excess states. This encompasses common disorders, such a polycystic ovary syndrome (PCOS) or acne and hirsutism, as well as rare conditions, such as congenital adrenal hyperplasia. Thousands of original scientific publications and hundreds of scholarly books have covered these conditions over the years. Why does the medical community need one more book on the subject?

In reviewing the available reference sources, we were struck that the underlying theme was focusing on the specific medical conditions affecting women, and working back to the underlying androgen dysregulation associated with the condition. This is an extremely valuable approach and has served the practitioner very well over the years. However, it is often helpful to consider a new perspective to stimulate new insights into a medical problem. How could we bring about an element of novelty?

What if we first focused on androgens and considered how they impact normal physiology, from the receptor all the way to the entire body? What if we focused on a variety of androgens, at the biochemical level and their respective physiological impact? What if we considered androgen effects not just in women but also in men? What about seeing how much we can learn from the endocrinology of androgens in rare androgen-linked conditions that we can apply to our understanding in the care of all women? What can medical practitioners with a wide range of specialty and clinical backgrounds learn from one another in a scientific forum? We thought such an approach may help reframe many of the issues of androgens in women's health.

This book represents an amazing collaboration of scientists that bought in to our novel approach, and agreed to participate in this experiment. We reached out to many specialists in the field. All of the contributing authors valued the opportunity and provided their unique and crisp insights in a way that we hope will open up new ideas, and hopefully drive further scientific progress.

At the same time, we wanted to make sure this book would be of practical value for every clinician. The clinical chapters provide a clear framework to work-up and manage patients. They are complemented by many chapters on the intricacies of making an accurate diagnosis, also essential for the clinician. And last but not least, the remaining chapters provide highly relevant insights into the scientific underpinnings of these complex conditions.

We hope you will find this book valuable and consult it frequently. Above all, we hope it will help you take even better care of your patients, and for some of you, trigger new research ideas.

Good reading.
Leo Plouffe Jr
Botros R. M. B. Rizk

Hirsutism: managing the basics

Héctor F. Escobar-Morreale

Introduction

The term "hirsutism" means presence of excessive terminal hair in androgen-dependent areas of the female body. Hirsutism is a frequent medical complaint that usually results from relatively benign functional disorders albeit, rarely, hirsutism may be the presenting sign of a life-threatening disorder. The present chapter aims to provide a comprehensive review on the diagnostic and therapeutic management of hirsutism based almost entirely on the author's experience and usual practices. The reader expecting evidence-based guidelines is kindly referred to a recent publication sponsored by the Androgen Excess and Polycystic Ovary Syndrome Society [1].

Pathophysiology of hirsutism

The pilosebaceous unit is a highly dynamic system that changes throughout the lifespan. Before puberty, the pilosebaceous unit includes a vellus hair – which is soft, short, and has no medulla – and a small sebaceous gland. In androgen-responsive areas, the increase in androgens characteristic of puberty induces the pilosebaceous unit to produce terminal hair – which is coarse, pigmented, and has a medulla – and the size of the sebaceous unit increases markedly.

Hair grows in asynchronous cycles that comprise three phases consisting of: (i) an anagen or growing phase that accounts for 85–90% of the duration of the hair cycle and may last for a few months in the case of terminal hair; (ii) a catagen or rapid involution phase that accounts for 2–3% of the hair cycle; and (iii) a telogen or resting phase accounting for 10–15% of cycle at which the end hair is ejected and anagen starts again (Figure 1.1) [2]. Because androgens stimulate the growth of terminal hair by prolonging its anagen

phase, the clinical effects of androgen excess and its amelioration necessarily take months to be apparent to both the patient and her physician. Being aware of this fact is of capital importance for the correct management of hirsutism.

Because androgens play a definite role in the transformation of vellus into terminal hair during puberty, and in the growth of terminal hair in androgen-dependent areas of the female body, hirsutism is considered to be a clinical marker of androgen excess [3]. However, some hirsute patients do not show any other evidence of androgen excess, such as hyperandrogenemia or ovarian dysfunction, and often receive a diagnosis of "idiopathic" hirsutism [4].

To understand this apparent paradox, it is important to know some facts about female androgen metabolism. In women, the adrenals and the ovaries secrete androgens into the circulation, because these are the only organs in the female body expressing the biosynthetic enzymes needed for the synthesis of androgens. Peripheral tissues, such as fat, also contribute to circulating androgen levels by converting other steroid precursors. Testosterone is the most important androgen and circulates mostly bound to serum albumin (low affinity, but large capacity) and to sex-hormone-binding globulin (SHBG) (high affinity, but small capacity). Given its high affinity for testosterone, SHBG actually regulates the amount of testosterone that reaches target tissues, even if its binding capacity is much less than that of albumin. Therefore, the lower the SHBG concentration, the larger the fraction of free or unbound testosterone that may reach target tissues.

However, testosterone is a prohormone that undergoes conversion into dihydrotestosterone in target cells before entering the cell nucleus and binding the androgen receptor. Both the conversion rate of

Androgens in Gynecological Practice, ed. Leo Plouffe and Botros Rizk. Published by Cambridge University Press. © Cambridge University Press 2015.

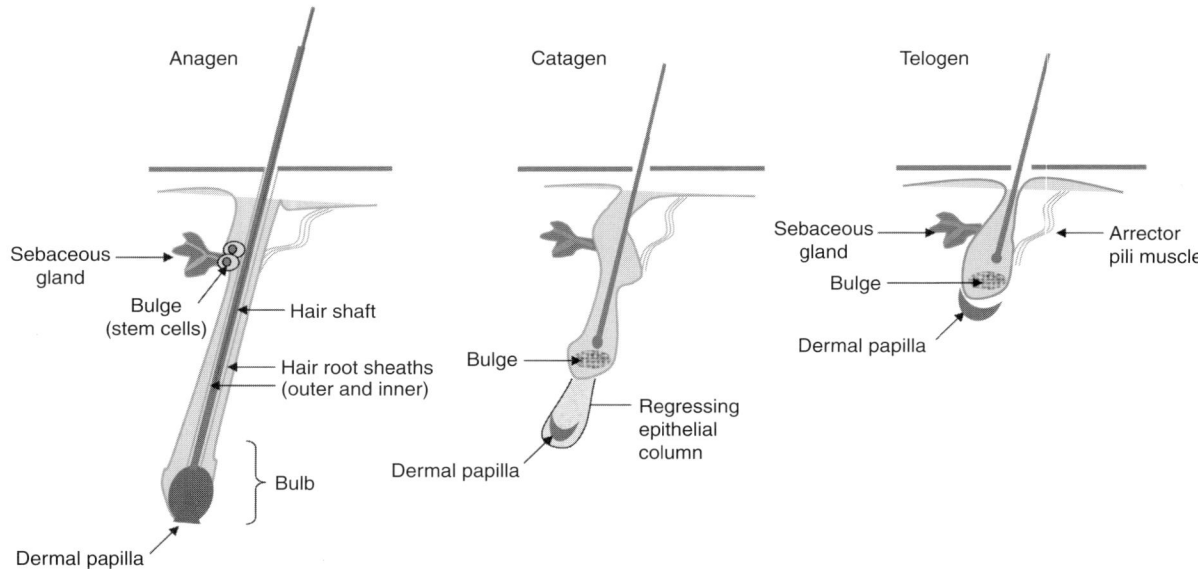

Figure 1.1 The hair follicle growth cycle. Hair follicles possess the ability, unique among mammalian tissues, to be partially regenerated in a process termed the hair follicle growth cycle. Hair follicles go through repeated cycles of development and growth (anagen), regression (catagen), and rest (telogen) to enable the replacement of hairs. Reproduced from Escobar-Morreale et al. [1], by permission of Oxford University Press.

testosterone – mediated by 5α-reductase – and the binding of dihydrotestosterone to the androgen receptor are subject to individual variation, and current hypotheses explain idiopathic hirsutism as the result of increased 5α-reductase activity and/or increased sensitivity of the androgen receptor to normal amounts of testosterone [4]. An alternate hypothesis is that women with idiopathic hirsutism have mild steroidogenic abnormalities that go undetected by the relatively insensitive biochemical tests applied in routine clinical practice [5]. Considering the well-known limitations of the assays of serum androgens currently available for clinical practice [6], it is my personal opinion that the presence of hirsutism should be considered an accurate marker of androgen excess, irrespective of serum androgen concentrations. In other words, most if not all hirsute patients are hyperandrogenic, and our limitation in detecting androgen excess is the actual culprit that we cannot confirm this diagnosis by analytical tools.

Quantification and epidemiology of hirsutism

The definition of hirsutism implies that the amount of terminal hair must be quantified before establishing such a diagnosis. After the initial attempts to standardize the quantification of body hair made by S. M. Garn (who developed his score to assess male hairiness) [7], D. Ferriman and J. D. Gallwey [8], and E. Moncada Lorenzo [9], in 1981, Hatch and colleagues [10] published the modification of the original Ferriman–Gallwey score that is currently the "gold-standard" for the quantification of hirsutism. The modified Ferriman–Gallwey score (mFG) estimates the presence of terminal hair in nine areas of the female body – upper lip, chin, chest, abdominal region above and below the navel, upper and lower back, arms and thighs – and assigns a score from 0 (absent) to 4 (complete cover) to each of these areas for a total score ranging from 0 to 36 [10]. Most clinicians and researchers use a cut-off value of 8 or above to diagnose hirsutism, and grade it as mild up to a score of 15, moderate from 16 to 25, and severe above 25.

The broad application of this scoring system provided researchers with a common language for the definition of hirsutism and was followed by significant advances in the study of hirsutism and related conditions, such as polycystic ovary syndrome (PCOS). However, the mFG score has evident limitations, the most notable being the subjective nature of the assessment – yet, it appears that inter-observer variation is acceptable [11] – the possibility that substantial terminal hair in one or

Table 1.1 Summary of studies addressing the prevalence of hirsutism in women

Author, year	Country	Race	Ethnicity	Score	Cut-off	Method of sample selection	Sample size	Prevalence (95% CI)	Comments
Diamanti-Kandarakis et al., 1999 [12]	Greece	White	Mediterranean	FG	≥6	Invitation of free medical examination	192	38% (31–45)	Possible selection self-referred bias[a]
Asuncion et al., 2000 [13]	Spain	White	Mediterranean	mFG	≥8	Unselected female blood donors from general population	154	7.1% (3.0–11.1)	
Zargar et al., 2002 [14]	India	Asian	Kashmir Dardic	FG	≥6	Hospital outpatient clinic	4780	10.5% (9.6–11.4)	Includes postmenopausal women
Sagsoz et al., 2004 [15]	Turkey	White	Middle Eastern	mFG	≥8	Regular check-up in outpatient clinic	204	8.3% (4.5–12.1)	
Cheewadhanaraks et al., 2004 [16]	Thailand	Asian	Thai and Chinese	mFG	≥3	Regular cervical smear check	531	2% (0.8–3.2)	
DeUgarte et al., 2006 [17]	USA	White Black	Caucasian and Hispanic African American	mFG	≥8	Pre-employment physical exam	293 350	5.4% (2.8–8.0) 4.3% (2.2–6.4)	Possible selection self-referred bias[b] 97.5% reproductive age
Noorbala and Kefaie, 2010 [18]	Iran	White	Middle Eastern	mFG	≥8	Randomized cluster sampling proportionate to population size	900	10.8% (8.8–12.8)	Included only teenagers
March et al., 2010 [19]	Australia	White	Caucasian	mFG	8	Unselected population cohort	728	21.2% (18.2–24.2)	Possible selection self-referred bias[c] 3% were not white
Sanchón et al., 2012 [20]	Spain	White	Mediterranean	mFG	≥8	Unselected female blood donors from general population	393	11.7% (8.5–14.9)	
Sanchón et al., 2012 [20]	Italy	White	Mediterranean	mFG	≥8	Unselected female blood donors from general population	199	13.1% (8.4–17.7)	
Gabrielli and Aquino, 2012 [21]	Brazil	Mixed	Mixed	mFG	≥6	Premenopausal women during cervical cancer screening	859	12.5% (10.4–14.8)	88.5% were black

FG, Ferriman–Gallwey score; mFG, modified Ferriman–Gallwey score.
[a] Invitation of free medical examination.
[b] Only 66% of invited women participated.
[c] Only 53% of invited women participated, and patients self-assessed their hirsutism scores.
Modified from Escobar-Morreale et al. [1], by permission of Oxford University Press.

Table 1.2 Suggested cut-offs for the modified Ferriman–Gallwey hirsutism score (mFG) according to the 95th percentile in different unselected populations of premenopausal women

Author, year	Year	Country	Race	Ethnicity	Sample size	Suggested mFG cut-off[a]
Asuncion et al., 2000 [13]	2000	Spain	White	Mediterranean	154	≥8
Sagsoz et al., 2004 [15]	2004	Turkey	White	Middle Eastern	204	≥9
Cheewadhanaraks et al., 2004 [16]	2004	Thailand	Asian	Thai and Chinese	531	≥3
Tellez and Frenkel, 1995 [23]	2005	Chile	White	Hispanic	236	≥6
DeUgarte et al., 2006 [17]	2006	United States	White	Caucasian and Hispanic	283	≥8
			Black	African American	350	≥8
Zhao et al., 2007 [24]	2007	China	Asian	Chinese Han	623	≥2
Api et al., 2009 [11]	2009	Turkey	White	Middle Eastern	121	≥11
Moran et al., 2010 [25]	2010	Mexico	White	Hispanic	150	≥10
Noorbala and Kefaie, 2010 [18]	2010	Iran	White	Middle Eastern	900	≥10
Kim et al., 2011 [26]	2011	Korea	Asian	Korean	1010	≥6
Sanchón et al., 2012 [20]	2011	Spain and Italy	White	Mediterranean	592	≥10

[a] As defined by the 95th percentile of an unselected population of premenopausal women.
Modified from Escobar-Morreale et al. [1], by permission of Oxford University Press.

two areas may yield total normal scores, and the lack of population-based and uniform cut-off values.

The prevalence of hirsutism varies according to the mFG score cut-off value and the population under study [12–21]. This prevalence is relatively homogeneous across the world with the exception of women of Asian ancestry, in whom hirsutism is much less frequent (Table 1.1). In American women, 7.6%, 4.6%, and 1.9% demonstrated a score of 6 or more, 8 or 10, and there was no significant racial difference, with hirsutism prevalences of 8.0%, 2.8%, and 1.6% in white women, and 7.1%, 6.1%, and 2.1% in black women, respectively, according to the chosen cut-off [22]. Similarly, we found that 7.1% of unselected blood donors in Spain had hirsutism as defined by an mFG score above 7 [13]. These and other studies addressing the prevalence of hirsutism, as defined by a pre-defined mFG score cut-off value in different populations according to their race and ethnicity, are summarized in Table 1.1. However, because race and ethnicity greatly influence the amount of body hair, ideally the cut-off values of the mFG score should be obtained from the particular population under study. Table 1.2 includes proposed mFG score cut-off values based on the 95th percentile of selected female populations of fertile age [11,13,15–18,20,23–26]. Broad application of these cut-off values would render a uniform 5% worldwide prevalence of hirsutism.

Diagnosis of hirsutism

After establishing the presence of hirsutism by an increased mFG score, or if a history of hirsutism is strongly suggested by the finding of some evidence of terminal hair in androgen-dependent areas in women successfully treated for this condition, the most likely etiology should be established in all patients.

Functional causes account for most cases [27–31]: PCOS, as defined by the combination of hyperandrogenism with ovarian dysfunction (oligo-ovulation or polycystic ovarian morphology), is the most frequent diagnosis, accounting for approximately 60% of cases, followed by idiopathic hyperandrogenism (when there is no evidence of ovarian dysfunction) in approximately 25% of cases, idiopathic hirsutism (when there is no evidence of hyperandrogenemia or ovarian dysfunction) in approximately 10% of cases, and nonclassic congenital adrenal hyperplasia in approximately 3–5% of cases (Table 1.3). Exceptionally, hirsutism derives from benign or malignant adrenal or ovarian tumors, from hyperplasia of ovarian cells, from androgenic medications or drugs that

Table 1.3 Frequencies of the etiologies of androgen excess in large clinical series

Author, year	Sample size (n)	PCOS (n)	Idiopathic hyperandrogenism (n)	Idiopathic hirsutism (n)	NCCAH (n)	Tumors (n)	Miscellaneous (n)
Azziz et al., 2004 [27]	873	749[a]	59[b]	39	18	2	6
Glintborg et al., 2004 [28]	340	134	86[b]	115	2	1	2
Unluhizarci et al., 2004 [29]	168	96	29[b]	27	12	3	1
Carmina et al., 2006 [30]	950	685[c]	150	72	41	2	0
Escobar-Morreale et al., 2008 [31]	270	171	61[b]	24	6	0	8
Total no. (%)	2601 (100)	1835 (71)	385 (15)	277 (10)	79 (3)	8 (0.3)	17 (0.7)

NCCAH, nonclassic congenital adrenal hyperplasia; PCOS, polycystic ovary syndrome.

[a] The original article considered 33 patients with the hyperandrogenism, insulin resistance, and acanthosis nigricans (HAIR-AN) syndrome as a separate disorder from PCOS, reducing this figure to 716.

[b] Polycystic ovarian morphology was not considered for the diagnosis of PCOS in these studies, which relied on the 1990 National Institute of Child Health and Human Development criteria [32]. Some of these patients might have been diagnosed with PCOS if ovarian morphology had been considered.

[c] This study considered polycystic ovarian morphology for PCOS diagnosis, and 147 of the 685 PCOS patients who had regular ovulatory cycles would have been included in the idiopathic hyperandrogenism subgroup if ovarian morphology had not been considered.

Reproduced from Escobar-Morreale et al. [1], by permission of Oxford University Press.

interfere with ovarian steroidogenesis such as valproate, or from gestational hyperandrogenism secondary to placental aromatase deficiency or Krukenberg tumors.

By and large, the most important tool for the etiological diagnosis of hirsutism is a detailed clinical history and a complete physical examination. Functional causes almost always show a peripubertal onset and a slow progression over years, a family history of hyperandrogenism is frequent, and signs of virilization, such as clitoromegaly or balding, or of defeminization, such as mammary atrophy, are extremely rare. In contrast, androgen-secreting tumors usually show a sudden onset – rarely coincidental with puberty – and a rapid progression with severe virilization and defeminization, usually accompanying hirsutism.

Moreover, clinical evaluation provides valuable clues to discriminate between functional causes of hirsutism. Oligo- or amenorrhea and infertility may indicate the ovarian dysfunction associated with PCOS. Because this disorder is frequently associated with insulin resistance, its diagnosis is also suggested by the presence of abdominal obesity or acanthosis nigricans.

However, certain tests are needed in order to properly ascertain the etiology of hirsutism. It always is my practice to obtain blood samples to measure serum androgens, SHBG, and other hormones to rule out secondary causes of androgen excess, although this practice is debated when dealing with mild cases of hirsutism [1,33]. In my opinion, a correct etiological diagnosis is essential, because of the lifelong consequences of some of the disorders associated with hirsutism, such as polycystic ovary syndrome and nonclassic congenital adrenal hyperplasia. These disorders cannot be reliably ruled out simply because hirsutism is mild and menstrual periods are normal; in as many as 30% of hirsute patients, regular menstrual cycles are actually anovulatory [34], and nonclassic congenital hyperplasia may even be asymptomatic in some women – the so-called "cryptic" cases – who carry one severe allele needing genetic counseling.

Therefore, I measure serum total testosterone and SHBG during the follicular phase of the menstrual cycle to calculate free testosterone concentrations in every woman with hirsutism and monitor ovulation by measuring luteal phase progesterone concentrations and/or body temperature. In those patients presenting with normal free testosterone concentrations and regular ovulatory menstrual cycles, I obtain an ovarian ultrasound scan to rule out ovulatory PCOS. In addition, I always measure serum thyrotropin, prolactin, and basal 17-hydroxyprogesterone levels, followed by a 1–24 adrenocorticotropin stimulation test when basal 17-hydroxyprogesterone levels are above 1.7 ng/mL (5.1 nmol/L) [31] to definitively rule out 21-hydroxylase deficiency (nonclassic 11β-hydroxylase deficiency is extremely rare in Spain).

In those patients whose clinical evaluation induces me to suspect PCOS, or when this diagnosis is confirmed after the initial evaluation, I order an oral glucose tolerance test for glucose and insulin to establish their glucose tolerance and to obtain a dynamic estimate of their insulin sensitivity. I also obtain a lipid profile, including cholesterol fractions and triglycerides.

In the rare instance that a patient presents with a clinical history suggestive of an androgen-secreting tumor, I start evaluation by ordering an adrenal computed tomography scan and a transvaginal ultrasound of the ovaries, because, in my limited experience with this kind of tumors, imaging, and not serum androgen measurements, is the most effective technique in these cases. In the rare cases when these techniques show negative results, or when in doubt, simultaneous selective venous sampling of adrenals and ovaries may be needed to ascertain the non-functional source of androgen excess [35].

Management of hirsutism

The goals of the correct management of hirsutism are ameliorating hirsutism and reproductive complaints, preventing and/or treating the possible metabolic derangements associated and, if possible, treating the underlying cause. To be truly effective, the management of hirsutism must follow a few, but quite important principles: (i) treatment must be chronic; (ii) the effects of drugs are not evident before 6 to 12 months of administration; (iii) treatment should change depending on the characteristics and expectations of the individual patient; and (iv) treatment must be monitored by an expert [1]. Unfortunately, during the past decade there have been very few, if any, advances in the tools available to the practitioner to accomplish these goals. These tools include cosmetic procedures and drugs.

Cosmetic procedures are essential for the correct management of hirsutism, and may be used alone in mild cases or in combination with drug treatment. These procedures include temporary and "permanent" methods of hair reduction and topical eflornithine, but I rarely recommend the latter because its effects take weeks to be apparent and its economic cost is high.

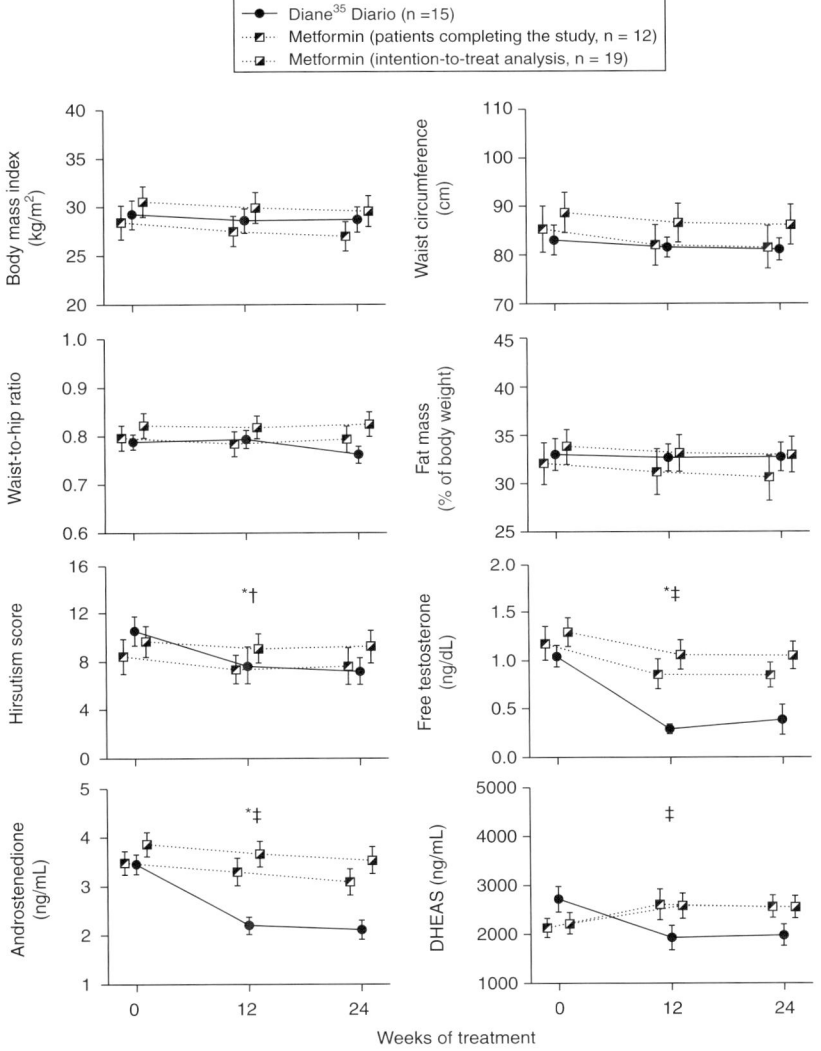

Figure 1.2 Effects on hirsutism score and serum androgen levels of an oral contraceptive containing 35 μg of ethinylestradiol and 2 mg of cyproterone acetate compared with the insulin sensitizer metformin, 850 mg twice a day. Data are means ± SEM. Because of the large drop-out rate in the metformin arm, the figures show the analysis of the patients completing the study (n = 12) and also an intention-to-treat analysis, including all the patients treated with metformin (n = 19). It was assumed in patients who did not complete the study that the dependent variables did not change at the missing visits with respect to the values observed in the previous visit. DHEAS, dehydroepiandrosterone sulfate. *$P < 0.05$ compared with baseline values in the whole group of patients, irrespective of the arm of treatment. †$P < 0.05$ for the differences in the changes of each variable depending on the arm of treatment, only in the intention-to-treat analysis. ‡$P < 0.05$ for the differences in the changes of each variable depending on the arm of treatment, both by analysis of patients who completed the study and by intention-to-treat analysis. Reproduced from Luque-Ramirez et al. [37] with permission. Copyright 2007, The Endocrine Society.

Bleaching and temporary methods of hair removal such as shaving, plucking, waxing, or the use of chemical depilatory agents are invaluable tools in the first months of treatment while waiting for drug treatment to be noticed, or may be used even as single procedures in mild cases [1]. The patient must be assured that even shaving does not increase the growth and thickness of hair, a common and incorrect belief among women, because the blunt tip of shaved hair is more visible than the tapered tip of uncut hair. Local discomfort is the common drawback of these procedures.

Among the methods of permanent hair reduction, I favor a competently performed galvanic electrolysis for small areas, such as the face, and laser or intense pulsed light (IPL) photoepilation for larger areas of the body, because, although real scientific evidence is lacking [33], I have the suspicion that galvanic electrolysis actually prevents hair regrowth in treated areas, whereas methods based on thermal destruction of the hair follicle, such as thermolysis or photoepilation, do not. Nevertheless, inexperienced electrolysis may cause considerable local side effects and even scarring, and this possibility should be weighed carefully. In some countries, including Spain, many women consider the presence of terminal hair, even in normal areas like the axilla and the pubic region, unacceptable. Photoepilation is the only realistic choice for these women.

Drug treatment of hirsutism includes oral contraceptive pills (OCP), antiandrogens, and the 5α-reductase inhibitor finasteride. Insulin sensitizers, on the contrary, are not effective for hair reduction, although they may be useful in hirsute patients with PCOS to ameliorate insulin resistance and improve the metabolic conditions frequently associated with this disorder [36]. However, there is no sound scientific evidence to recommend insulin sensitizers solely for hirsutism [1].

For most hirsute patients, an OCP containing a neutral (gestodene or desogestrel) or antiandrogenic (cyproterone or chlormadinone acetate, or drosperinone) progestin is the drug of choice. OCPs not only decrease circulating testosterone levels by reversibly suppressing gonadotropin secretion, but also induce a marked increase in SHBG concentrations, thereby decreasing free testosterone concentrations to levels below those observed in healthy women (Figure 1.2). Additional advantages are the regularization of menstrual bleeding in women with menstrual dysfunction, avoiding endometrial hyperplasia, and the safe contraception needed for the combination with antiandrogens.

Drawbacks of these third generation OCPs include their deleterious effects on coagulation, especially among smokers, and an increased risk of non-fatal venous thromboembolism when compared with second generation OCPs containing the androgenic progestin levonorgestrel [38]. Although the increased risk of venous thromboembolism with third generation OCPs is small, it must be noted that even older formulations containing androgenic progestins may also ameliorate hirsutism [39]. However, these older OCPs may be less effective on hirsutism and might increase body mass index, compared with OCPs containing neutral progestins [40]. Therefore, the choice of an OCP for the treatment of hirsutism must carefully balance the greater efficacy of third generation OCPs against the safer coagulation profile of second generation OCPs, especially in adolescents, hypertensive women, and smokers.

Because of my clinical and research experience [37,41–44], I am particularly fond of a 2 mg cyproterone acetate plus 35 μg ethinylestradiol combination (CPA + EE) that has been available for decades in Europe, and I am still to be convinced that newer formulations containing drosperinone are of similar efficacy. The CPA + EE OCP ameliorates hirsutism and suppresses hyperandrogenemia (Figure 1.2), regularizes menstrual bleeding, and has an overall safe metabolic profile in hyperandrogenic women, showing a neutral effect on glucose tolerance and insulin sensitivity and a beneficial effect on the lipid profile consisting of an increase in Apo-A and high-density lipoprotein phospholipids [37]. In women presenting with moderate to severe hirsutism, I add 50 mg of CPA during the first 10 days of each CPA + EE cycle to obtain a faster amelioration of hirsutism. After attaining a satisfactory amelioration of hirsutism, I withdraw the additional CPA and continue maintenance therapy with the CPA + EE pill for years.

Because the CPA + EE combination may induce a small, but significant increase in blood pressure in PCOS patients [41], in hypertensive women I am currently recommending combinations of ethinylestradiol with a neutral progestin plus spironolactone as antiandrogen, with excellent tolerability and clinical results comparable to those of the CPA + EE pill.

Only in those hirsute patients in whom oral estrogens are contraindicated do I favor the use of finasteride over other antiandrogens, and only after ensuring safe non-hormonal contraception to avoid fetal male pseudohermaphroditism in case of unadverted pregnancy. The efficacy of finasteride is similar to that of CPA, spironolactone, and flutamide [45], but lacks the menstrual disturbances associated with CPA and spironolactone when used as single agents [46,47], and the potential for severe hepatotoxicity of flutamide [48].

Treatment has to be monitored at least annually, and must always consider additional measures related to the etiology of hirsutism, such as diet and lifestyle recommendations in women with PCOS or mineralo- and/or glucocorticoid replacement therapy in women with classic congenital adrenal hyperplasia. Of note, quitting smoking must be strongly advised because this habit worsens the undesirable effects of almost all the drugs available for hirsutism and related conditions [43]. Drug treatment for hirsutism may be stopped in patients seeking fertility and reinstated after delivery and lactation.

Conclusions

Hirsutism is a frequent complaint in women of fertile age and usually results from functional disorders of androgen excess. The diagnosis of hirsutism requires quantifying the amount and distribution of terminal hair and establishing the most likely underlying etiology. Treatment of hirsutism is effective, but should be chronic and multidisciplinary. Cosmetic measures can be used in every case and, while treatment with an OCP is indicated for most patients, an antiandrogen may be added in moderate or severe cases. Finally, successful management of hirsutism should also consider the treatment of any reproductive or metabolic comorbidity.

Acknowledgments

Professor Michael Pugeat, Department of Medicine, Université Claude Bernard Lyon 1, Lyon, France, is the author of the original drawings in Figure 1.1. This work was supported by the Spanish Ministry of Economy and Competitiveness, Instituto de Investigación Carlos III, grants FIS PI080944 and PI1100357. CIBERDEM is also an initiative of Instituto de Investigación Carlos III.

References

1. Escobar-Morreale HF, Carmina E, Dewailly D, et al. Epidemiology, diagnosis and management of hirsutism: a consensus statement by the Androgen Excess and Polycystic Ovary Syndrome Society. *Hum Reprod Update* 2012; 18: 146–70.

2. Olsen EA. Methods of hair removal. *J Am Acad Dermatol* 1999; 40: 143–55.

3. Azziz R, Carmina E, Dewailly D, et al. Position statement: criteria for defining polycystic ovary syndrome as a predominantly hyperandrogenic syndrome: an Androgen Excess Society guideline. *J Clin Endocrinol Metab* 2006; 91: 4237–45.

4. Azziz R, Carmina E, Sawaya ME. Idiopathic hirsutism. *Endocr Rev* 2000; 21: 347–62.

5. Escobar-Morreale HF, Serrano Gotarredona J, García Robles R, et al. Mild adrenal and ovarian steroidogenic abnormalities in hirsute women without hyperandrogenemia: does idiopathic hirsutism exist? *Metabolism* 1997; 46: 902–7.

6. Rosner W, Auchus RJ, Azziz R, et al. Position statement: Utility, limitations, and pitfalls in measuring testosterone: an Endocrine Society position statement. *J Clin Endocrinol Metab* 2007; 92: 405–13.

7. Garn SM. Types and distribution of the hair in man. *Ann N Y Acad Sci* 1951; 53: 498–507.

8. Ferriman D, Gallwey JD. Clinical assessment of body hair growth in women. *J Clin Endocrinol Metab* 1961; 21: 1440–7.

9. Lorenzo EM. Familial study of hirsutism. *J Clin Endocrinol* 1970; 31: 556–64.

10. Hatch R, Rosenfield RL, Kim MH, Tredway D. Hirsutism: implications, etiology, and management. *Am J Obstet Gynecol* 1981; 140: 815–30.

11. Api M, Badoglu B, Akca A, et al. Interobserver variability of modified Ferriman-Gallwey hirsutism score in a Turkish population. *Arch Gynecol Obstet* 2009; 279: 473–9.

12. Diamanti-Kandarakis E, Kouli CR, Bergiele AT, et al. A survey of the polycystic ovary syndrome in the Greek island of Lesbos: hormonal and metabolic profile. *J Clin Endocrinol Metab* 1999; 84: 4006–11.

13. Asuncion M, Calvo RM, San Millan JL, et al. A prospective study of the prevalence of the polycystic ovary syndrome in unselected Caucasian women from Spain. *J Clin Endocrinol Metab* 2000; 85: 2434–8.

14. Zargar AH, Wani AI, Masoodi SR, et al. Epidemiologic and etiologic aspects of hirsutism in Kashmiri women in the Indian subcontinent. *Fertil Steril* 2002; 77: 674–8.

15. Sagsoz N, Kamaci M, Orbak Z. Body hair scores and total hair diameters in healthy women in the Kirikkale Region of Turkey. *Yonsei Med J* 2004; 45: 483–91.

16. Cheewadhanaraks S, Peeyananjarassri K, Choksuchat C. Clinical diagnosis of hirsutism in Thai women. *J Med Assoc Thail* 2004; 87: 459–63.

17. DeUgarte CM, Woods KS, Bartolucci AA, Azziz R. Degree of facial and body terminal hair growth in unselected black and white women: toward a populational definition of hirsutism. *J Clin Endocrinol Metab* 2006; 91: 1345–50.

18. Noorbala MT, Kefaie P. The prevalence of hirsutism in adolescent girls in Yazd, Central Iran. *Iran Red Crescent Med J* 2010; 12: 111–17.

19. March WA, Moore VM, Willson KJ, et al. The prevalence of polycystic ovary syndrome in a community sample assessed under contrasting diagnostic criteria. *Hum Reprod* 2010; 25: 544–51.

20. Sanchón R, Gambineri A, Alpanes M, et al. Prevalence of functional disorders of androgen excess in unselected premenopausal women: a study in blood donors. *Hum Reprod* 2012; 27: 1209–16.

21. Gabrielli L, Aquino EM. Polycystic ovary syndrome in Salvador, Brazil: a prevalence study in primary healthcare. *Reprod Biol Endocrinol* 2012; 10: 96.

22. Knochenhauer ES, Key TJ, Kashar-Miller M, et al. Prevalence of the polycystic ovary syndrome in unselected black and white women of the Southeastern United States: a prospective study. *J Clin Endocrinol Metab* 1998; 83: 3078–82.

23. Tellez R, Frenkel J. [Clinical evaluation of body hair in healthy women]. *Rev Med Chil* 1995; 123: 1349–54.

24. Zhao JL, Chen ZJ, Shi YH, et al. [Investigation of body hair assessment of Chinese women in Shandong region and its preliminary application in polycystic ovary syndrome patients]. *Zhonghua Fu Chan Ke Za Zhi* 2007; 42: 590–4.

25. Moran C, Tena G, Moran S, et al. Prevalence of polycystic ovary syndrome and related disorders in Mexican women. *Gynecol Obstet Invest* 2010; 69: 274–80.

26. Kim JJ, Chae SJ, Choi YM, et al. Assessment of hirsutism among Korean women: results of a randomly selected sample of women seeking pre-employment physical check-up. *Hum Reprod* 2011; 26: 214–20.

27. Azziz R, Woods KS, Reyna R, et al. The prevalence and features of the polycystic ovary syndrome in an unselected population. *J Clin Endocrinol Metab* 2004; 89: 2745–9.

28. Glintborg D, Henriksen JE, Andersen M, et al. Prevalence of endocrine diseases and abnormal glucose tolerance tests in 340 Caucasian premenopausal women with hirsutism as the referral diagnosis. *Fertil Steril* 2004; 82: 1570–9.

29. Unluhizarci K, Gokce C, Atmaca H, et al. A detailed investigation of hirsutism in a Turkish population: idiopathic hyperandrogenemia as a perplexing issue. *Exp Clin Endocrinol Diabetes* 2004; 112: 504–9.

30. Carmina E, Rosato F, Janni A, et al. Extensive clinical experience: relative prevalence of different androgen excess disorders in 950 women referred because of clinical hyperandrogenism. *J Clin Endocrinol Metab* 2006; 91: 2–6.

31. Escobar-Morreale HF, Sanchon R, San Millan JL. A prospective study of the prevalence of nonclassical congenital adrenal hyperplasia among women presenting with hyperandrogenic symptoms and signs. *J Clin Endocrinol Metab* 2008; 93: 527–33.

32. Zawadzki JK, Dunaif A. Diagnostic criteria for polycystic ovary syndrome: towards a rational approach. In: Dunaif A, Givens JR, Haseltine FP, Merriam GR, editors. *Polycystic Ovary Syndrome*. Boston: Blackwell Scientific Publications; 1992. pp. 377–84.

33. Martin KA, Chang RJ, Ehrmann DA, et al. Evaluation and treatment of hirsutism in premenopausal women: an Endocrine Society clinical practice guideline. *J Clin Endocrinol Metab* 2008; 93: 1105–20.

34. Azziz R, Waggoner WT, Ochoa T, et al. Idiopathic hirsutism: an uncommon cause of hirsutism in Alabama. *Fertil Steril* 1998; 70: 274–8.

35. Alpanes M, Gonzalez-Casbas JM, Sanchez J, et al. Management of postmenopausal virilization. *J Clin Endocrinol Metab* 2012; 97: 2584–8.

36. Lord JM, Flight IH, Norman RJ. Insulin-sensitising drugs (metformin, troglitazone, rosiglitazone, pioglitazone, D-chiro-inositol) for polycystic ovary syndrome. *Cochrane Database Syst Rev* 2003: CD003053.

37. Luque-Ramirez M, Alvarez-Blasco F, Botella-Carretero JI, et al. Comparison of ethinyl-estradiol plus cyproterone acetate versus metformin effects on classic metabolic cardiovascular risk factors in women with the polycystic ovary syndrome. *J Clin Endocrinol Metab* 2007; 92: 2453–61.

38. Parkin L, Sharples K, Hernandez RK, Jick SS. Risk of venous thromboembolism in users of oral contraceptives containing drospirenone or levonorgestrel: nested case-control study based on UK General Practice Research Database. *BMJ* 2011; 342: d2139.

39. Breitkopf DM, Rosen MP, Young SL, Nagamani M. Efficacy of second versus third generation oral contraceptives in the treatment of hirsutism. *Contraception* 2003; 67: 349–53.

40. Sanam M, Ziba O. Desogestrel+ethinylestradiol versus levonorgestrel+ethinylestradiol. Which one has better effect on acne, hirsutism and weight change? *Saudi Med J* 2011; 32: 23–6.

41. Luque-Ramirez M, Mendieta-Azcona C, Alvarez-Blasco F, Escobar-Morreale HF. Effects of metformin versus ethinyl-estradiol plus cyproterone acetate on ambulatory blood pressure monitoring and carotid intima media thickness in women with the polycystic ovary syndrome. *Fertil Steril* 2009; 91: 2527–36.

42. Luque-Ramirez M, Alvarez-Blasco F, Escobar-Morreale HF. Antiandrogenic contraceptives increase serum adiponectin in obese polycystic ovary syndrome patients. *Obesity* 2009; 17: 3–9.

43. Luque-Ramirez M, Mendieta-Azcona C, del Rey Sanchez JM, et al. Effects of an antiandrogenic oral contraceptive pill compared with metformin on blood coagulation tests and endothelial function in women with the polycystic ovary syndrome: influence of obesity and smoking. *Eur J Endocrinol* 2009; 160: 469–80.

44. Luque-Ramirez M, Escobar-Morreale HF. Treatment of polycystic ovary syndrome (PCOS) with metformin ameliorates insulin resistance in parallel with the decrease of serum interleukin-6 concentrations. *Horm Metab Res* 2010; 42: 815–20.

45. Moghetti P, Tosi F, Tosti A, et al. Comparison of spironolactone, flutamide, and finasteride efficacy in the treatment of hirsutism: a randomized, double blind, placebo-controlled trial. *J Clin Endocrinol Metab* 2000; 85: 89–94.

46. Levitt JI. Spironolactone therapy and amenorrhea. *JAMA* 1970; 211: 2014–15.

47. van Wayjen RG, van den Ende A. Clinical-pharmacological investigation of cyproterone acetate. *Gynecol Invest* 1971; 2: 282–9.

48. Wallace C, Lalor EA, Chik CL. Hepatotoxicity complicating flutamide treatment of hirsutism. *Ann Intern Med* 1993; 119: 1977.

Chapter

2

Alopecia: managing the basics

Thamer Mubki and Jerry Shapiro

The relationship between sexual maturity and hair loss was recognized for the first time by Aristotle. Since then, many steps have been undertaken to convert that observation to scientific fact. Androgens, which were found to be responsible for sexual maturity, also have a major role in normal hair growth and induction of certain hair diseases. The hair follicle as a skin appendage is considered as a target tissue for androgen actions [1]. This chapter will discuss the relationship between androgens and hair.

Physiology of hair growth

Hair can be classified structurally into three types: lanugo, vellus, and terminal hair. Lanugo is the soft hair that covers the infant's body and is usually shed shortly before birth. Vellus hairs are the short, fair, and non-pigmented hair that covers the apparently hairless parts of the skin. The longer, pigmented, and hence more visible hair is called terminal hair. The latter normally presents on the scalp, eyebrows, eyelashes, axillae, and pubic areas in both sexes. Moreover, it can also be seen on the face, back, and abdomen in males and hirsute females. Hair can also be classified into sexual and nonsexual depending on its sensitivity to sex steroids. Sexual hair usually responds to androgens and grows on the face, lower abdomen, anterior thighs, chest, breast, pubic area, and axillae.

Hair growth is cyclic and is characterized by three distinct phases: telogen (resting phase), anagen (growth phase), and catagen (involution phase) [2]. In telogen, the hair is loosely attached to the base of the largely inactive hair follicle. As hair growth (anagen) begins, cells in the epithelial matrix at the base of the hair follicle begin to proliferate and continue to do so for a period of time. Completion of anagen is followed by catagen, during which a programmed cell death

(apoptosis) and regression of the hair follicle occurs back to a resting state. The length of the hair is determined primarily by the duration of the anagen phase. Scalp hair remains in anagen for 2–5 years and spends only a relatively short time in telogen (about 3 months), leading to a ratio of anagen to telogen hairs of approximately 9:1 [3]. Elsewhere, such as on the forearm, the hair cycle has a shorter anagen duration and a longer telogen phase, yielding a shorter hair of relatively stable length.

Effects of sex steroids on hair growth

Sex steroids include androgens, estrogens, and progestogens. The androgens are the main normal regulators of human hair growth [4]. They regulate the hair's characteristics and distribution pattern [5]. Onset of puberty is associated with gradual replacement of tiny vellus hairs with larger, more pigmented terminal hairs in the pubis and later in the axillae [6,7]. Later, similar changes occur on the male face, chest, and limbs. These changes essentially correspond to the pubertal rise in plasma androgens.

Dehydroepiandrosterone sulfate (DHEA-S), dehydroepiandrosterone (DHEA), and androstenedione are the major circulating androgens in women. All of these androgens have little or no direct intrinsic androgenic activity and need to be converted to testosterone and dihydrotestosterone (DHT) to exert androgenic activity [8]. Androgens bind to specific intracellular androgen receptors in the hair follicles. The resulting hormone–receptor complex then binds to specific hormone response elements in DNA, promoting the expression of specific hormone-regulated genes [9]. Androgens are required to irreversibly trigger a follicle's specific genetic programming [9]. For instance, androgens will trigger the beard hair follicles

Androgens in Gynecological Practice, ed. Leo Plouffe and Botros Rizk. Published by Cambridge University Press. © Cambridge University Press 2015.

in men and hirsute women to irreversibly change from vellus to terminal hair follicles. Thereafter, androgens are needed to maintain the production of good terminal hair quality. Once the hair follicle has been triggered to produce terminal hair, any future lack of significant levels of androgens will not cause terminal follicles to regress back to the vellus type; but rather, will reduce the speed and the quality by which the terminal hair is produced [10]. This is particularly important to know when dealing with hirsutism, as correcting the underlying hormonal abnormality might not be sufficient to produce cosmetically acceptable results and various hair depilation methods may be needed.

Androgens, in particular testosterone, stimulate growth and increase the diameter and pigmentation of non-scalp hair. They achieve this by increasing the proportion of time hairs spend in anagen [4]. Paradoxically, testosterone affects the hair follicles on the scalp of genetically predisposed individuals, leading to a shorter duration of anagen and hence, converting large terminal scalp follicles into vellus ones and producing androgenetic alopecia (pattern hair loss) [4,11,12]. In other words, the response to androgens is variable between individual follicles, ranging from stimulation to inhibition, depending on the body site and sensitivity of individual follicles to the hormone. For example, female levels of circulating androgens may be high enough to promote axillary hair and the female pubic pattern of terminal hair, but male patterns of body hair normally require the higher male level. On the other hand, estrogens exert opposite effects on the hair follicles, producing thinner and less pigmented hairs [5]. This is clearly noticed in pregnant women who usually display a slower rate of replacement of spontaneous hair loss or plucked hairs. Progestogens have variable effects depending on their androgenic activity [5].

Although testosterone is the major circulating androgen, it is not the most potent one. DHT is regarded as the most potent nuclear androgen in many androgen-sensitive tissues, including the hair follicles [8,13]. In the skin, DHT has a major role in the pathogenesis of skin diseases, namely hirsutism, acne, and androgenetic alopecia [14]. The 5α-reductase enzyme is responsible for converting testosterone to DHT. This conversion takes place peripherally in the prostate, sebaceous glands, and the hair follicles. The resulting DHT binds to the same androgen receptors as the parent compound, but with a fivefold greater affinity [15]. Two isoenzymes of 5α-reductase have been described in the skin; 5α-reductase type I in sebaceous glands and 5α-reductase type II in dermal papillae and the outer and inner root sheaths of hair follicles [16]. Developments of male body hair patterns do not usually occur in 5α-reductase type II-deficient individuals. Instead, female patterns of pubic and axillary hair will develop. Androgenetic patterns of alopecia do not develop in such people [4]. Therefore, it seems that the intrafollicular binding of DHT to the androgen receptors is a prerequisite for developing secondary sexual hair.

Female pattern hair loss (androgenetic alopecia)

The term female pattern hair loss (FPHL) is commonly used synonymously with female androgenetic alopecia. However, since the role of androgens and androgen receptor genes in the pathogenesis of FPHL is uncertain as compared with their well-defined role in male androgenetic alopecia (AGA-M), the term FPHL should probably be used instead [17,18]. In this chapter the term FPHL will be used to refer to women with progressive thinning of scalp hair that follows a pattern distribution. FPHL is common with a prevalence ranging from 6% of women under the age of 50 to 38% of women over the age of 70 [19,20]. Although biochemical and clinical evidence of hyperandrogenism has been reported in 30–40% of cases of FPHL in one prospective study [21], 60–70% of women with FPHL usually have no underlying endocrinological disturbances [22].

Pathogenesis of FPHL

The progressive thinning of scalp hair basically results from two processes. First is gradual miniaturization of scalp hair follicles and second is alteration in the hair cycle dynamics. Miniaturization is a process in which the hair follicle transforms gradually from a terminal follicle to a vellus-like follicle. In addition, shortening of anagen and stabilization or increase in the duration of the telogen usually occurs. These changes in hair cycle dynamics will eventually reduce the anagen/telogen ratio to less than the normal of 9:1 [23].

While DHT–androgen receptor complex overactivity has a central role in the pathogenesis of AGA-M, its role is controversial in FPHL. The proposed role of the overactive hormone–receptor complex may result from local or systemic factors [24]. Increased androgen receptor numbers, over-sensitive receptors produced by functional polymorphism, or increased local production of DHT can be regarded as local factors. On the other hand, various systemic factors can result in

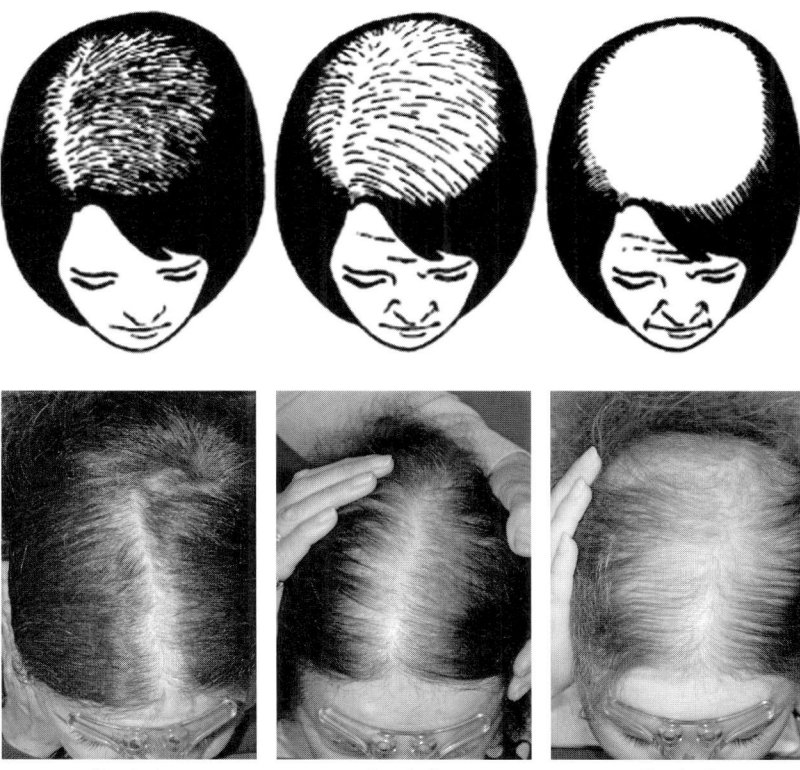

Figure 2.1 Ludwig's classification of FPHL. Adapted with permission from Ludwig [30], John Wiley and Sons ©.

Grade 1   Grade 2   Grade 3

systemic increase in circulating androgens [25]. One study has shown that 30–40% of females with FPHL have an underlying systemic endocrine alteration, with polycystic ovary syndrome (PCOS) being the most common cause of androgen excess in such females, followed by other less common causes like late-onset congenital adrenal hyperplasia (CAH), and even more rarely an androgen-secreting tumor [25]. However, the majority (60–70%) of FPHL occurs in patients with no clinical or biochemical evidence of androgen excess (idiopathic FPHL) [22]. Alterations of androgen receptor sensitivity may be responsible for the latter condition [22].

The role of genetics in FPHL is complex. A polygenic pattern of inheritance is probably present. Several different genes are involved, including those involved in androgen production, conversion of androgens to DHT, and controlling the sensitivity of the receptors [26]. Moreover, the family history in women with FPHL is not as clear as it is in AGA-M [27].

Clinical features and classification

FPHL has a variable postpubertal age of onset, with most cases presenting clinically at an age of around 40 years [28]. It has a specific pattern of scalp involvement. The central portion of the scalp is usually affected, sparing the frontal hairline, in contrast to AGA-M in which the frontal hairline is mostly affected and progressively receding. Classically, the disease starts with a wider midline part of the crown as compared with the occipital region. This might progress eventually to a more diffuse involvement, resulting in an oval-shaped alopecia surrounded by a rim of normal hair density (Ludwig's classification) [29– 31]. Ludwig has proposed three grades of FPHL starting with grade 1 (minimal), grade 2 (moderate), and grade 3 (intense) (Figure 2.1). Another possible presentation of FPHL is thinning hair that is more pronounced on the frontal portion of the scalp, resulting in a part resembling a fir tree (Christmas tree) behind the frontal hairline [31]. This has been referred to as frontal accentuation by Olsen, who proposed a classification with three triangular stages of hair loss [31]. On the other hand, it is not uncommon to have a woman presenting with a hair loss pattern not only confined to the crown portion of the scalp, but in addition, involving the frontal and fronto-parietal hair line causing these to recede in a pattern more or less similar to AGA-M (Hamilton–Norwood classification) [31]. The latter pattern was regarded at one time as a marker for androgen excess; however, it was reported as a normal finding in 13% and 37% of premenopausal and postmenopausal women, respectively [25,29,30]. Other signs of androgen excess such as acne, seborrhea, hirsutism (SAHA: seborrhea, acne, hirsutism, and alopecia), obesity, menstrual cycle irregularities, and low fertility can be associated with FPHL with underlying androgen excess.

Evaluation

Diagnosis of FPHL is usually made clinically. However, histological evaluation using a 4 mm punch skin biopsy can be performed very rarely to confirm the diagnosis or to rule out other differential diagnoses of non-scarring alopecia, such as acute/chronic telogen effluvium (resting hair loss) and diffuse alopecia areata or early scarring hair loss, such as early central centrifugal cicatricial alopecia and fibrosing alopecia with a patterned distribution [28]. When needed, the skin biopsy should be taken from a representative site within the central scalp. Both temporal regions should be avoided as miniaturized hair follicles can present there independent of FPHL [28]. A punch biopsy of at least 4 mm size should be taken following the direction of the hair growth and reaching deeply into the level of subcutaneous tissue where the bulbs of anagen hair follicles usually reside. Horizontal sectioning of the biopsy is the standard technique used by many dermatopathologists as it allows a more accurate count of terminal and vellus-like hair follicles. A reduction of the terminal to vellus-like hair ratio from the normal of 9:1 to 1.9:1 – 2.2:1 is considered the hallmark of androgenetic alopecia. This might be associated with a reduction in the number of hairs per unit area to less than the normal of 240–400 hairs/cm^2 or 30–50 hairs per 4 mm punch biopsy, a change that might not occur in the early stages of FPHL. A perifollicular lymphohistiocytic infiltrate and perifollicular fibrosis might occasionally be seen in androgenetic alopecia [28,32].

A screening blood work is usually needed to rule out the presence of other abnormalities that might cause telogen effluvium (resting hair loss) and unmask FPHL. These include serum ferritin, thyroid-stimulating hormone, vitamin D, and zinc. Iron deficiency might cause the hair follicle to enter telogen, causing telogen effluvium. In addition, it might also interfere with the efficacy of treatments used for FPHL [28]. Low serum ferritin is diagnostic

for iron deficiency. However, the sensitivity of serum ferritin as a tool to diagnose low iron stores in patients with chronic inflammatory diseases, malignancy, or infections can be an issue, as ferritin is considered an acute phase reactant [28]. Measurement of both serum iron and total iron-binding capacity can be used as an alternative. Low serum iron and a high total iron-binding capacity can be expected in cases of iron deficiency. Hypothyroidism may also cause hair loss by inducing telogen effluvium. Moreover, zinc deficiency can cause hair loss, and vitamin D receptor-mutated animals and humans usually have alopecia [28,33].

Most women with FPHL have no underlying evidence of androgen excess [22,25]. Therefore, the decision to evaluate the patient's serum androgen levels should be guided by the presence of relevant symptoms and signs. An early onset of FPHL can be a marker of a carrier state of a gene causing PCOS [34]. Moreover, the presence of concomitant signs and symptoms of hyperandrogenism such as hirsutism, moderate to severe or refractory resistant adult acne, irregular menses, infertility, or signs of insulin resistance, such as acanthosis nigricans and obesity, should warrant screening the patient for biochemical evidence of androgen excess. The most useful initial test is serum total testosterone level. Total testosterone assays measure free testosterone, albumin bound, and sex hormone-binding globulin (SHBG) bound testosterone. Although assessment of the free testosterone level has a more diagnostic yield as compared with that of total testosterone, its measurement is time consuming and costly [8]. Moreover, an excellent correlation between total and free testosterone has been described; and therefore, bioavailable testosterone can usually be predicted from the total testosterone level [8]. Measurement of DHEA-S should be ordered in cases where an androgen adrenal secreting tumor is suspected [8]. Testosterone levels greater than 2.5× normal or >150–200 ng/dL, or DHEA-S greater than 2× normal or >700 μg/dL in premenopausal or >400 μg/dL in postmenopausal women, should necessitate a radiological workup to rule out the presence of androgen-secreting tumors of ovaries or adrenals [35]. An early morning serum 17-OH progesterone during the follicular phase of the cycle (days 1–14) can be used as a screening test for nonclassic CAH [36]. Prolactin level can be ordered in cases of galactorrhea or increased serum testosterone level [8]. Further evaluation by an endocrinologist will be needed in patients with high values of testosterone, DHEA-S, prolactin, or 17-OH progesterone [28].

Treatment

Successful treatment of FPHL depends on meeting realistic expectations made by patients, as almost all available treatment modalities are aimed at slowing down the hair loss and at most arresting it. Hence, no dramatic changes in hair thickness should be expected. However, stimulating growth and thickening of pre-existing hair can be achieved in some patients using drug treatments. Starting medications early and combining more than one medication together might be the key for better results. On the other hand, hair transplantation surgery can be very effective, which, when performed and combined with the use of appropriate medicines, can result in superior results compared with using medications alone.

Considering the mode of action, treatments used for FPHL can be divided into two categories: androgen independent and androgen dependent. Other miscellaneous groups of treatments will also be discussed.

Androgen-independent therapies

Minoxidil

There are two different concentrations for topical minoxidil solution commercially available: 2% and 5%. This is in addition to the recently approved 5% foam preparation. Minoxidil 2% solution (twice daily) and 5% foam are the only concentrations approved by the United States Food and Drug Administration and other countries' regulatory equivalents for treating FPHL. However, the 5% solution is increasingly being used off-label for FPHL. Minoxidil sulfate, minoxidil's active metabolite, works by opening the cellular potassium channels, leading to impaired entry of calcium into the hair follicle cells, thus decreasing epidermal growth factors and enhancing hair growth. Minoxidil was also found to prolong the anagen phase of hair growth, transforming vellus and vellus-like hairs into terminal hairs [1,37]. Minoxidil can either arrest hair loss or induce mild to moderate hair regrowth in 60% of women [28,38]. A randomized, placebo-controlled trial comparing the efficacy of topical 5% solution with topical 2% and placebo in FPHL patients demonstrated statistically significant increased hair growth in both 5% and 2% groups over the placebo group. However, although not statistically significant, the 5% solution seemed to work better than the 2% solution [38]. Side effects include contact dermatitis in 6.5% of patients and facial hypertrichosis in 3–5% of patients [37].

Tachycardia and hypotension can also be seen in 1 out of every 1000 patients [1]. Switching to the lower concentration of 2% solution or to the 5% foam preparation can help to alleviate most of the side effects.

Minoxidil should be avoided during pregnancy or breast feeding. Minoxidil is considered as pregnancy category C by the US Food and Drug Administration (FDA). However, a one-year prospective study showed no increase in cardiovascular events or adverse pregnancy outcomes among patients on topical minoxidil versus controls, but there have been scattered reports of fetal malformations [37]. Minoxidil can be used for FPHL with or without hyperandrogenism. It can also be used alone or in combination with other medications, such as antiandrogens. One mL of minoxidil should be applied and spread over the frontal and crown region of a dry scalp twice daily for at least 6 months before judging the efficacy. When effective, minoxidil should be used for life, as discontinuing minoxidil will cause the responsive hair follicles to fall out. It is the author's opinion that minoxidil should be included in any treatment regimen of FPHL and should be considered the first-line treatment.

Androgen-dependent therapies

Two groups of antiandrogen agents are recognized. The first group comprises the classic antiandrogen agents, including spironolactone, cyproterone acetate, and flutamide, which are capable of blocking cytoplasmic androgen receptors. The second group includes peripheral antiandrogen agents, such as finasteride and dutasteride, which work by blocking the conversion of testosterone to DHT in the hair follicles. None of the above groups are approved for use in FPHL and they are uncommonly used in North America, unlike in Europe where they are used more commonly.

Spironolactone

Spironolactone is a potassium-sparing diuretic and a synthetic steroid structurally related to aldosterone. It exerts its antiandrogen effects by competitively blocking the DHT cytoplasmic receptors and in addition it has a weak inhibitory effect on androgen synthesis. Although many studies have demonstrated its efficacy in the treatment of FPHL, especially when associated with other signs of hyperandrogenism, such as hirsutism and acne, spironolactone is being used off-label to treat FPHL [39,40]. Most women will require a minimum of 200 mg per day for at least 6 months [29].

Tolerable side effects include lethargy, nausea, breast engorgement, and menorrhagia. It can also induce hyperkalemia and hypotension, warranting a periodic monitoring of potassium levels [25]. Spironolactone is in pregnancy category D and might cause feminization of a male fetus. Therefore, it is contraindicated for women of childbearing age without the use of appropriate contraception methods. Low-dose oral contraceptive pills (OCPs) should be helpful to prevent pregnancy and to reduce the menorrhagia.

Cyproterone acetate

Cyproterone acetate is an androgen receptor blocker. It can also decrease testosterone levels by suppressing luteinizing hormone and follicle-stimulating hormone release [29]. Improvement of FPHL was demonstrated when cyproterone acetate was used alone or in combination with ethinylestradiol or spironolactone [37]. It can be used for FPHL with or without hyperandrogenism. The most useful dose is 100 mg on days 5–15 combined with 50 µg ethinylestradiol on days 5–25 of the menstrual cycle [28]. The hair loss requires a higher dose than the 2 mg daily dose found in the combined estrogen–progestin OCP called Diane (2 mg cyproterone acetate days 5–15 and 50 µg of ethinylestradiol days 5–25 of menstrual cycle) that can be used for hirsutism [25]. Fatigue, edema, loss of libido, weight gain, nausea, and depression are the most recognized side effects. Moreover, higher doses have been reported to induce hepatitis [25]. Cyproterone acetate can cause feminization of the male fetus. Therefore, its use should be accompanied by the use of appropriate OCPs.

Flutamide

Flutamide is a nonsteroidal androgen receptor blocker used primarily for prostate cancer. A low dose of the drug (62.5 mg per day) was found to improve FPHL in hyperandrogenic women and to be well tolerated at the same time. A dose-dependent risk of severe hepatotoxicity has made this medication less attractive in dermatology [40,41].

Finasteride

Finasteride works by inhibiting type II 5α-reductase. This lowers serum and scalp levels of DHT [37]. Although one double-blind controlled trial failed to show any statistical difference in efficacy in a group of postmenopausal women who used finasteride 1 mg for one year compared with a placebo, another study has shown the efficacy of oral daily dosage of finasteride 5 mg

in treating a group of normoandrogenic pre- and postmenopausal women with FPHL [42]. Moreover, finasteride 2.5 mg per day was effective in 62% of a group of premenopausal normoandrogenic females with FPHL while taking OCPs containing drospirenone and ethinylestradiol [31]. Finasteride 1.25 mg per day was also beneficial in stabilizing or improving FPHL in one case series of four premenopausal women with hyperandrogenism [43]. Therefore, considering the uncertain role of increased androgens versus excessive activity of 5α-reductase enzyme in the pathogenesis of FPHL as compared with AGA-M, there will be some variability in the proportion of female patients who would respond to finasteride and in the minimum effective dose that is needed to induce a clinical improvement. It seems that a higher dose of finasteride as compared with the regular 1–1.25 mg daily dose for men will be needed to manage FPHL [37]. Finasteride is in pregnancy category X. Therefore, its use is contraindicated in premenopausal women without appropriate contraceptive methods as it can cause feminization of a male fetus. Pregnant women are even instructed not to handle crushed or broken pills given that possible risk [37]. Finasteride is metabolized in the liver, but it does not affect the cytochrome P450-linked drug metabolizing enzymes. Thus, no clinically recognized drug interactions should be expected. A safer approach is to avoid using it for patients with known liver abnormalities [37]. Longer and better controlled trials may be needed to confirm finasteride's safety and efficacy in women.

Dutasteride

Dutasteride exerts a dual action by blocking both types of 5α-reductase enzyme, I and II. It is also one hundred times and three times as potent as finasteride at inhibiting type I and II 5α-reductase enzymes, respectively. Given these enhanced effects on 5α-reductase enzymes, it was found that dutasteride can reduce serum DHT level by 90% as compared with the 70% reduction rate seen with finasteride [37]. One case report showed that dutasteride 0.5 mg per day was effective at reversing FPHL in a woman who failed to respond to finasteride [37]. Dutasteride is not yet approved for use in pattern hair loss in either males or females.

Miscellaneous therapies

Hair transplantation

Hair transplantation surgery is considered to be an important option for females in whom medical therapy alone has failed or induced only a partial response. It is an outpatient procedure which is performed under local anesthesia. It classically involves dividing a harvested occipital strip into individual hair follicles in a process called follicular unit transplantation (FUT). The resulting individual hair follicles are then strategically placed in the recipient's sites, which are usually located in the frontal scalp [31].

Camouflage

A variety of camouflaging topical sprays, powders, or keratin fibers are available. Their use can be helpful for patients as they usually give a temporary denser look to the scalp hair by sticking to pre-existing hair fibers.

Cranial prostheses (wigs, hairpieces)

Different types of cranial prosthesis are available, ranging from off-the-shelf synthetic wigs to custom-made real hair wigs that are designed to cover the entire scalp. Localized areas of hair loss can be covered using specially designed hairpieces. Moreover, hair extensions are an alternative option to add more volume or length to the hair.

Hirsutism

Hirsutism is defined as excessive terminal hair growth in females growing in typical male distribution. It only affects the androgen-dependent hair follicles. Hirsutism should be differentiated from hypertrichosis in which there is excessive growth of androgen-independent vellus hair in nonsexual areas.

Hirsutism affects 5–10% of women of reproductive age [5]. The modified Ferriman–Gallwey scoring system is regarded as the standard for assessing severity of hirsutism in clinical investigations [8]. In this scoring system, each of nine areas most sensitive to androgens is assigned a score from 0 (no hair) to 4 (frankly virile), and then the sum of the scores provides a hirsutism score. Although no value is defined as a threshold to diagnose hirsutism, a score of 6–8 is probably an acceptable value [44]. Mild hirsutism is defined as a score of 8 to 15. Moderate to severe hirsutism is associated with a score of >15 [11] (Figure 2.2).

Androgens and mainly testosterone, as discussed earlier, play a major role in inducing hair follicles to grow thicker, longer, and darker hairs. This is invariably true in androgen-sensitive areas such as the face, chest, lower abdomen, pubis, and axillae. Moreover, hirsutism is regarded as the most common complaint associated with androgen excess, with most women who have at least twice the normal upper limit of androgens developing some degree of hirsutism [8]. Despite that, severity of hirsutism is usually determined by the subject's individual follicular androgen sensitivity rather than the actual androgen level [11].

Causes of hirsutism

Hirsutism has always been considered as a sign of androgen excess. Therefore, different possible causes of hyperandrogenism will be outlined here. Causes will be discussed under four headings: ovarian, adrenal, iatrogenic, and idiopathic.

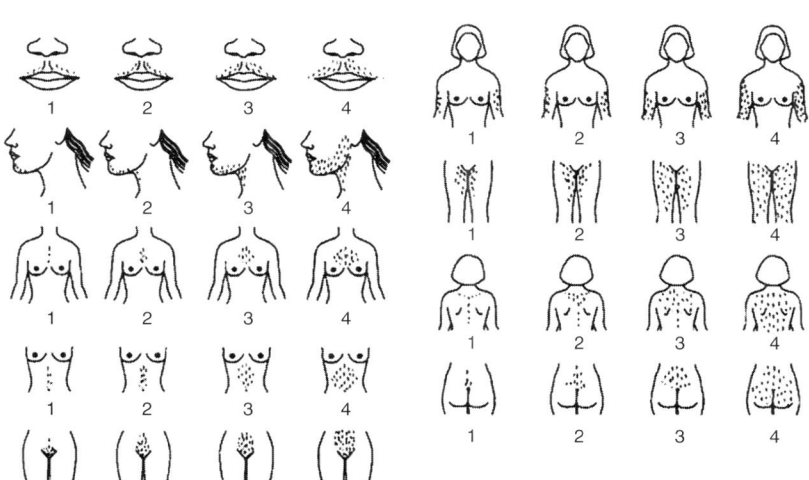

Figure 2.2 Modified Ferriman–Gallwey scoring system. Each one of the nine body areas most sensitive to androgens is assigned a score from 0 (no hair) to 4 (frankly virile). These scores are summed to give a hirsutism score. Reprinted with permission from R. Azziz (Yildiz BO, Bolour S, Woods K, Moore A, Azziz R. Visually scoring hirsutism. *Hum Reprod Update* 2010;16:51–64.). Copyright Oxford University Press, 2010.

Ovarian causes

Polycystic ovary syndrome (PCOS)

PCOS is regarded as the most common cause of hirsutism in woman presenting with signs of androgen excess, with a prevalence ranging between 57% and 82% [8,45,46]. Hirsutism secondary to PCOS usually starts around puberty and progressively and slowly worsens over time. Other signs of androgen excess can accompany the hirsutism, including irregular cycles, infertility, weight gain, acne, and hair loss in a pattern distribution. Associated insulin resistance can give rise to acanthosis nigricans.

Features of PCOS are believed to occur as a consequence of chronic anovulation caused by a wide variety of factors rather than being attributed to a specific endocrinological disorder [8]. Therefore, a history of ovulatory dysfunction (such as amenorrhea and oligomenorrhea) of pubertal onset is considered as the most prominent feature of PCOS. In fact, the history of cyclic predictable menses makes the diagnosis of PCOS unlikely. The Androgen Excess Society (AES) has suggested criteria for the diagnosis of PCOS. Clinical or biochemical signs of hyperandrogenism, such as hirsutism and/or hyperandrogenemia, evidence of ovarian dysfunction either clinically in a form of oligo-anovulation or radiological evidence of polycystic ovaries, and lastly the exclusion of other causes of androgen excess or related disorders were all included in the AES diagnostic criteria of PCOS [47].

Hyperthecosis

This refers to the hyperplasia of luteinized theca cells in the ovarian stroma. Apart from a greater degree of hyperandrogenism, the symptoms of hyperthecosis are very similar to those of PCOS [8]. Hyperthecosis is regarded as a severe variant of PCOS [47].

Ovarian neoplasms

Rapidly progressing hirsutism, which might be accompanied by other signs of virilization, should warrant searching for androgen-secreting neoplasms. Androgen-secreting tumors of ovaries and adrenals are the cause of androgen excess in 0.2% of cases [47]. While most androgen-secreting neoplasms arise from ovaries, less than 1% of ovarian neoplasms secrete androgens. Sertoli–Leydig and hilus cell tumors are the most common testosterone-secreting ovarian tumors. The former tumor occurs during the reproductive years, while the latter affects postmenopausal women. These ovarian tumors secrete large quantities of testosterone or its precursor androstenedione, resulting in a total testosterone values exceeding 200 ng/dL [35,47].

Adrenal causes

Congenital adrenal hyperplasia (CAH)

Females with classic CAH are usually recognized at birth with ambiguous genitalia. On the other hand, the nonclassic form (late onset) CAH presents usually at or after puberty with hirsutism and menstrual irregularities. CAH is responsible for 1–2% of hirsutism cases [45]. 21-Hydroxylase deficiency is the most common underlying cause in both types of CAH. The 21-hydroxylase enzyme is responsible for converting 17α-hydroxyprogesterone to 11-deoxycortisol. The resulting low glucocorticoid concentration stimulates adrenocorticotropic hormone (ACTH) production, which in turn causes adrenal hyperplasia and an increase in androgen production [5].

Cushing's syndrome

Elevated ACTH concentration can result from a pituitary gland tumor/hyperplasia (Cushing's disease) or from ectopic ACTH secretion by a non-pituitary tumor (Cushing's syndrome). However, exogenous administration of glucocorticoids remains the most common cause of Cushing's syndrome. Cushing's syndrome is a very rare cause of hirsutism [5].

Adrenal tumors

Adrenal tumors are a very rare cause of hirsutism. Testosterone, DHEA-S, DHEA, and cortisol can be secreted by a variety of adrenal carcinomas and adenomas [8]]

Iatrogenic hirsutism

Hirsutism can be induced by some medications, such as danazol, glucocorticoids, testosterone, and OCP, especially those containing androgenic progesterone. Hence, a thorough evaluation of drug history is very important in assessing hirsute patients [5].

Idiopathic hirsutism

Idiopathic hirsutism is a diagnosis to be made in a hirsute woman who has no clinical or biochemical evidence of androgen excess. It occurs in less than 20% of cases of hirsutism [5]. Increased peripheral 5α-reductase activity was proposed as the most acceptable explanation [5]. Therefore, even a normal androgen level can induce hirsutism in such patients.

Miscellaneous causes

Hyperprolactinemia

Elevated serum prolactin level can induce a state of hyperandrogenism either directly, by stimulating adrenals to produce DHEA-S which in turn undergoes peripheral conversion to testosterone and then to DHT, or indirectly, by inducing a state of chronic anovulation [8].

Evaluation

Successful evaluation of hirsute women starts with a detailed history. Onset and progression of the hirsutism are of great importance, as the condition caused by PCOS or CAH usually starts at puberty and progresses very gradually. A later age of onset or a rapid progression over a period of months favors the diagnosis of Cushing's or androgen-secreting neoplasms. Moreover, regularity of menses can give a clue to the most likely cause of hirsutism. PCOS and CAH patients typically report irregular menses starting at or around menarche. Although PCOS and CAH tend to have almost the same clinical presentation, hirsutism is more severe in CAH [8]. Other symptoms of androgen excess, such as acne and symptoms of virilization (clitoromegaly, deepening of voice, increase in muscle mass, increased libido, breast atrophy, and hair loss in a pattern distribution) should also be sought. Moreover, galactorrhea or symptoms of thyroid disease should be documented, given the causative role of these two factors in inducing a state of hyperandrogenism [5,45].

A familial predisposition of hirsutism, menstrual irregularities, infertility, and obesity can be suggestive of PCOS or occasionally late-onset CAH. Lastly, a detailed history of drug intake should be obtained to rule out iatrogenic causes of hirsutism and hypertrichosis. Whereas danazol, glucocorticoids, testosterone, and androgenic progesterone containing OCP can cause hirsutism, other non-androgenic medications, such as phenytoin, cyclosporine, and minoxidil, can result in the development of hypertrichosis [8].

Physical examination should include calculation of body mass index (BMI), as a high BMI can be associated with PCOS. Signs of virilization, such as breast atrophy, increased muscle mass, clitoromegaly, and hair loss in a pattern distribution, in addition to acne, are usually associated with androgen excess. The presence of acanthosis nigricans, which presents as brownish or grayish velvety discoloration of posterior neck or underarm skin, is highly suggestive of insulin resistance. On the other hand, Cushing's syndrome can be associated with remarkable striae, central obesity, a moon face, and buffalo hump. A palpable pelvic mass can be a sign of an androgen-secreting ovarian tumor.

Lastly, clinical implementation of the modified Ferriman–Gallwey scoring system may carry some limitations. First, it is unreliable in assessing hirsutism in ethnic group patients who might have sparse body hair, but may have other signs of hyperandrogenism [8]. Second, the scoring system does not take into account any hair that was removed by different depilating methods. Therefore, it is probably better to use other means of assessment, such as determining the method (e.g., plucking and waxing or shaving) and the frequency of hair removal, to provide a clinically relevant measure for assessing severity as well as response to the treatment. Assessment using these measures was found to correlate very well with the Ferriman–Gallwey scores [8].

Laboratory investigations

The aim of evaluating hirsute women is to identify any underlying endocrinological condition that requires specific treatment. However, the threshold for investigating patients with hirsutism should be guided by the severity of hirsutism, rate of progression, and presence of associated clinical signs and symptoms. Mild hirsutism (Ferriman–Gallwey score of 8–15), which is not associated with any other signs of virilization or anovulation (such as menstrual irregularities), should not be investigated, as most of these cases turn out to be an idiopathic hirsutism [11]. On the other hand, appropriate laboratory evaluation should be done for moderate to severe hirsutism (Ferriman–Gallwey score >15) or any hirsutism that has an onset at older age, is rapidly progressing, or is associated with other symptoms and signs of virilization [48].

The serum total testosterone is the only hormone that needs to be measured in every woman with hirsutism necessitating laboratory evaluation. It provides the best screening tool for overall androgen production. Although measurement of free serum testosterone concentration is more sensitive than measurement of total testosterone, given the lack of uniform laboratory standard and the excellent correlation between total and free testosterone, bioavailable testosterone can usually be predicted from the total testosterone level

[8,11]. A low level of SHBG, although not specific for a single disease, is usually inversely related to the free testosterone level [5].

DHEA-S concentration can be used to identify adrenal causes of hyperandrogenism. However, the evidence for measuring DHEA-S levels in every hirsute patient is conflicting for two reasons. First, the diagnosis of adrenal disorders, such as nonclassic CAH and Cushing's syndrome, usually requires other specific tests like performing ACTH hormone levels, basal and post corticotropin-releasing hormone injection, and low and high dexamethasone suppression tests for the former and 17α-hydroxyprogesterone for the latter. Second, moderately elevated levels of DHEA-S can be found in PCOS patients [8]. These two reasons, in addition to the fact that DHEA-S usually undergoes a peripheral conversion to testosterone, resulting in elevated serum testosterone concentrations, would limit the utility of this test to only evaluating hirsute women with highly suspected adrenal tumors.

Radiological evaluations of ovarian and adrenal tumors using transvaginal ultrasound and CT scan are warranted in patients with a testosterone level >150–200 ng/dL or DHEA-S >700 µg/dL in premenopausal women, or >100 ng/dL or >400 µ g/dL respectively in postmenopausal women [35].

Further evaluation by an endocrinologist is reasonable in patients with high values of testosterone, DHEA-S, prolactin, or with existing signs of Cushing's syndrome or nonclassic CAH.

Treatment

Cosmetic measures and medical therapies are considered together as integral parts of hirsutism treatment. Whereas the medical therapy takes care of the underlying androgen-related disturbances and reduces the chance of further progression of hirsutism, the cosmetic measures usually restore the feminine look and produce a quick, but not necessarily long-lasting improvement. Treatment of hirsutism should be offered to every hirsute woman, regardless of the condition's severity.

Medical therapies

Medical therapies consist mainly of estrogen-progestin OCPs and antiandrogens, including both classic androgen receptor blockers, such as spironolactone, cyproterone acetate, and flutamide and peripheral antiandrogens, such as finasteride and dutasteride. The miscellaneous group of therapies include insulin-sensitizing drugs and topical eflornithine cream.

Estrogen-progestin contraceptives

Combination OCPs are usually composed of ethinylestradiol, in doses ranging from 20 µg to 50 µg daily in addition to one progestational agent. They mainly exert their antiandrogenic effects by suppressing pituitary luteinizing hormone secretion, which eventually suppresses ovarian androgen production. Moreover, the high level of estrogen in the combination OCPs stimulates hepatic SHBG production, which in turn reduces the circulating bioavailable testosterone [8,11]. The overall end result is the reduction of circulating free testosterone. Other antiandrogenic effects of combination OCPs have been proposed, including suppression of adrenal DHEA-S production and inhibition of 5α-reductase activity in the skin [8,44].

Some of the progestational agents found in OCPs have antiandrogenic properties. Drospirenone, a derivative of spironolactone, is an example. However, it is unlikely that the very small usual dose of drospirenone (3 mg dose equivalent to approximately 25 mg of spironolactone) mixed with 30 µg ethinylestradiol in Yasmin, has any superior impact on hirsutism when compared to other contraceptives [8,48]. Cyproterone acetate is an androgen receptor blocker exerting antiandrogenic properties. Diane 35 contains 35 µg ethinylestradiol and 2 mg cyproterone acetate.

Other forms of combination hormonal contraceptives, including transdermal contraceptive patches and vaginal contraceptive rings, can induce a reduction in androgen levels comparable to that induced by oral OCPs, and hence, can be used to treat hirsutism. Progestin-only contraceptives, such as medroxyprogesterone acetate in both the 3-monthly intramuscular 150 mg dose (Depo) and the 10–20 mg oral daily dose, can also reduce the free testosterone level [8].

OCPs arrest or decrease new terminal hair growth and induce the pre-existing hairs to grow finer and more slowly, which will eventually reduce the need for shaving by 50% [11]. Moreover, 60–100% of women report improvement in their hirsutism during treatment with OCPs [8]. The beneficial effects are most obvious after six cycles and are continuous thereafter at a slower rate [5]. Despite all the treatments mentioned above, cosmetic measures to remove the hair are also usually needed, as OCPs will stop the progression of hirsutism for various reasons, but will not reverse it [11]. OCPs can also improve acne and seborrhea, symptoms that can accompany hirsutism in PCOS [8]. Although adding an antiandrogen to OCPs failed to show any statistically significant difference in changing hirsutism scores when compared to the use of OCPs alone, it is probably more practical to do so in selected non-responders to OCPs [5].

Spironolactone

The action and safety of spironolactone has been discussed earlier. It is a testosterone receptor blocker; therefore, it can be effective in treating hirsutism, including the idiopathic cases regardless of the cause of hyperandrogenism [11]. Spironolactone 100 mg per day was statistically superior to placebo or other antiandrogens in improving hirsutism [5]. A reduction in the Ferriman–Gallwey score by 15–40% can be expected within 6 months of starting spironolactone or cyproterone acetate [11]. However, subjective variations between women can be expected and 9–12 months of continuous use should be enough to induce the maximum benefit [39]. Use of combination OCPs with spironolactone is highly recommended to prevent pregnancy and normalize the menstrual irregularities that commonly accompany spironolactone use, in addition to the possible antiandrogenic synergistic effect.

Cyproterone acetate

The action and safety of cyproterone acetate as an androgen receptor blocker was discussed earlier. Cyproterone acetate is the progestin which, in a 2 mg dose, is usually combined with 50–35 µg of ethinylestradiol to produce the combination OCPs (Diane, Diane 35). This combination was more effective than placebo for reducing hair growth in hirsute women. Moreover, higher doses of cyproterone acetate added to the standard cyproterone acetate containing OCPs showed no additional benefit over the combination OCP alone [5,49].

Flutamide

Flutamide is an androgen receptor blocker that is used in doses of 250–750 mg per day to inhibit hair growth. It was no more effective than spironolactone in the treatment of hirsutism. Therefore, it is probably not a good option here, given its high chance of inducing fatal hepatitis [8,5].

Finasteride

Finasteride 5 mg per day can be prescribed for hirsutism. It was found to be as effective as cyproterone acetate containing combination OCPs, but on the other hand, it was less effective than spironolactone for treatment of hirsutism [5,49].

Insulin sensitizers

Insulin resistance is a common feature of PCOS, the condition regarded as the most common cause of hirsutism. Metformin and thiazolidinediones (rosiglitazone and pioglitazone) can reduce circulating insulin, leading to a lower androgen biosynthesis and a higher level of SHBG. Despite their ability to reduce the androgen levels, insulin sensitizers were not found to be beneficial when used alone for the treatment of hirsutism [11,49].

Eflornithine

Eflornithine is an irreversible inhibitor of ornithine decarboxylase, an enzyme that is needed for the synthesis of polyamines, the critical building blocks for hair. Eflornithine hydrochloride 13.9% cream is approved by the United States Food and Drug Administration as well as by many other countries for treatment of unwanted facial hair. The cream significantly reduces the rate of hair growth and improves the appearance of facial hair in 90% of patients after 24 weeks of twice daily use; however, the earliest results should be seen within 8 weeks of its use [49]. Moreover, combining the cream with photoepilation (laser hair reduction) resulted in a more rapid and complete reduction of unwanted facial hair when compared with using laser hair reduction alone [50]. Its relatively high price and the need for continuous use in order to maintain the good results are the only drawbacks of the drug.

Cosmetic measures

Various depilatory methods should be considered in combination with medical treatments for the management of hirsutism. They can provide a quick improvement in quality of life and help to alleviate the psychological distress caused by unwanted hair growth in women. Depilatory methods can be classified as: physical and chemical depilation, electroepilation, and photoepilation.

Physical and chemical depilation

Shaving and hair plucking along with waxing, sugaring, or electrical rotating depilation devices are considered the most commonly used epilating methods. They are simple, effective, quick, and not expensive. Skin irritation and folliculitis may be considered as the main disadvantages.

Chemical depilatory creams work by breaking disulfide bonds of hair keratins, leading to dissolution of the hair shaft. These chemicals can cause skin irritation. Therefore, their use should be restricted to legs, avoiding skin folds. Bleaching of hair using 12% hydrogen peroxide in ammonia-based preparations can be used to make the hair less visible. This approach is commonly used for facial and arm hair. Skin irritation can result.

Electroepilation

Electroepilation is aimed at destroying the hair follicle by heat produced by either a direct current or a high-frequency alternating current delivered by a special probe inserted into each individual hair follicle. Although this procedure carries potential disadvantages including pain, follicular hyperpigmentation, infections, and scarring, it is considered to be the only method that results in permanent hair follicle destruction [49].

Photoepilation

The physics behind laser hair reduction involves two principles. First, selective photothermolysis is based on targeting a specific part of the hair follicle without damaging the surrounding tissues. Second, the presence of melanin pigment in the hair follicles serves as a chromophore for a specific laser wavelength. Therefore, white and lightly pigmented hair should not be affected by laser hair reduction. Many different wavelength lasers including 755 nm (Alexandrite), 800 nm (diode), and 1064 nm (Nd:Yag) have been used. A single laser hair-reduction session can result in a 10–40% reduction in hair density [49]. The final results will depend on the number of sessions, fluence used, and intensity of hair color. Although Nd:Yag is less effective than Alexandrite or diode lasers, it is safer to use on darker skin [49].

Conclusion

In conclusion, recent advances in our understanding of the role of androgens in the pathogenesis of hair disorders in females, namely FPHL and hirsutism, have resulted in more accurate diagnoses and treatment of these disorders. Moreover, changes in hair growth patterns in women, especially when accompanied by other signs of irregular ovulation or virilization, should warrant the search for any underlying cause of androgen excess. However, a subset of FPHL and hirsutism can occur in women with no underlying androgen excess.

While antiandrogens are generally effective in treating most of the types of hirsutism, including the idiopathic type, their effectiveness in treating FPHL is less recognized in the medical literature. This can be explained by the lack of a standard female optimal dosage.

Acknowledgment

The authors acknowledge the valuable comments from Professor Kevin McElwee in the preparation of the manuscript.

References

1. Otberg N, Finner AM, Shapiro J. Androgenetic alopecia. *Endocrinol Metab Clin North Am* 2007; 36(2): 379–98.

2. Paus R, Cotsarelis G. The biology of hair follicles. *N Engl J Med* 1999; 341: 491–7.

3. Trüeb RM. Molecular mechanisms of androgenetic alopecia. *Exp Gerontol* 2002; 37: 981–90.

4. Randall VA. Androgens: the main regulator of human hair growth. In: Camacho FM, Randall VA, Price VH, eds. *Hair and Its Disorders: Biology, Pathology, and Management*. London: Martin Dunitz Ltd, 2000: 69–82.

5. Shah D, Patel S. Hirsutism. *Gynecol Endocrinol* 2009; 25(3): 140–8.

6. Marshall WA, Tanner JM. Variations in pattern of pubertal changes in girls. *Arch Dis Child* 1969; 44: 291–303.

7. Marshall WA, Tanner JM. Variations in the pattern of pubertal changes in boys. *Arch Dis Child* 1970; 45: 13–23.

8. Fritz MA, Speroff L. Hirsutism. In: *Clinical Gynecologic Endocrinology and Infertility*. Philadelphia: Lippincott Williams & Wilkins, 2011: 533–63.

9. Randall VA. Androgens and human hair growth. *Clin Endocrinol* 1994; 40: 439–57.

10. Hamilton JB. Male hormone stimulation is a prerequisite and an incitant in common baldness. *Am J Anat* 1942; 71: 451–80.

11. Rosenfield RL. Clinical practice: Hirsutism. *N Engl J Med* 2005; 353(24): 2578–88.

12. Messenger AG. The control of hair growth: an overview. *J Invest Dermatol* 1993; 101: 4S–9S.

13. Hanneken S, Ritzmann S, Nothen M, et al. Androgenetic alopecia: current aspects of a common phenotype. *Hautarzt* 2003; 54(8): 703–12.

14. Sawaya ME. Differences in the mechanisms of androgen action in hair follicles from women and men with androgenetic alopecia. In: Camacho FM, Randall VA, Price VH, eds. *Hair and Its Disorders: Biology,*

Pathology, and Management. London: Martin Dunitz Ltd, 2000: 153–8.

15. Kaufman KD. Androgen metabolism as it affects hair growth in androgenetic alopecia. *Dermatol Clin* 1996; 14: 697–711.

16. Bayne EK, Flanagan B, Einstein M, et al. Immunohistochemical localisation of types 1 and 2 5a reductase in human scalp. *Br J Dermatol* 1999; 141: 481–91.

17. Richards KN, Rashid RM. Problems in pattern alopecia. *J Cosmet Dermatol* 2012; 11(2): 131–3.

18. El-Samahy MH, Shaheen MA, Saddik DE, et al. Evaluation of androgen receptor gene as a candidate gene in female androgenetic alopecia. *Int J Dermatol* 2009; 48: 584–7.

19. Norwood OT. Incidence of female androgenetic alopecia (female pattern alopecia). *Dermatol Surg* 2001; 27: 53–4.

20. Birch MP, Messenger JF, Messenger AG. Hair density, hair diameter and the prevalence of female pattern hair loss. *Br J Dermatol* 2001; 144: 297–304.

21. O'Driscoll JB, Mamtora H, Higginson J, et al. A prospective study of the prevalence of clear-cut endocrine disorders and polycystic ovaries in 350 patients presenting with hirsutism or androgenic alopecia. *Clin Endocrinol (Oxf)* 1994; 41: 231–6.

22. Sawaya ME. Novel agents for the treatment of alopecia. *Semin Cutan Med Surg* 1998; 17: 276–83.

23. Curtois M, Loussouarn G, Horseau C, et al. Hair cycle and alopecia. *Skin Pharm* 1994; 7: 84–9.

24. Grino PB. Testosterone at high concentrations interacts with the human androgen receptor similarly to dihydrotestosterone. *Endocrinology* 1990; 126(2): 1165–72.

25. Sinclair RD, Dawber RP. Androgenetic alopecia in men and women. *Clin Dermatol* 2001; 19: 167–78.

26. Kuster W, Happle R. The inheritance of common baldness: two B or not two B. *J Am Acad Dermatol* 1984; 11: 921–6.

27. Norwood OT, Lehr B. Female androgenetic alopecia: a separate entity. *Dermatol Surg* 2000; 26: 679–82.

28. Olsen EA, Messenger AG, Shapiro J, et al. Evaluation and treatment of male and female pattern hair loss. *J Am Acad Dermatol* 2005; 52: 301–11.

29. Camacho-Martínez FM. Hair loss in women. *Semin Cutan Med Surg*. 2009; 28: 19–32.

30. Ludwig E. Classification of the types of androgenetic alopecia (common baldness) occurring in the female sex. *Br J Dermatol* 1977; 97: 247–54.

31. Shapiro J. Clinical practice: hair loss in women. *N Engl J Med* 2007; 357(16): 1620–30.

32. Whiting DA, Waldstreicher J, Sanchez M, et al. Measuring reversal of hair miniaturization in androgenetic alopecia by follicular counts in horizontal sections of serial scalp biopsies. Results of finasteride 1 mg treatment of men and postmenopausal women. *J Invest Dermatol Symp Proc* 1999; 4: 282–4.

33. Bikle D. Nonclassic actions of vitamin D. *J Clin Endocrinol Metab* 2009; 94(1): 26–34.

34. Carey AH, Chan KL, Short F, et al. Evidence for a single gene effect causing polycystic ovaries and male pattern baldness. *Clin Endocrinol* 1993; 38: 653–8.

35. Pittaway DE. Neoplastic causes of hyperandrogenism. *Infertil Reprod Med Clin North Am* 1991; 2: 479–94.

36. Kuttenn F, Couillin P, Girard F, et al. Late onset adrenal hyperplasia in hirsutism. *N Engl J Med* 1985; 313: 224–31.

37. Rogers NE, Avram MR. Medical treatments for male and female pattern hair loss. *J Am Acad Dermatol* 2008; 59(4): 547–66.

38. Lucky AW, Piacquadio DJ, Ditre CM, et al. A randomized, placebo-controlled trial of 5% and 2% topical minoxidil solutions in the treatment of female pattern hair loss. *J Am Acad Dermatol* 2004; 50(4): 541–53.

39. Burke BM, Cunliffe WJ. Oral spironolactone for female patients with acne, hirsutism or androgenetic alopecia. *Br J Dermatol* 1985; 112: 124–5.

40. Sinclair R, Wewerike M, Jolly D. Treatment of female pattern hair loss with oral antiandrogens. *Br J Dermatol* 2005; 152: 466–73.

41. Young R, Sinclair RD. Continuing medical education: Hirsutes part 2. *Australas J Dermatol* 1998; 39: 151–67.

42. Yeon JH, Jung JY, Choi JW, et al. 5 mg/day finasteride treatment for normoandrogenic Asian women with female pattern hair loss. *J Eur Acad Dermatol Venereol* 2011; 25: 211–14.

43. Shum KW, Cullen DR, Messenger AG. Hair loss in women with hyperandrogenism: four cases responding to finasteride. *J Am Acad Dermatol* 2002; 47: 733–9.

44. Escobar-Morreale HF, Lasuncion MA, Sancho J. Treatment of hirsutism with ethinyl estradiol-desogestrel contraceptive pills has beneficial effects on the lipid profile and improves insulin sensitivity. *Fertil Steril* 2000; 74: 816–19.

45. Nikolaous D, Gilling-Smith C. Hirsutism. *Curr Obstet Gynaecol* 2005; 15: 174–82.

46. O'Driscoll JB, Mamtora H, Higginson J, et al. A prospective study of the prevalence of clear-cut endocrine disorders and polycystic ovaries in 350 patients presenting with hirsutism or androgenic alopecia. *Clin Endocrinol (Oxf)* 1994; 41: 231–6.

47. Bulun SE. Physiology and pathology of the female reproductive axis. In: Melmed S, Polonsky KS, Larsen PR, et al, eds. *Williams Textbook of Endocrinology*. Philadelphia: Elsevier Saunders, 2011: 581–660.

48. Martin KA, Chang RJ, Ehrmann DA, et al. Evaluation and treatment of hirsutism in premenopausal women: An Endocrine Society clinical practice guideline. *J Clin Endocrinol Metab* 2008; 93: 1105–20.

49. Blume-Peytavi U, Hahn S. Medical treatment of hirsutism. *Dermatol Ther* 2008; 21(5): 329–39.

50. Hamzavi I, Tan E, Shapiro J, et al. A randomized bilateral vehicle-controlled study of eflornithine cream combined with laser treatment versus laser treatment alone for facial hirsutism in women. *J Am Acad Dermatol* 2007; 57: 54–9.

Polycystic ovary syndrome: managing the basics

Keith A. Hansen

Introduction

The less we know about something, the greater the need to constantly redefine it. (P. G. McDonough [1])

Polycystic ovary syndrome (PCOS) affects 6–10% of reproductive age females, making it the most common metabolic endocrine condition in this population [2]. PCOS was first officially described in 1935 by Drs. Stein and Leventhal, resulting in its eponymous name, the Stein–Leventhal syndrome. The syndrome they described were patients who presented with amenorrhea, hirsutism, polycystic-appearing ovaries, and often obesity. They described the operation known as a "wedge resection," which was often followed by regular menses and sometimes pregnancy. On pathological examination of tissue from the ovarian "wedge resection," they noted multiple subcapsular cysts associated with a thickening of the tunica [3].

Later, this constellation of signs and symptoms became known as PCOS, which reflected the multiple subcapsular follicular cysts. Since 1935, our understanding of this syndrome has expanded and in 2012 an evidence-based methodology workshop was sponsored by the National Institutes of Health. At this workshop one recommendation was to change the name from PCOS to more closely reflect the complex pathophysiology of this condition, which would improve accurate communication and progress in both the clinical and research arenas [4].

Nature vs. nurture

PCOS is a complex condition characterized by anovulation and androgen overproduction affecting reproductive age women. This phenotype can result in infertility due to ovulatory dysfunction, irregular menses, endometrial hyperplasia, endometrial carcinoma, iron deficiency anemia due to heavy menses, hirsutism, virilization, miscarriage, insulin resistance, glucose intolerance, gestational and type 2 diabetes mellitus, hyperlipidemia, and the potential for increased risk of cardiovascular disease. Other conditions associated with PCOS include obesity, sleep apnea, anxiety, depression, and possible reduction in quality of life.

Studies have demonstrated a strong genetic component, supported by family and twin studies, with an important environmental contribution. Recent studies, especially in the rhesus monkey and sheep, suggest that exposure to elevated androgen levels in utero may predispose a woman to the development of a polycystic ovary type of phenotype [5]. In humans, it is possible that exposure to excess androgen at any time during development prior to the onset of puberty could increase the risk of hyperandrogenic, chronic anovulation [6]. This reflects the developmental plasticity of the hypothalamic–pituitary ovarian and adrenal axes.

PCOS is often accompanied by the metabolic syndrome, with an estimated >50% of patients who are overweight or obese. With the current epidemic of obesity these numbers will most likely continue to increase. The metabolic syndrome is characterized by central obesity, insulin resistance, and increased risk of developing type 2 diabetes, dyslipidemia, and hypertension. In 1992, David J. Barker proposed the thrifty phenotype theory. This theory proposes that intrauterine growth restriction accompanied by reduced fetal growth places the individual at higher risk of obesity, hypertension, type 2 diabetes, and cardiovascular disease. This theory proposes that alterations in fetal nutritional supply result in limited fetal growth or that fetal glucocorticoid exposure as a stress response may play a critical role in the pathogenesis [7]. As the metabolic

Androgens in Gynecological Practice, ed. Leo Plouffe and Botros Rizk. Published by Cambridge University Press. © Cambridge University Press 2015.

syndrome often accompanies PCOS, it is possible that understanding the effects of fetal exposures, including adrenal androgens and glucocorticoids, may help unravel the enigmatic origin of PCOS.

Diagnosis

The diagnosis of PCOS has been imprecise and has varied over time, which has resulted in difficulty for research as well as clinical management of this condition. These issues have resulted in three consensus conferences to arrive at diagnostic criteria to improve diagnosis and communication. In 2003, the Rotterdam consensus conference published their criteria, which included two out of the following three: oligo- or anovulation, clinical or biochemical evidence of androgen excess, or polycystic appearing ovaries. "Polycystic ovaries" were defined as the presence of 12 or more follicles measuring 2–9 mm in diameter or a total ovarian volume of >10 mL. There are a couple of important caveats about the ultrasound appearance of polycystic ovaries. First, this definition only requires one ovary to meet the criteria. The patient should not be on oral contraceptive pills. If there is a dominant follicle or corpus luteum cyst, then the scan needs to be repeated in a future menstrual cycle, or if there is a cystic mass on the ovary further investigation is necessary. Additionally, the endocrine phenocopies of PCOS, such as androgen-producing ovarian or adrenal tumors, exogenous androgen administration, congenital adrenal hyperplasia especially late-onset 21-hydroxylase deficiency, hyperprolactinemia, thyroid dysfunction, Cushing's syndrome, or acromegaly, must be ruled out [8,9].

In 2006, the Androgen Excess-PCOS Society published criteria for the diagnosis of PCOS to highlight the importance of androgen excess and ovulatory dysfunction as central and critical to this syndrome. In this consensus definition, the patient must have evidence of androgen excess, either clinically or biochemically, accompanied by ovulatory dysfunction and/or polycystic ovaries, while excluding other androgen excess disorders [10]. The recent National Institutes of Health Evidence-Based Methodology Workshop on Polycystic Ovary recommended maintaining the Rotterdam 2003 diagnostic criteria for PCOS as well as explicitly identifying the PCOS phenotype: androgen excess with ovulatory dysfunction, androgen excess with polycystic ovary morphology, ovulatory dysfunction with polycystic ovary morphology, and androgen excess

with ovulatory dysfunction and polycystic-appearing ovaries [4]. The diagnosis of PCOS in adolescents is particularly challenging, with the suggestion that the diagnosis only be made when all three of the Rotterdam criteria are present: irregular menses or amenorrhea for at least 2 years after menarche, documented hyperandrogenemia, and that ultrasound should document increased ovarian volume >10 cm^3 [11].

Metabolic syndrome is characterized by central obesity, hypertension, dyslipidemia, insulin resistance, and a proinflammatory and prothrombotic state, which increases the risk for developing cardiovascular disease and type 2 diabetes mellitus. The National Cholesterol Education Program's Adult Treatment Panel III identified the following criteria, and required three of the five to make the diagnosis of metabolic syndrome in women: central obesity with a waist circumference of >88 cm or 35 inches, HDL cholesterol of <50 mg/dL, triglycerides of ≥150 mg/dL, blood pressure of ≥130/85 mmHg, and fasting glucose of ≥100 mg/dL [12].

Evaluation

The evaluation of patients for PCOS should be directed at confirming the diagnostic criteria are met, as well as ruling out PCOS phenocopies. The importance of a careful history and physical examination with supportive laboratory data needs to be emphasized. In the history and physical examination, one wants to evaluate for evidence of clinical hyperandrogenism, menstrual abnormalities, and any evidence of the metabolic syndrome (Table 3.1).

On physical examination, it is important to obtain height and weight, calculate the body mass index (BMI), measure the waist circumference, and obtain a resting blood pressure. The BMI is the mass in kilograms divided by the height in meters squared and is given in units of kg/m^2. Underweight is defined as a BMI <18.5, normal weight 18.5 to 25, overweight is 25 to 30, and obesity is >30 kg/m^2.

The skin examination can be supportive of a diagnosis of metabolic syndrome as well as PCOS by looking for evidence of hyperandrogenism, insulin resistance, and PCOS phenocopies, such as the stigmata of Cushing's syndrome. Signs and symptoms of hyperandrogenism include acne, hirsutism, and evidence of male pattern balding. The Ferriman–Gallwey score is a common method used to determine the presence and severity of excess hair growth in women. The modified Ferriman–Gallwey score determines the presence and

Table 3.1 Clinical diagnosis of PCOS: historical Information

Symptom	
Birth weight	Small for gestational age
	Large for gestational age: mother with gestational diabetes mellitus
Menarche	Premature pubarche
	Delayed puberty
Reproductive history	Infertility
	Previous pregnancies
	– ovulation induction
	– gestational diabetes mellitus
Menstrual irregularity	≤25 days or ≥35 days interval
	<10 menses per year
Clinical evidence of androgen excess	Hirsutism
	Acne
	Alopecia
	Virilizing signs
Polycystic ovaries on ultrasound	≥12 follicles 2 to 9 mm diameter
	Volume of 10 cm^3 or greater
	Only one ovary must meet criteria
	May not be on oral contraceptive pills
	If ovarian cyst: re-evaluate
Metabolic syndrome	Overweight/obese: body mass index
	Hypertension
	Insulin resistance
	– acanthosis nigricans
	– skin tags
	– glucose intolerance/type 2 diabetes
	Dyslipidemia
	Sleep apnea
Family history	PCOS
	Type 2 diabetes/metabolic syndrome
	Infertility
	Endometrial cancer/cancer syndrome
	Congenital adrenal hyperplasia
Medications	Androgenic medications
	Partner on androgens, especially gels
	History of glucocorticoid use
	Oral contraceptive pills or other hormones
Phenocopies	Galactorrhea
	Thyroid symptoms
	Vasomotor symptoms
	Symptoms of acromegaly and Cushing's syndrome
	Congenital adrenal hyperplasia

amount of hair growth in nine different areas including the upper lip, chin, chest, upper and lower back, upper and lower abdomen, upper arms, and thighs. Each area is given a score from 0, with no growth of terminal hair to 4, which is extensive growth of terminal hair. In Caucasian women a score ≥ 8 is consistent with excess androgen. For other ethnic groups one should use an ethnically appropriate standardized scale [13]. The presence and rapid progression of virilizing signs should alert one to the possibility of more severe androgen excess: androgen-producing ovarian and adrenal neoplasms, cases of more severe insulin resistance, such as hyperandrogenism (HA), insulin resistance (IR) and acanthosis nigricans (AN), or HAIR-AN syndrome, or exogenous androgen exposure as with inadvertent exposure to testosterone gel [14]. Virilization is defined by the presence of increased muscle mass, clitoromegaly, lowering of pitch of the voice, hirsutism, frontal balding, and ovulatory dysfunction.

The skin is also a good site for evidence of insulin resistance. Patients with insulin resistance may have skin tags, acanthosis nigricans, and varying degrees of lipodystrophy. Skin tags or acrochordons are benign lesions which may be sessile or pedunculated, smooth or irregular, and their color may vary from skin-color to dark. Acanthosis nigricans is a darkening, thickening of the skin usually in intertriginous zones like the nape of the neck, axilla, and medial thigh. Lipodystrophy is a metabolic abnormality in fat metabolism, which results in loss of subcutaneous fat and this can be a localized or more generalized phenomenon.

Tanner staging of the breasts can help determine whether there was normal puberty with adequate levels of estrogen to induce breast development. Tanner staging of pubic and axillary hair can also determine whether there is an adult pattern and extent of hair distribution. The pelvic examination can provide additional information, including evidence of clitoromegaly or an ovarian mass, which would suggest an androgen-producing ovarian neoplasm. The pelvic examination can give evidence of adequate estrogenization. In a female with PCOS and unopposed estrogen production, one would expect to discover normal external genitalia with normal vaginal rugae and thick stratified squamous, vaginal epithelium. In the presence of unopposed estrogen there will be abundant, clear cervical mucus with spinnbarkeit. Under progesterone's influence, the cervical mucus becomes scant, thick, and tacky. In the absence of estrogen, the vaginal walls shrink and become flat-surfaced and pale pink in color. In the presence of estrogen, vaginal pH is usually less than 4.5 because of glycogen metabolism by vaginal bacteria, while in the absence of estrogen the pH is usually >4.5 due to loss of this lactic acid production. A careful physical examination should be able to determine whether the patient has been exposed to estrogen during puberty (Tanner stages) and whether there is current estrogen production by the ovaries.

The physical examination must also evaluate for PCOS mimics: including thyroid dysfunction, galactorrhea, late-onset or nonclassic 21-hydroxylase deficiency (congenital adrenal hyperplasia), acromegaly, and Cushing's syndrome.

Laboratory evaluation

After a thorough history and physical examination, the physician will have developed a differential diagnosis, which helps to focus appropriate laboratory testing in the patient with suspected PCOS. Many of these patients will present with absent or irregular menses. The initial evaluation of the patient with absent or irregular menses includes a pregnancy test, thyroid function tests, and a serum prolactin. This is accompanied by evaluation of the patient's estrogen status. As noted above, physical examination can often give insight into whether estrogen is present or not. One laboratory test that can help determine estrogen status is the Vaginal Maturation Index. The Vaginal Maturation Index is performed by obtaining a scraping from the lateral vaginal wall, and microscopically determining the percentage of basal, para-basal, and superficial epithelial cells. Under the influence of estrogen, there will be normal maturation of the vagina and one will find more superficial cells, while in the absence of estrogen there will be primarily basal and para-basal cells.

The most common method for determining estrogen status is using the patient as her own in vivo biological assay, the progesterone challenge, in which progesterone is administered and she is observed for the onset of a withdrawal bleed. In the patient with chronic, hyperandrogenic anovulation (PCOS), the administration of progesterone is expected to induce a withdrawal bleed. If the patient does not have a withdrawal bleed, then the differential diagnosis includes primary ovarian insufficiency, müllerian agenesis and other absent or obstructive abnormalities of the uterus and outflow tract, hypogonadotropic hypogonadism, pregnancy, congenital adrenal hyperplasia with elevated progestin levels (17-hydroxyprogesterone), and occasionally with markedly elevated testosterone with its accompanying hypoestrogenism.

The presence of a progestin withdrawal bleed in this setting is consistent with normogonadotropic anovulation (WHO group II) [15]. In this case, further testing is aimed at discovering the etiology of this normogonadotropic, anovulatory state. Thyroid function tests are obtained in the initial part of the evaluation, because hypothyroidism can cause ovarian dysfunction, but rarely hyperandrogenism.

Prolactin is an important hormone in human reproduction and abnormally elevated levels can disturb ovulation, often inducing hypogonadotropic hypogonadism. There are a number of situations which can elevate prolactin levels, including stress, diet, chronic chest stimulation, drugs, and a prolactin-secreting, pituitary adenoma. As prolactin is often elevated by stress and diet it is important to draw the level in the fasting, relaxed state. In a patient discovered to have an elevated prolactin, repeating the blood draw in the fasting state, after 30 minutes of rest following placement of an intravenous heparin lock to allow for an atraumatic blood draw, will assist in determining whether the elevation is due to stress or diet. One must also remember the high-dose hook effect and macroprolactinemia as causes of spurious results. The high-dose hook effect can occur in a one-step, immunometric, sandwich assay when there is a large amount of analyte present in the unknown sample. In this situation there is such a large amount of analyte that one molecule binds to the labeled antibody and a different molecule binds to the solid phase antibody, which results in no sandwich forming and no signal. This high-dose hook effect can be detected by a two-step sandwich assay, in which there is a washing step prior to addition of the second antibody. In this case, one would detect saturation of the assay, where all of the initial antibodies are bound by analyte, then any free analyte is washed off, followed by signal due to formation of multiple sandwiches in the assay. The high-dose hook effect can also be detected by diluting the unknown specimen and noting that the results are not linear in relation to the dilution. Macroprolactinemia occurs when prolactin molecules are bound together by endogenous antibodies to form big prolactin and big-big prolactin. These macrocomplexes still have antigenic determinants, so they can be detected in the immunoassay, but have limited biological activity. Such macrocomplexes can be precipitated by addition of polyethylene glycol to the sample, and if >50% of the prolactin activity precipitates out, it is consistent with macroprolactinemia.

The most important circulating androgen is testosterone. The adrenal androgens, dehydroepiandrosterone sulfate (DHEA-S) and androstenedione, primarily function as pre-hormones, which have to undergo conversion to testosterone prior to biological activity. Testosterone is highly protein bound with only 1–2% in a free state and 60–70% tightly bound to the high-affinity sex hormone-binding globulin, and the rest loosely associated with albumin. In adult females, most testosterone assays can differentiate between the normal range and hyperandrogenism. However, these same assays are inaccurate in quantifying the degree of hyperandrogenism. In women, free testosterone levels, including the free androgen index (FAI), correlate well with the degree of hyperandrogenism. One should suspect an androgen-producing ovarian tumor in women with markedly elevated testosterone levels (defined as >2.5 to 3 standard deviations above the upper limit of the normal range or a concentration of total testosterone >200 ng/dL in a premenopausal woman), especially if present in women with new-onset, rapidly progressing hirsutism or virilization [16]. Most androgen-producing ovarian tumors are palpable on pelvic examination. Pelvic ultrasound remains the gold standard for imaging the ovaries for the presence of an ovarian neoplasm. In androgen-producing adrenal tumors, there are usually marked elevations in DHEA-S (8 µg/mL) and other adrenal androgen precursors [17].

There are some important points in evaluating the patient with hyperandrogenism. First, one needs to have a high index of suspicion for an androgen-producing neoplasm in the patient with new-onset, rapidly progressing hirsutism or virilization. Second, androgen-producing neoplasms may episodically secrete androgens, so multiple sampling times may be necessary to detect abnormally high androgens. Third, other precursor hormones may be secreted by the neoplasm and be reflected in elevated circulating levels. An example is androstenedione, which may be markedly elevated in androgen-producing neoplasms. Lastly, some androgen-producing tumors will present with lower levels of androgens, which is especially important to remember in evaluating the postmenopausal patient who presents with new-onset, rapidly progressing hirsutism or virilization. On the other hand, there are patients with PCOS and hyperthecosis with markedly elevated androgens, and no evidence of an androgen-producing neoplasm.

Conditions that can mimic PCOS besides androgen-producing neoplasms include late-onset congenital adrenal hyperplasia due to 21-hydroxylase deficiency, Cushing's syndrome, and acromegaly. Late-onset, or nonclassic, 21-hydroxylase deficiency is due to a partial deficiency in 21-hydroxylation resulting in hyperandrogenism in female patients, and can be tested for by drawing an early morning serum

17-hydroxyprogesterone level. A follicular phase, early morning, basal 17-hydroxyprogesterone level of <3 ng/mL is normal. If the 17-hydroxyprogesterone level is ≥ 3 ng/mL then further testing with an ACTH stimulation test is indicated. In the ACTH stimulation test, ACTH 250 μg is given intravenously and blood drawn for 17-hydroxyprogesterone levels at 0 and 60 minutes. These blood levels are then compared to nomograms developed by Dr. New to determine the existence and degree of 21-hydroxylase blockade [18]. A common cause of an elevated 17-hydroxyprogesterone level is the inadvertent drawing of the sample during the luteal phase, as this hormone is made by the corpus luteum.

Cushing's syndrome can be screened for with a 24-hour urine free cortisol or by administering an overnight dexamethasone suppression test. The 24-hour urine free cortisol is an excellent screening tool, but is sometimes difficult to obtain and may be incomplete. Determination of 24-hour urine creatinine can help determine whether the collection is a complete specimen. For the overnight dexamethasone suppression test, dexamethasone 1 mg is given at 11 p.m. with a serum cortisol drawn the next morning at 8 a.m. In a normal individual, one expects the a.m. cortisol level to be <1.8 μg/dL. An elevated 24-hour urine free cortisol and elevated cortisol after overnight dexamethasone suppression can be due to Cushing's or pseudo-Cushing's syndrome, which requires further differentiation. Pseudo-Cushing's syndrome is a heterogeneous group of conditions including obesity, depression, anorexia nervosa, alcohol withdrawal, and poorly controlled diabetes. One method that may discern between Cushing's and pseudo-Cushing's syndrome is the maintenance of diurnal cortisol's variation in the latter condition. In the patient with clinical signs and symptoms as well as supporting laboratory evidence for Cushing's syndrome, further testing is necessary to confirm and determine the source of excess cortisol [19].

Acromegaly is an often insidious, slowly progressing condition with common signs and symptoms making it difficult to diagnose until it has significantly progressed. Symptoms include a coarsening of facial features, enlarged hands and feet, excessive perspiration, muscle weakness often accompanied by fatigue, acrochordon, snoring and sleep apnea, deepening of the voice, headaches, visual changes, and irregular menstruation in women. These patients have an increased frequency of insulin resistance and type 2 diabetes mellitus [20]. Insulin-like growth factor-1 (IGF-1) is often used as a screening test for the presence of acromegaly [21]. In acromegaly, the IGF-1 level is often elevated, but it can also be raised in pregnancy. IGF-1 levels decline with age and in patients with poorly controlled disease of the liver and kidney, and diabetes. When a patient has an elevated IGF-1 level, a more provocative test is performed to confirm the diagnosis of acromegaly. The glucose tolerance test, in which 75–100 g of glucose is given to the patient, should suppress growth hormone secretion to less than 1 ng/mL in a healthy individual. In patients with acromegaly the growth hormone will not suppress during the oral glucose tolerance test [22].

In the evaluation of irregular menses, follicle-stimulating hormone (FSH) and luteinizing hormone (LH) levels will often be drawn. FSH levels are often drawn to rule out primary ovarian insufficiency. In PCOS, one may find an inverted LH:FSH ratio frequently being >2:1 or 3:1 in about 50% of patients. In the past, the inverted LH/FSH ratio was often used as one of the diagnostic markers of PCOS [23–25].

Anti-müllerian hormone (AMH) is a glycoprotein of the transforming growth factor beta (TGF-β) family made by the granulosa cells of the primary, secondary, preantral, and small antral follicles, which are still gonadotropin-releasing hormone (GnRH) independent. The concentration of AMH has been correlated with ovarian reserve. Recent studies suggest that elevated levels of AMH may be predictive of PCOS [26].

Pelvic ultrasound, preferably with a vaginal probe transducer, is important in hyperandrogenic patients to exclude ovarian neoplasms as well as determine the presence of polycystic appearing ovaries. The Rotterdam consensus conference defined a polycystic ovary as one having ≥12 follicles of 2 to 9 mm in diameter or an ovarian volume of >10 cm³. The ovarian volume is calculated using the formula for a prolate ellipse: volume = 0.5 × length × width × thickness. These standards require that only one ovary has to meet the criteria, that the patient should be scanned in the follicular phase, and not on oral contraceptive pills (OCPs). It is also important to use state-of-the-art equipment [27,28].

Should women with PCOS be screened with an oral glucose tolerance test and lipid panel? In women with PCOS, as many as 40% will have evidence of glucose intolerance and by the fourth decade 10% will have type 2 diabetes mellitus [29]. Some recommend that we should only screen those at high risk for developing diabetes: increased BMI, history of gestational diabetes, family history of diabetes, or membership in a high-risk ethnic group [30]. The oral glucose tolerance test is a 75 g glucose load, followed by blood draws in the fasting state and 2 hours after ingestion of the glucose. Normal levels are fasting plasma glucose less than 100 mg/dL and 2-hour glucose level below 140 mg/dL. Fasting glucose levels between 100 and 125 mg/dL are diagnosed as having impaired fasting glycemia, while 2-hour levels between 140 and 200 mg/dL are diagnosed as having impaired glucose tolerance. Diabetes is diagnosed when the fasting level is ≥126 mg/dL and 2-hour glucose levels are ≥200 mg/dL [31]. A fasting lipid panel can also be helpful in diagnosing the metabolic syndrome (HDL cholesterol of <50 mg/dL, triglycerides of ≥150 mg/dL) [11].

Women with PCOS and ovulatory dysfunction have a higher frequency of developing endometrial cancer. Endometrial cancer is the most common gynecologic cancer in the USA, with increasing frequency with advancing age, but it has recently been seen increasingly in younger women [30]. In those 13 to 18 years of age, the development of endometrial cancer is rare, but one should consider further evaluation with an endometrial biopsy in the patient with prolonged abnormal bleeding, obesity, no other etiology for the bleeding, and failed medical treatment. For those between the ages of 19 and 39, endometrial cancer is a relatively unusual diagnosis, but consider an endometrial biopsy if there has been a prolonged period of unopposed estrogen, failed medical therapy, or a family history of endometrial cancer or applicable cancer syndrome. If the endometrial biopsy is non-diagnostic, further testing with hysteroscopy and possible dilatation and curettage may be necessary. In women over the age of 40, especially those over age 45, an endometrial biopsy should be performed for abnormal uterine bleeding [32–34].

Sleep apnea is a common complaint in PCOS and does not appear to be explained solely by obesity, but may be more correlated with insulin resistance. Common signs and symptoms of sleep apnea include daytime sleepiness, long pauses in breathing, loud and chronic snoring, as well as choking or gasping during sleep. Other minor symptoms include waking with a headache, dry or sore throat, insomnia, waking up feeling short of breath, and moodiness, irritability or depression [35,36]. If the patient has symptoms suggestive of sleep apnea, then a formal sleep study with polysomnography can assist in making the diagnosis.

Recent studies have demonstrated an increased incidence of depression and anxiety in women with PCOS. Deeks et al. showed that in women with PCOS, 29% met criteria for depression, especially if infertile, and 57% met the criteria for anxiety. Of these women, few were diagnosed and treated for these conditions, highlighting the importance of screening for anxiety and depression [37]. To evaluate their psychological condition, the Beck Depression Inventory or the Hospital Anxiety and Depression Scale can be administered to patients with PCOS [38,39].

Treatment

As there is a large variability in the phenotype and the basic pathophysiology of PCOS remains enigmatic, our treatment is directed at short- and long-term signs and symptoms of the condition, and not at an underlying abnormality. Hence, as our understanding of the pathophysiology grows in this age of molecular discovery, we can expect to see more directed treatments, and more personalized therapy based on the individual's genome.

Ovulatory dysfunction: infertility, irregular menses, endometrial cancer

Common presenting complaints for women with PCOS include abnormal uterine bleeding and infertility. The treatment of ovulatory dysfunction in PCOS depends on the patient's fertility desires. In the PCOS patient with infertility, ovulation induction is often necessary to achieve her goal.

The first step in inducing ovulation in the PCOS patient is to evaluate for the metabolic syndrome, and if the patient is overweight or obese have them intensify weight management by increasing physical activity and diet. Previous studies have demonstrated that a 2–7% loss of body weight can reduce hyperandrogenemia and restore ovulatory cycles. The addition of metformin or other insulin-sensitizing agents can also lower insulin levels, decrease androgen levels, and improve ovulation [40,41].

Historically, the first ovulation induction arose from the pioneering work of Drs. Stein and Leventhal, who serendipitously discovered that many patients ovulated following an ovarian wedge resection [3]. In the 1960s, Greenblatt et al. introduced the world to a new medication termed MRL-41, which later became known as clomiphene citrate [42]. In the late 1950s and 1960s, a number of investigators studied ovulation

Table 3.2 Options for ovulation induction in clomiphene-resistant PCOS

Dexamethasone or prednisone with clomiphene	1. Elevated DHEA-S: continuously 2. Normal DHEA-S: continuously or 10-day course during ovulation induction

Oral medications	Aromatase inhibitors
	   1. Letrozole
	   2. Anastrozole
	SERM
	   1. Tamoxifen

Parenteral	Human menopausal gonadotropins GnRH pump
Surgery	Laparoscopy ovarian drilling Wedge resection

Metabolic syndrome	Metformin
	Thiazolidinediones
	   1. Pioglitazone
	   2. Rosiglitazone
	   3. Troglitazone: no longer available: liver toxicity
	Weight loss
	   1. Diet
	   2. Exercise
	   3. Bariatric surgery

induction with extracted human pituitary gonadotropins. Lunenfeld was able to extract gonadotropins from the urine of menopausal women and use this material (human menopausal gonadotropins) to induce ovulation [43]. In the 1970s, GnRH and dopamine agonists were added to available ovulation induction regimens.

Clomiphene citrate is a racemic mixture of two isomers, enclomiphene (trans-) and zuclomophene (cis-). Zuclomophene has the longer half-life and is the more potent of the two isomers. Clomiphene is a selective estrogen receptor modulator (SERM), which binds to the estrogen receptor with varying estrogenic effects depending on the endogenous endocrine milieu. In the patient with WHO group II anovulation, clomiphene citrate functions by binding to the estrogen receptor in the hypothalamic–pituitary axis and increasing gonadotropin secretion, which simulates ovulation. Clomiphene citrate is administered at a dose of 50–150 mg daily for 5 days, starting on days 3, 4, or 5 of a spontaneous menses or progestin-induced withdrawal bleed. The initial cycle starts at 50 mg per day for 5 days and increases by 50 mg a day per cycle until ovulation is achieved. Once ovulation is achieved, one may continue at the ovulatory dose for up to six cycles. In women with PCOS, 73% will ovulate in response to clomiphene citrate, but only 36% will become pregnant. There are a number of factors that may explain the discrepancy between ovulation and pregnancy rates, including antiestrogen effects of clomiphene citrate on the cervical mucus and endometrium or coexistent infertility issues, such as tubal disease or male factor. There are a number of potential risks with clomiphene citrate use, including multiple pregnancy, ovarian cysts, vasomotor symptoms, and occasionally visual changes. As clomiphene citrate is a SERM, it does have estrogenic activity and has been associated with deep venous thrombosis and central retinal vein occlusion. If a patient develops visual disturbances while on clomiphene citrate, it should be stopped with appropriate referral. If taken for greater than one year, clomiphene may increase the risk of ovarian cancer [44].

Approximately one-third of patients with PCOS are resistant to ovulation induction with clomiphene citrate. This resistance to ovulation induction with clomiphene citrate is associated with obesity, hyperandrogenism, and insulin resistance. There are a number of options available for ovulation induction in women resistant to clomiphene citrate (Table 3.2).

Letrozole and anastrozole are third generation aromatase inhibitors that can be used to induce ovulation in WHO group II patients. Letrozole has been studied more than anastrozole in terms of ovulation induction. The aromatase inhibitors work by inhibiting the conversion of androgens to estrogens, which lowers circulating estrogens, resulting in the release of gonadotropins by the pituitary gland so that folliculogenesis occurs. Because the aromatase inhibitors do not bind to the estrogen receptor, rising estradiol levels during folliculogenesis can stimulate production of cervical mucus and thicken the endometrial lining. Studies are very promising with letrozole, suggesting that this may replace clomiphene citrate as the first-line agent for ovulation induction in PCOS. Complications with letrozole include multiple gestation, ovarian cysts, and vasomotor symptoms. The optimal starting dose of letrozole has not been determined, but most start with 2.5 mg daily for 5 days starting on day 2, 3, or 4 of menses. The dose can be increased to 5 mg or to 7.5 mg a day for the 5 days to achieve ovulation. Concerns about possible birth defects resulted in a black box warning from the FDA about the use of letrozole during pregnancy. Subsequent studies have demonstrated letrozole's safety for ovulation induction, as it has a relatively short half-life (about 2 days), resulting in rapid clearance from the body following ovulation induction, prior to pregnancy [45].

Human menopausal gonadotropins (HMG) are reasonable options for patients who are clomiphene citrate resistant or do not conceive with clomiphene citrate. The largest risks in patients with PCOS undergoing ovulation induction with HMG are high-order multiple pregnancies and the ovarian hyperstimulation syndrome. A number of different protocols have been developed to try to minimize the risk of multiple pregnancy and ovarian hyperstimulation syndrome: clomiphene citrate followed by HMG, step-down protocols that lower the dose of HMG during the stimulation protocol, and step-up protocols that slowly increase the dose with a goal to achieve three or fewer follicles. During HMG ovulation induction, if the patient has multiple follicles, the cycle can be cancelled or the patient switched over to in vitro fertilization (IVF) [46]. IVF remains an option for patients with clomiphene citrate-resistant ovaries or those who do not conceive after ovulation with clomiphene citrate. One clear advantage of IVF over HMG is that one can control the number of embryos replaced with IVF and reduce the risk of high-order multiple pregnancy.

Stein and Leventhal described one of the first techniques, ovarian wedge resection, for inducing ovulation in women with PCOS. This procedure was complicated by adhesion formation with resultant reduced fertility. Advances in minimally invasive surgery have resulted in resurgence of interest in surgical treatment of PCOS. Following laparoscopic ovarian drilling, regular menses and ovulation occurred in about 80% of subjects, and some maintain ovarian function over a long period of time [47]. A recent prospective trial comparing metformin with laparoscopic ovarian drilling demonstrated that metformin was more helpful in terms of reproductive and overall health [48].

The chronic anovulation of PCOS increases the risk of developing endometrial hyperplasia or carcinoma. An endometrial biopsy should be considered in women with abnormal uterine bleeding who are older (>40), with a long history of irregular menses, >1 year of amenorrhea, or unresponsive to medical therapy. Endometrial thickness on ultrasound may assist in determining the need for a biopsy. Endometrial hyperplasia and cancer can be prevented by the use of progestins: cyclically or continuously, as well as with the return of regular ovulation with improvement in the metabolic syndrome.

Skin: hirsutism and acne

In PCOS, the excess production of androgens can cause hirsutism, worsening acne, male pattern balding, and virilization. The treatment for hirsutism involves suppressing androgen production, blocking androgen action, and then either temporally or permanently removing unwanted hair.

OCPs remain a mainstay in the treatment of hirsutism due to the hyperandrogenism of PCOS. The estrogenic component of the OCP suppresses FSH, while the progestin suppresses LH. By this combined suppression of LH and FSH, androgen production and folliculogenesis are inhibited. The estrogenic component of the OCP also increases hepatic production of binding proteins, specifically sex hormone-binding globulin (SHBG). The combined effect of reducing androgen production and increasing SHBG reduces free testosterone, which reduces the amount of testosterone that is available to bind to the androgen receptor. The progestins in the OCP are important, because they can have variable androgenic activity. The progestins norgestimate and desogestrel have very little androgenic activity, while drospirenone has some antiandrogen activity. For PCOS, OCPs are a good therapy for the signs and symptoms of hyperandrogenism. OCPs also prevent unopposed estrogen stimulation of the endometrium, preventing the development of endometrial hyperplasia and cancer [49]. One important relative contraindication for the use of OCPs, which is increased in the metabolic syndrome, is hypertriglyceridemia. Estrogen can increase triglyceride levels, so for those with triglyceride levels >250 mg/dL it is important to follow their lipid panels monthly for a few months until reassured that they are stable. If the patient has a triglyceride level >750 mg/dL, a coexistent history of cardiovascular disease, or other risk factors like smoking, then OCPs should not be prescribed.

There are a number of antiandrogens which function by blocking the action of testosterone, usually by interfering with its ability to bind to the androgen receptor. Spironolactone is an aldosterone antagonist that inhibits androgen binding to the androgen receptor and also weakly inhibits testosterone synthesis. Spironolactone is a moderately effective antiandrogen, which is often used at doses between 100 and 200 mg per day.

Flutamide is a nonsteroidal, potent androgen receptor antagonist which is effectively used in the treatment of hirsutism in women. Because of its association with hepatotoxicity, it is not often used to treat hirsutism. Cyproterone acetate is a progestin with antiandrogen activity that can be used to treat hirsutism and acne. Cyproterone exerts its antiandrogen action by being a competitive antagonist for the androgen receptor and partially blocking enzymes in androgen biosynthesis. Because it is a progestin, cyproterone can be used as a component of OCPs, termed Dianette in the United Kingdom, Diane-35 in Canada, and Dixi-35 in Chile. Cyproterone acetate is not approved for use in the USA [50].

Finasteride is a competitive antagonist of the 5α-reductase enzyme, primarily the type II enzyme. Finasteride blocks the conversion of testosterone to dihydrotestosterone. It was approved for use in the male to treat benign prostatic hypertrophy and male pattern baldness. It has also been discovered to be effective in the treatment of hirsutism in women [51].

When using an antiandrogen like spironolactone, flutamide, or finasteride, it is important to make sure the patient is using adequate contraception, as these medications have been associated with birth defects. Because of their effects on androgens, OCPs are an excellent contraceptive choice to combine with antiandrogens. Antiandrogens can also induce irregular menses, which OCPs will regulate.

In patients with severe hyperandrogenism, such as seen with hyperthecosis or in those who do not respond to typical therapies, GnRH agonists can be used effectively. The GnRH agonists function by inducing a hypogonadotropic, hypogonadal state with resultant decrease in androgen levels. One of the adverse effects of the GnRH agonists is that they induce a hypoestrogenic state, which may be accompanied by vasomotor symptoms, vaginal dryness, and bone loss. The addition of add-back therapy with estrogen and progestin can reduce these adverse effects of the GnRH agonists. The addition of estrogen will stimulate SHBG production and further reduce free testosterone, which has been found to have minimal clinical response compared with GnRH alone.

When treating hirsutism, it is important to inform the patient that long-term treatment is necessary to see significant changes. The therapy should be used for a minimum of 6 months before evaluating its effectiveness. It is also important to instruct the patient that terminal hairs will become lighter, thinner and will grow more slowly but will not completely stop growing or disappear.

There are a number of cosmetic therapies which can be used in conjunction with medical therapy to improve results. Eflornithine hydrochloride is an inhibitor of ornithine decarboxylase, which is an enzyme important for hair growth. Eflornithine hydrochloride is a cream that is applied locally twice a day and reduces facial hair growth. If it is discontinued the hair growth will resume and by 8 weeks returns to pre-treatment levels.

Temporary methods used to remove hair include plucking, shaving, waxing, and use of depilatory agents. One method of measuring response to antiandrogen therapy is to have the patient keep track of the frequency of needing to use a temporary method of hair removal. More permanent methods of removing hair include electrolysis and laser hair removal. These methods work best when combined with medical therapy, which not only slows the growth of current terminal hairs, but helps prevent conversion of existing vellus hairs into terminal hairs.

Metabolic syndrome

Around 60–70% of women with PCOS are overweight or obese and this percentage continues to rise with the increasing obesity epidemic. The primary interventions should be increased weight management by increasing physical activity and appropriate diet. The addition of metformin has been demonstrated to decrease development of the metabolic syndrome [52]. The Diabetes Prevention Trial has shown that exercise, diet, and metformin are effective in decreasing development of diabetes in subjects with glucose intolerance. Although bariatric surgery has been suggested as an alternative treatment for those with metabolic syndrome and PCOS, it should be considered only after other therapies have been exhausted due to the risks of the surgical procedure [53].

In patients with metabolic syndrome, it is important to monitor for development of hypertension, dyslipidemia, glucose intolerance, type 2 diabetes, and cardiovascular disease. Hopefully, with aggressive weight management through diet and exercise, the development of the metabolic syndrome and its complications can be prevented. It is reasonable to consider having these patients take 81 mg of aspirin daily. If they develop evidence of hypertension, then appropriate antihypertensive medications should be initiated to maintain normal blood pressure. If the patient develops dyslipidemia, then appropriate addition of lipid-lowering medications is indicated and if they develop diabetes, then appropriate oral hypoglycemics or insulin should be used.

Conclusion

PCOS is a complex condition characterized by hyperandrogenic anovulation, resulting in a number of characteristic signs and symptoms. A careful history and physical examination with appropriate laboratory testing is necessary to confirm the diagnosis of PCOS and rule out PCOS mimics. Women with PCOS deserve careful evaluation and management of both the short- and long-term consequences of the condition. Continued research on PCOS will help determine the etiology and arrive at better treatments for the most common metabolic condition in women.

References

1. McDonough PG. How many of the items in the polycystic ovary syndrome can be validated statistically? *Fertil Steril* 2006;85:530–1.

2. Azziz R, Woods KS, Reyna R, et al. The prevalence and features of the polycystic ovary syndrome in an unselected population. *JCEM* 2004;89:2745–9.

3. Stein LF, Leventhal ML. Amenorrhea associated with bilateral polycystic ovaries. *Am J Obstet Gynecol* 1935;77:826.

4. National Institutes of Health Evidence-based Methodology Workshop on Polycystic Ovary Syndrome, December 2012; pages 1–14 at http://prevention.nih.gov/workshops/2012/pcos/.

5. Eisner JR, Barnett MA, Dumesic DA, Abbott DH. Ovarian hyperandrogenism is adult female rhesus monkeys exposed to prenatal androgen excess. *Fertil Steril* 2002;77:167–72.

6. Franks S, McCarthy MI, Hardy K. Development of polycystic ovary syndrome: involvement of genetic and environmental factors. *Int J Androl* 2006;29:278–85.

7. De Boo HA, Harding JE. The developmental origins of adult disease (Barker) hypothesis. *Aust N Z J Obstet Gynaecol* 2006;46:4–14.

8. The Rotterdam ESHRE/ASRM-sponsored PCOS consensus workshop group. Revised 2003 consensus on diagnostic criteria and long-term health risks related to polycystic ovary syndrome (PCOS). *Hum Reprod* 2004;19:41–7.

9. Kaltsas GA, Mukherjee JJ, Jenkins PJ, et al. Menstrual irregularity in women with acromegaly. *J Clin Endocrinol Metab* 1999;84:2731–5.

10. Azziz R, Carmina E, Dewailly D, et al. The Androgen Excess and PCOS Society criteria for the polycystic ovary syndrome: the complete task force report. *Fertil Steril* 2009;91:456–88.

11. Wiksten-Almstromer M, Hirschberg AL, Hagenfeldt K. Prospective follow-up of menstrual disorders in adolescence and prognostic factors. *Acta Obstet Gynecol Scand* 2008;87:1162–8.

12. Third report of the National Cholesterol Education Program (NCEP) expert panel on detection, evaluation, and treatment of high blood cholesterol in adults (Adult Treatment Panel III). Final report. *Circulation* 2002;10:3143–421.

13. Ferriman D, Gallwey JD. Clinical assessment of body hair growth in women. *J Clin Endocrinol* 1961;21:1440–7.

14. De Ronde W. Hyperandrogenism after transfer of topical testosterone gel: case report and review of published and unpublished studies. *Hum Reprod* 2009;24:425–8.

15. Dhont M. WHO-classification of anovulation: background, evidence and problems. *Gynaecology, Obstetrics, and Reproductive Medicine in Daily Practice: International Congress Series* 2005;1279:3–9.

16. Friedman CI, Schmidt GE, Kim MH, et al: Serum testosterone concentrations in the evaluation of androgen-producing tumors. *Am J Obstet Gynecol* 1985;153:44–9.

17. Derksen J, Nagesser SK, Meinders AE, et al: Identification of virilizing adrenal tumors in hirsute women. *N Engl J Med* 1994;331:968–73.

18. New MI. Extensive clinical experience: nonclassical 21-hydroxylase deficiency. *J Clin Endocrinol Metab* 2006;91:4205–14.

19. Papanicolaou DA, Yanovski JA, Cutler GB, et al. A single midnight serum cortisol measurement distinguishes Cushing's syndrome from Pseudo-Cushing states. *J Clin Endocrinol Metab* 1998;83:1163–7.

20. American Association of Clinical Endocrinologists Medical Guidelines for Clinical Practice for the Diagnosis and Treatment of Acromegaly. *Endocr Pract* 2004;10(3):213–25.

21. Melmed S. Medical progress: acromegaly. *N Engl J Med* 2006;355:2558–73.

22. Freda PU, Post KD, Powell JS, Wardlaw SL. Evaluation of disease status with sensitive measures of GH secretion in 60 postoperative patients with acromegaly. *J Clin Endocrinol Metab* 1998;83:3808–16.

23. Carmina E, Rosato F, Jannì A, et al. Relative prevalence of different androgen excess disorders in 950 women referred because of clinical hyperandrogenism. *J Clin Endocrinol Metab* 2006;91:2–6.

24. Lobo RA, Carmina E. The importance of diagnosing the polycystic ovary syndrome. *Ann Intern Med* 2000;132:989–93.

25. Adams J, Polson DW, Franks S. Prevalence of polycystic ovaries in women with anovulation and idiopathic hirsutism. *Br Med J (Clin Res Ed)* 1986;293:355–9.

26. Pigny P, Merlen E, Robert Y, et al. Elevated serum level of anti-mullerian hormone in patients with polycystic ovary syndrome: relationship to the ovarian follicle excess and to the follicular arrest. *J Clin Endocrinol Metab* 2003;88:5957–62.

27. Jonard S, Robert Y, Cortet-Rudelli C, et al. Ultrasound examination of polycystic ovaries: is it worth counting the follicles? *Hum Reprod* 2003;18:598–603.

28. Jonard S, Robert Y, Dewailly D. Revisiting the ovarian volume as a diagnostic criterion for polycystic ovaries. *Hum Reprod* 2005;20:2893–8.

29. Ehrmann DA, Barnes RB, Rosenfield RL, et al. Prevalence of impaired glucose tolerance and diabetes in women with polycystic ovary syndrome. *Diabetes Care* 1999;22:141–6.

30. Ehrmann DA. Polycystic ovary syndrome. Medical progress. *N Engl J Med* 2005;352:1223–36.

31. Genuth S, Alberti KG, Bennett P, et al. Follow-up report on the diagnosis of diabetes mellitus. The Expert Committee on the Diagnosis and Classification of Diabetes Mellitus. *Diabetes Care* 2003;26:3160–7.

32. Management of abnormal uterine bleeding associated with ovulatory dysfunction. Practice Bulletin No. 136. American College of Obstetricians and Gynecologists. *Obstet Gynecol* 2013;122:176–85.

33. Haidopoulos D, Simou M, Akrivos N, et al. Risk factors in women 40 years of age and younger with endometrial carcinoma. *Acta Obstet Gynecol Scand* 2010;89:1326–30.

34. Pellerin GP, Finan MA. Endometrial cancer in women 45 years of age or younger: a clinicopathological analysis. *Am J Obstet Gynecol* 2005;193:1640–4.

35. Vgontzas AN, Legro RS, Bixler EO, et al. Polycystic ovary syndrome is associated with obstructive sleep apnea and daytime sleepiness: role of insulin resistance. *J Clin Endocrinol Metab* 2001;86:517–20.

36. Fogel RB, Malhotra A, Pillar G, et al. Increased prevalence of obstructive sleep apnea syndrome in obese women with polycystic ovary syndrome. *J Clin Endocrinol Metab* 2001;86:1175–80.

37. Deeks AA, Gibson-Helm ME, Teede HJ. Anxiety and depression in polycystic ovary syndrome: a comprehensive investigation. *Fertil Steril* 2010;93:2421–3.

38. Beck AT, Steer RA, Ball R, Ranieri WF. Comparison of Beck Depression Inventories-IA and –II in psychiatric outpatients. *J Pers Assess* 1996;67:588–97.

39. Zigmond A, Snaith R. The Hospital Anxiety and Depression Scale. *Acta Psychiatr Scand* 1983;67:361–70.

40. Huber-Buchholz MM, Carey DG, Norma RJ. Restoration of reproductive potential by lifestyle modification in obese polycystic ovary syndrome: role of insulin sensitivity and luteinizing hormone. *J Clin Endocrinol Metab* 1999;84:1470–4.

41. Nestler JE, Jakubowicz DJ, Evans WS, Pasquali R. Effects of metformin on spontaneous and clomiphene-induced ovulation in the polycystic ovary syndrome. *N Engl J Med* 1998;338:1876–80.

42. Greenblatt RB, Barfield WE, Jungck EC, Ray AW. Induction of ovulation with MRL. *JAMA* 1960;178:101–5.

43. Lunenfeld B. Treatment of anovulation by human gonadotropins. *Int J Obstet Gynecol* 1963;1:153–7.

44. Homburg R. Clomiphene citrate-end of an era? A mini-review. *Hum Reprod* 2005;20:2043–51.

45. Tulandi T, Martin J, Al-Fadhli R, et al. Congenital malformations among 911 newborns conceived after infertility treatment with letrozole or clomiphene citrate. *Fertil Steril* 2006;85:1761–5.

46. Fauser BC, van Heusden AM. Manipulation of human ovarian function: physiological concepts and clinical consequences. *Endocr Rev* 1997;18:71–106.

47. Gjonnaess H. Polycystic ovarian syndrome treated by ovarian electrocautery through the laparoscope. *Fertil Steril* 1984;41:20–5.

48. Palomba S, Orio F, Jr, Nardo LG, et al. Metformin administration versus laparoscopic ovarian diathermy in clomiphene-resistant women with polycystic ovarian syndrome: a prospective parallel randomized double-blind placebo-controlled trial. *J Clin Endocrinol Metab* 2004;89:4801–9.

49. Diamanti-Kandarakis E, Baillargeon J-P, Iuorno MJ, et al. A modern medical quandary: polycystic ovary syndrome, insulin resistance, and oral contraceptive pills. *J Clin Endocrinol Metab* 2003;88:1927–33.

50. Moghetti P, Tosi F, Tosti A, et al. Comparison of spironolactone, flutamide, and finasteride efficacy in the treatment of hirsutism: a randomized, double blind, placebo-controlled trial. *J Clin Endocrinol Metab* 2000;85:89–94.

51. Venturoli S, Marescalchi O, Colombo FM, et al. A prospective randomized trial comparing low dose flutamide, finasteride, ketoconazole, and cyproterone acetate-estrogen regimens in the

treatment of hirsutism. *J Clin Endocrinol Metab*, 1999;84:1304–10.

52. Orchard TJ, Temprosa M, Goldberg R, et al. Diabetes Prevention Program Research Group. The effect of metformin and intensive lifestyle intervention on the metabolic syndrome: the Diabetes Prevention

Program randomized trial. *Ann Intern Med* 2005;142:611–19.

53. Malik SM, Traub ML. Defining the role of bariatric surgery in polycystic ovarian syndrome patients. *World J Diabetes* 2012;3:71–9.

Chapter

4

Sexual dysfunction, including hypoactive sexual desire disorder: diagnosis and treatment recommendations

Ricki Pollycove and James A. Simon

Female sexual dysfunction and testosterone therapy: essentials for gynecologists

Throughout the reproductive life cycle, women experience varying degrees of sexual satisfaction as well as dysfunction. According to the Diagnostic and Statistical Manual of Mental Disorders (DSM-IV TR), there are six female sexual disorders: hypoactive sexual desire disorder, aversion disorder, sexual arousal disorder, orgasmic disorder, vaginismus, and dyspareunia [1]. The National Health and Social Life Survey found that 43% of American women experience lack of interest in sex, anorgasmia, and pain or decreased pleasure with sexual activity [2].

Female sexual dysfunction (FSD) is also a prevalent global problem for women aged 40–80, with decreased libido and inability to achieve orgasm the most common issue, 26–43% and 18–41% respectively in these age groups [3]. The largest US survey of FSD, the Prevalence of Female Sexual Problems Associated with Distress and Determinants of Treatment Seeking (PRESIDE), which included over 30 000 women who responded to standardized questionnaires, documented a 21% prevalence of orgasmic dysfunction [4].

Specific disease states and comorbid conditions are associated with higher rates of FSD. These include renal failure with chronic dialysis and diabetes [5]. Androgen insufficiency may play a role in the observed increased incidence of FSD in these women as well as those considered to be in good health [4]. Although there is no clear relationship between sexual function and androgen levels, a significant percentage of otherwise healthy women with FSD have lower testosterone levels and will experience benefit with supplementation

[6]. Currently available evidence suggests that women with significant depletion of endogenous androgens due to oophorectomy, ovarian and/or adrenal insufficiency, and clinical symptoms of impaired well-being or distressingly low libido are the most suitable candidates for androgen replacement.

Debates are not resolved regarding testosterone replacement in women rendered androgen deficient due to oophorectomy and iatrogenic medication effects, such as contraceptive hormone suppression of ovarian sex steroidogenesis. A significant issue in the controversy regarding testosterone therapy revolves around safety, as without an accurate assessment of safety, the risk/benefit ratio of androgen treatments is difficult to determine. Short-term studies, up to 2 years, where serum plasma testosterone levels are at the upper portion or slightly above the reference range for reproductive aged women, confirm that testosterone does not increase the risk of hepatotoxicity, endometrial hyperplasia, or behavioral hostility. No adverse cardiovascular effects, including changes in blood pressure, blood viscosity, arterial vascular reactivity, hypercoagulable states, and polycythemia, have been shown. Longer-term safety studies will provide conclusive evidence as to testosterone safety in women [7].

The gynecologist is frequently called upon to be the primary source of patient assessment as well as therapeutics when such issues exist. The etiology of FSD may relate to complex non-endocrine factors such as depression, anxiety, fatigue (both psychological and physical), relationship conflicts, past sexual abuse, lack of privacy, medication side effects, other medical diseases, physical disabilities, and body image (i.e., obesity has a strong association with FSD). The list of medical conditions associated with a higher observed incidence of FSD is long:

Androgens in Gynecological Practice, ed. Leo Plouffe and Botros Rizk. Published by Cambridge University Press. © Cambridge University Press 2015.

Table 4.1

Comorbid conditions associated with sexual dysfunction

Diabetes	Infection	Stress
Menopause	Endometriosis	Psychological issues
Arthritis	Surgery	History of sexual abuse
Incontinence	Chemotherapy	Relationship issues
Cancer	Radiation therapy	Gender identity conflicts
Fatigue	Heart, lung, liver, kidney, or thyroid disease	Medications
Alcohol or drug abuse	Depression	

While the gynecologist often makes the primary intervention, a collaborative care approach with a mental health practitioner and practitioners from other disciplines can be of significant value in addressing many of these concomitant comorbidities. For over 20 years female androgen insufficiency has been implicated in a significant percentage of female sexual dysfunctions. The "new tool in the tool box" for gynecologists is testosterone or its precursors. The purpose of this chapter is to review the basics of normal and abnormal female sexual function as it relates to androgens, identify specific types of FSD that may be treated successfully with testosterone, and provide guidelines for monitoring therapy.

Eliciting the history

Discussing sexual issues with a clinician is neither easy nor natural for most patients. Recent data suggest that clinicians of all specialties are not talking about sex with their patients. In a recent poll of adult women in the United States, only 9% of those aged 40 to 80 were asked about sexual concerns by their primary care clinician in the last 3 years, and a paltry 22% by their gynecologists [8].

When sexual dysfunction is present, patients rarely broach the topic with their clinicians. Reasons for this include embarrassment, fear that her concerns would be dismissed, anticipation that the clinician would be uncomfortable, and fear that there will be no medical treatment available for women. Since the widespread use of Viagra® (PDE5 inhibitors, in general), the absence of FDA-approved medical therapies for women further frustrates patients and clinicians alike.

Eliciting a sexual history is part of comprehensive gynecological care.

In the current practice environment, constraints placed on the amount of time that can be spent with each patient, coupled with psychosocial discomforts, make it unlikely that women will have their sexual dysfunction addressed during a standard medical encounter, even with their gynecologist. Current research implies most patients are currently treated by psychotherapists, even when significant contributions to FSD may be of medical/hormonal etiology and within the purview of the gynecologist. Clinicians are in a position to help fill this gap in the total care of the patient by educating themselves, recognizing opportunities to diagnose and treat FSD, becoming better informed, and becoming comfortable with the subject and initiating the conversation with patients. Waiting room or website questionnaires can be of great value in eliciting patient complaints efficiently and with fewer stigmas attached to broaching the topic [9].

Definition of FSD

The normal female sexual experience is commonly divided into one of three phases: desire or libido, excitement or arousal, and orgasm. A fourth aspect of this sequence is now simply referred to as "satisfaction": the woman's perception of how satisfying the overall sexual experience has been for her. FSD includes disorders of desire, arousal, orgasm, painful sex, and overall satisfaction. A significant percentage of desire, arousal and anorgasmia disorders may relate to testosterone insufficiency, with attention to the timing of onset of symptoms being of great importance.

Women with low desire throughout their reproductive lives (primary dysfunction) should likely be referred to a mental health practitioner with specialized expertise in sexuality. FSD problems that present with a specific time of onset following a period of normal functioning (secondary dysfunction/acquired dysfunction) may relate to changes in hormonal milieu as well as emotionally important, psychologically significant events. These distinctions may facilitate more expeditious, proper treatment.

Desire disorders

Hypoactive sexual desire disorder (HSDD) is a syndrome defined in the Diagnostic and Statistical Manual of the American Psychiatric Association (DSM-IV R) classification as "personal distress resulting from a diminished feeling of sexual interest, sexual thoughts and distress with the lack of one's responsive desire beyond what would be considered normal for time of life and duration of the relationship." According to the consensus of the Sexual Function Health Council of the American Foundation for Urologic Disease, HSDD is defined by persistent or recurrent deficiency of sexual fantasies, thoughts and/or desire for, or receptivity to, sexual activity resulting in personal distress [10]. While the actual prevalence of HSDD remains elusive, recent figures indicate about 50% of female sexual dysfunction is based on these criteria in a gynecology outpatient cohort, of which HSDD is the most prevalent [11].

HSDD is most often multifactorial in origin. When patients complain of low libido, their concerns must be validated, with an approach that invites dialogue, appreciating the multiple potential elements in her experience. Although androgen deficiency may be a contributing factor, important additional psychological factors may come to light during the course of evaluation. Should sexual abuse, trauma, or mood disorders be revealed, it is important to acknowledge them and refer the patient for appropriate psychological therapy (a wide variety of mental health professionals specialize in addressing such concerns, such as International Society for the Study of Women's Sexual Health [ISSWSH] or American Association of Sexuality Educators Counselors and Therapists [AASECT] certified practitioners).

A history of HSDD onset with anorexia, bulimia, severe anxiety, premenstrual depressive disorder, postpartum blues or depression, infertility, premature ovarian failure, postoperative surgical bilateral oophorectomy, or administration of contraceptive hormones, perimenopause or menopause may be particularly associated with low androgen levels [12,13]. Administration of many antidepressants, such as selective serotonin reuptake inhibitors (SSRIs) or serotonin–norepinephrine reuptake inhibitors (SNRIs), can create negative sexual side effects (particularly anorgasmia) and may require pharmacologic consultation with a medicating psychiatrist.

Anorgasmia

Anorgasmia is the absence of being able to achieve orgasm, defined as "a variable, transient peak sensation of intense pleasure creating an altered state of consciousness, usually accompanied by involuntary, rhythmic contractions of the pelvic striated circumvaginal musculature, often with concomitant uterine and anal contractions and myotonia that resolves the sexually-induced vasocongestion (sometimes only partially), usually with an induction of well-being and contentment" [14]. The presence of a normal sexual excitement phase is a prerequisite for female orgasmic disorder. In other words, if the absence of orgasm follows a time of decreased desire for sexual activity, an aversion to genital sexual contact, or a decreased lubrication-swelling response, diagnoses such as hypoactive sexual desire disorder, sexual aversion disorder, or female sexual arousal disorder, respectively, might be more appropriate, even if anorgasmia is the common final outcome.

Upon ruling out physiologic or biologic causes of anorgasmia, such as Turner's syndrome, primary depression, underlying psychological factors such SSRIs, substance abuse including alcohol, and a relationship problem, we are left with the diagnosis of primary anorgasmia. It is significantly less common than low libido or HSDD, and can result from insufficient levels of free, biologically available testosterone. An increased frequency of this less common FSD is observed in women in the lowest quartile of free testosterone as measured in serum. While systematic investigation of testosterone for the treatment of isolated orgasmic disorder is lacking, many studies document the improvement in orgasmic dysfunction in women with HSDD when given testosterone.

Painful intercourse

If intercourse is experienced as painful, it is likely to create negative associations and lead to decreased desire for sex. As clinicians, we must be systematic in elucidating the cause. It is exceedingly rare to remediate dyspareunia with correction of low testosterone levels alone. If pain is experienced upon entry/vaginal penetration, etiologic considerations include vulvar pathology (atrophy, infection, dystrophy, vestibulitis, or vulvodynia or provoked vestibulodynia) as well as urethral anomalies (diverticulum), or vaginal/genital/urethral atrophy. This is especially frequent in menopausal women not receiving estrogen therapy, local or systemic.

Although lichen sclerosis et atrophicus was formerly treated with topical testosterone preparations, testosterone deficiency per se does not cause this condition nor is it associated with this diagnosis, which is now uniformly treated with topical corticosteroids. Urethritis and cystitis (with urinary infections) can be a frequent source of acute onset painful intercourse and must be ruled out. This infectious urinary tract complication is associated with the effects of low estrogen concentrations in these tissues, not androgen deficiency per se. However, as the vestibule and vestibular glands (Bartholin's, Skene's, etc.) are rich in androgen receptors, and derived from embryonic endoderm, they are highly sensitive to androgens. As such, topical estrogen *and* and androgen may have a more effective outcome than estrogen therapy alone [15]. If pain is elicited deeper into the vagina or at its apex, other considerations may include endometriosis, ovarian cyst, pelvic inflammatory disease, malpositioned IUD, as well as vaginal canal atrophy associated with very low serum estrogen levels and should be investigated. Rarely, severe uterine retroversion can be associated with dyspareunia with deeply posterior penetration, which is not associated with androgen deficiency.

The examination

The gynecologic physical exam done with care may reveal problems of genital anatomy or personal perception. With increasing media attention to plastic surgical "vaginal rejuvenation," many women have become self-conscious and critical of the appearance of their labia and vaginal introitus. Labial asymmetry is common yet may now be a cause for impaired body image, reducing sexual confidence. Laxity of the introitus and ptosis of the labia minora and/or majora may have an association with undesirable age-related skin changes in other parts of the body. Graying or sparse pubic hair, sometimes referred to by patients with the unfortunate term "old lady vagina" carry negative stereotypes of aging, and potentially diminished sexual stimulation. Shaving and waxing fashions may reveal long-standing labial and vaginal anatomic asymmetries as well as variations of pigmentation that do not fit the patient's aesthetic expectations and can create negative impact on body image. Unrealistic genital expectations recently have been conditioned by misleading media and plastic surgical advertisements as well as remunerative motivation on the part of physicians [16].

Comorbidities on examination

Trauma

Evidence of sexual abuse may also be seen with scarring that is not attributable to obstetric history or prior gynecologic surgery. Healed lacerations from sexual trauma may appear irregular and thickened due to chronic inflammation and healing by secondary intent. Vaginal or labial hematoma can result from aggressive sexual play or abuse and must be sensitively evaluated. Asking women whether their sexual activity has ever been forced or resulted from intimidation can open the dialog to elicit a history of traumatic experiences.

Atrophy

Vaginal atrophy is seldom a feature of menstruating or perimenopausal women. Estrogen levels sufficient to menstruate amply support normal vaginal health. However, in premenopausal women on long-standing hormonal contraception, the potent progestogens present can actually cause atrophy, particularly of the vestibule. In such women, switching to a non-hormonal form of contraception or adding vaginal atrophy treatment with estrogen or estrogen and androgen can be helpful. In postmenopausal women, vaginal atrophy due to low estrogen levels affects an estimated 70–80% of women over time, and can result in dyspareunia, with concomitant dryness further decreasing sexual pleasure [17]. Adequate estrogenization of the vagina can be achieved with local tablets, creams, or an estradiol releasing ring, widely available FDA-approved products. However, systemic estrogen levels may be required for the creation of an appropriate central nervous system hormonal milieu to support acceptable levels of libido, mood, and overall sexual satisfaction [18,19].

Androgen insufficiency – local

Having addressed the many factors of vaginal health and emotional/psychological effects of adequate estrogen, one then considers the potential role of androgen insufficiency. Estrogens modulate genital sensitivity and maintain function of sexual organs. They act via vasodilatory pathways to increase vaginal, clitoral, and urethral blood flow, with resultant genital congestion and increased vaginal lubrication. Testosterone has similar effects when administered directly to vaginal tissues, enhancing vaginal blood flow and lubrication with sexual arousal [20]. These effects appear to be attributable to testosterone directly, as trials show that women on aromatase inhibitors (which block the conversion of testosterone to estradiol) exhibit similar salutary genital/sexual effects [21]. Further, topical estrogens and androgens may have synergistic effects [15].

Systemic testosterone treatments

Background and hormonal measurements

HSDD is not highly correlated with androgen levels as measured in serum, supported by data from the USA and other countries. A recently published cross-sectional study in a cohort of 1423 Australian women found no significant correlation of circulating androgen levels with self-reported perception of sexual desire and sexual satisfaction [6]. In spite of this lack of correlation, it is unfortunate that commercially available testosterone assays are notoriously unreliable and imprecise in women. Using total testosterone or bio-available testosterone may be better than measured or calculated free testosterone, as the normal ranges are greater in menstruating and perimenopausal women. Individual perimenopausal patients may exhibit wide variations, depending on the cycle in which testosterone is measured as levels fluctuate rapidly.

While the authors wish to dissuade the reader from relying upon testosterone assays in the assessment of therapeutic response, there are several approaches which can help to reduce the variability of such assay measurements. The suggested timing for minimizing cyclic variation in menstruating women (i.e., perimenopause) is cycle days 7–10, avoiding early menstrual cycle levels when free and total testosterone are at their nadir. Blood sampling at this time also avoids the pre-ovulatory peak in testosterone unless women have less than a 25-day cycle length. Data support trials of testosterone supplementation in women whose testosterone levels are in the lower 25–33% of normal range [22].

Pre- and perimenopause

Actual androgen deficiency is difficult to explain in normal premenopausal women with regular menses, as androgens are requisite for normal menstrual function. The benefits of testosterone therapy in perimenopausal women with low sexual desire have been demonstrated in clinical trials by Goldstat et al. [23]. Other comorbidities can cause androgen deficiency in these women (such as hypothalamic amenorrhea, adrenal insufficiency, etc.). Functional hypoandrogenism could be the result of hormonal contraception with suppression of both granulosa (estrogen-producing) and theca (testosterone-secreting) cells in the ovary. Decreased sexual desire is a relatively common complaint for perimenopausal women in spite of mostly adequate androgen production prior to full menopause

Table 4.2 Princeton Consensus Statement

2001 Princeton Consensus Statement
Clinical symptoms
Diminished sense of well-being, dysphoric mood, and/or blunted motivation
Persistent, unexplained fatigue
Sexual function changes, including decreased libido, sexual receptivity, and pleasure
Other potential signs or symptoms
Bone loss, decreased muscle strength, changes in cognition and memory
As the most common presenting symptoms are nonspecific, symptoms alone are not sufficient for the diagnosis of androgen insufficiency

[24]. As previously stated, there can be measurable benefits of testosterone therapy in HSDD perimenopausal women with low serum testosterone levels. Although hypoandrogenemia is less common, it can occur in pre- and perimenopausal women [23].

Thus a strictly biologic explanation for this phenomenon in women who continue to maintain functional ovulatory cycles is not substantiated, as there is no concomitant fall in serum testosterone levels. One hypothesis in this subgroup suggests that the evolutionarily beneficial midcycle surge in testosterone and its associated increase in libido may be absent in perimenopausal women [25]. Both ovarian testosterone and adrenal androgens (DHEA, DHEA-S, and androstenedione) fall gradually from the early twenties, dropping to about 50% of their peak adult values by the onset of menopause, which might offer an explanation for this phenomenon [26,27]. The Princeton Consensus Statement (see Table 4.2) published in 2002 by US and Australian experts defined female androgen deficiency syndrome (FADS) as the presence of decreased well-being and libido, occurring in a woman with androgen levels in the lower quartile of the normal female range and adequate levels of estrogen [28]. However controversial the diagnosis of true androgen deficiency may be in menstruating or perimenopausal women, correcting low testosterone or DHEA levels can help eliminate these "endocrinopathies" as a potential source of HSDD.

Postmenopause

HSDD (estimated as low desire with distress) occurs in 9–14% of postmenopausal women, a large number as greater longevity has increased the population of older women [4]. Numerous studies validate the therapeutic sexual benefits of testosterone in postmenopausal women who experienced loss of libido. Clinical data support the usefulness of transdermal testosterone administration for women with both low sexual desire and difficulties with arousal [29]. Clinically, such low-libido patients are common. They may not present primarily with sexual complaints, which often need to be elicited by more thorough questioning in the clinical setting. Persistent and unexplained fatigue without symptoms of depression or hypothyroidism may be the presentation, with concomitant complaints of muscle weakness and/or lack of muscle definition with previously adequate exercise efforts. These women are more likely to be surgically menopausal as removal of ovaries further decreases low androgen concentrations in older women, even though sexual ideation in some studies did not differ in oophorectomized older women [28,30].

Testosterone treatment

The route of testosterone administration is important. Oral testosterone results in deleterious effects on lipids in women due to first-pass hepatic metabolism (decreased HDL cholesterol primarily). Replacement via transdermal or subcutaneous routes, with serum levels within the physiologic range, is the desired goal. The strongest evidence for use of testosterone replacement for HSDD was generated with the development of the transdermal testosterone patch (TTP) (Intrinsa; Procter & Gamble/Warner Chilcott). Initially, 75 surgically menopausal women on oral estrogen, conjugated equine estrogen, were randomized to placebo or TTP. Those on TTP demonstrated improved sexual function [29]. Subsequently, several other studies have demonstrated the efficacy and safety of this form of treatment [31,32; see below].

Route of delivery of androgen replacement in women

Choosing both a convenient and efficient mode of androgen administration in women remains a challenge, especially with no FDA-approved preparations in this category for women. When given oral testosterone preparations (methyltestosterone and testosterone undecanoate) women exhibit wide variability due to intestinal absorption, resulting in widely fluctuating levels of circulating testosterone and short half-lives due to first-pass metabolism in the liver.

Non-FDA-approved compounded subcutaneous testosterone implants are used by some clinicians, require insertion every 4–6 months, and result in abnormally elevated serum levels, often reaching the normal male range. Although enthusiastic anecdotes abound, this "compounded implant method" is not recommended. Testosterone pellets are irretrievable in the face of complications; frequently they induce supraphysiological concentrations over weeks and months, which can result in uncontrollable acne, hirsutism, clitoromegaly, voice changes, and complications of pellet extrusion and insertion site infections.

TTP (Intrinsa) are applied twice a week and are more convenient to use. Of note, transdermal delivery bypasses first-pass metabolism and can provide consistent levels of hormone over time. Testosterone patches for women had been available in the EU for years, but were never approved for sale in the USA. These patches were recently withdrawn from the market in the EU for business reasons. Testosterone patches, like all patches, may cause skin irritation or fail to adhere, although these problems were rare in clinical trials [33]. Patch problems are minimized by the use of transdermal testosterone gel, which has been shown to provide higher and more stable testosterone bioavailability (and has been FDA approved for and studied in men) [34]. Several testosterone gels or gel-like transdermals are FDA approved for men.

In women as compared with men, much lower circulating testosterone concentrations are desired in order to achieve concentrations in the premenopausal female reference range, lower than the male range by approximately a factor of 10. Thus, the use of available testosterone products that are FDA approved for men can be treacherous, and result in serum androgen levels that far exceed the desired female range. A paucity of published data on androgen delivery by transdermal testosterone gel in women exists, but they have shown a dose-dependent increase of bioavailable testosterone in treated women [35].

Treatment options

Currently, there are no available FDA-approved testosterone products for women, whether pre- or postmenopausal. However, off-label use of the AndroGel 1% pump delivery system (this product is also available in 1.62% pump and packaged in sachets) dosage form can achieve therapeutic levels in women. One pump depression of AndroGel 1% yields 12.5 mg transdermal testosterone, which if administered 2–3 times per week is in a similar range to the 10 mg dose used daily in the sexual function study of testosterone therapy in premenopausal women. In this trial, 31 women were evaluated with baseline testosterone deficiency (total testosterone less than 2.2 nmol/L) and complaints of low libido. This placebo-controlled, crossover design double-blinded, with two 12-week treatment periods separated by a 4-week washout interval, showed significant benefit to well-being, mood, and sexual function with skin application of 10 mg testosterone gel every morning [23]. The mean level of serum total testosterone increased by 1.54 nmol/L, starting from baselines in the lower third of normal female reproductive range. The free androgen index (FAI), a ratio of testosterone to SHBG, increased by 3.6 with 10 mg testosterone applied in a transdermal cream. Serum estradiol levels were unchanged. Forty-six percent of the treatment group exhibited a 50% or higher increase in their sexual self-rating score compared with only 19% improvement in the placebo group. None of the trial participants reported virilizing side effects as a result of testosterone therapy. Thus, an empirical trial of testosterone therapy may be warranted in premenopausal women for whom decreased sexual desire causes distress.

An alternative delivery approach utilizes Testim 1% transdermal gel, also FDA approved for use in hypogonadal men. This product is supplied in re-cappable tubes. As each tube provides 50 mg, a standard daily dose for a man, and consists of 60 drops of liquid-like testosterone gel, a dose equivalent to one-tenth the standard male dose is about 6 drops daily. In either case (AndroGel 1% or Testim 1%), the application should be used on the leg or other easily shaved area, as a small percentage of women will actually grow excess hair at the application site. Additionally, as neither of these products is FDA approved for women, insurance coverage is often difficult to obtain. While other FDA-approved male therapies can be modified for use in women, they are more difficult to modify for use in women and are more likely to result in overdosing.

No US FDA approval of available preparations: a significant challenge

At present, no preparation for testosterone replacement therapy in women has been licensed by the US FDA. Previous testosterone regulatory applications aimed at the treatment of HSDD have been rejected, primarily due to concerns with the lack of long-term safety data, and whether the improvement satisfied "meaningful clinical benefits." Several testosterone products for women are licensed by the European Medicines Agency (EMA). They include the testosterone patches Intrinsa (Procter & Gamble, OH, USA) and Livensa (Warner Chilcott, Dublin, Ireland). AndroFeme (Lawley Pharmaceuticals, Perth, Australia) is a 1% testosterone cream licensed by the state of Western Australia. These products are approved for the treatment of HSDD in surgically menopausal women or similar indications. Due to lack of profitability, Warner Chilcott has suspended distribution of the Intrinsa testosterone patch in the EU.

Bioidentical compounded testosterone cream

A robust "compounded bioidentical testosterone cream" industry has sprung forth in the USA. The quality of such products, clinical parameters of absorption, and distribution and metabolism remain highly variable. Individual compounding pharmacies ideally obtain Pharmacy Compounding Accreditation Board (PCAB) certification of purity and dose reliability of their products to optimize quality and clinical outcomes for both prescriber and consumer [36]. Disappointingly, only a small percentage of compounding pharmacies belong to PCAB and achieve the quality assurance PCAB certification indicates.

Typical starting prescriptions for correcting androgen deficiency include: 0.5 mg/0.5 mL of testosterone cream. Some women will have satisfactory results with this low dose, and others may require 1–2 mg twice daily application in order to achieve therapeutic efficacy with serum testosterone levels in the mid-to-upper third of the normal reproductive female range. Morning application is encouraged to mimic the natural peak testosterone diurnal maximum concentration in the morning, nadir at night.

Serum testosterone testing in this setting is encouraged with the aforementioned provisos, as variability and potential for overdosage should be avoided. As serum testosterone concentrations do not typically require fasting, morning administration with late afternoon phlebotomy to approximate average to nadir testosterone serum levels (between 4 and 6 p.m.) make monitoring easier. If higher doses of testosterone are needed to achieve clinical improvement and optimize patient satisfaction, minimal volumes can be delivered with an increase in testosterone concentration rather than by prescribing large volumes of cream. Common doses ranges are 1.0 mg/0.5 mL to 2 mg/mL for long-term maintenance of clinically satisfactory testosterone concentrations.

For consumer safety, unsubstantiated "take it on faith" creams and gels are to be avoided. Patients may bring such unlabeled pump vials to the office requesting replacement. Serum testing is necessary to assess levels achieved with "mystery products" as well as to prescribe appropriate dose ranges to achieve desired clinical effects and avoid undesired adverse events as well as potential future litigation.

Long-term side effect profile of androgen therapy

Most of the studies concerning androgen replacement in women have concentrated on the potential effects on female libido and well-being and recorded androgenic effects on skin (acne, hirsutism, and alopecia) and metabolic endpoints, like the lipid profile, glucose tolerance, and polycythemia. However, detailed data on potential androgen effects on insulin sensitivity, body composition, bone mineral density, cardiovascular events, and cancer are more scarce and often preliminary (see Table 4.3). A recently published, 4-year follow-up study in 967 oophorectomized patients on estrogen replacement therapy, who received at least one application of 300-µg TTP, showed no meaningful changes in the safety or tolerability profile of the TTP. Also observed in this trial were three cases of invasive breast cancer, consistent with age-expected rates [32]. Monitoring of standard metabolic chemistries and blood count is recommended in all testosterone-treated women.

Oncologic concerns and androgen therapy

Controversies continue regarding an association with estrogens in the development of breast, ovarian, and endometrial cancers. Despite lower rates of invasive breast cancer in conjugated equine estrogen users compared with placebo (the estrogen-only arm of the Women's Health Initiative [WHI] after 10.7 years of follow-up), these data have not yet changed the clinical paradigm of fear regarding cancer risks [37]. Similarly, several prospective studies have linked circulating androgen levels to the risk of postmenopausal breast cancer, though data are also conflicting due to the association of higher endogenous androgen levels with greater body mass index (BMI) and the association of obesity with increased rates of breast cancer.

Data from the Nurses' Health Study and the WHI Observational Study concluded that naturally menopausal women using hormonal replacement therapy containing both estrogens and androgens (Estratest) were at greater risk of developing breast cancer compared with never-users of postmenopausal hormone therapy [38]. Further analysis of these same data revealed that androgen levels were related to other risk factors for cancer, such as elevated BMI, higher levels of alcohol intake and cigarette smoking, making exact cause–effect association difficult [39].

Reassuringly with physiologic levels of androgen therapy, Davis et al. reported no changes in mammographic density, a more recently identified risk factor for breast cancer [40]. Davis and colleagues, following a small group of postmenopausal women using 300 µg/day TTP for 52 weeks, found no significant change in digital mammographic density. Additionally, a large UK study in postmenopausal women confirmed the absence of mammographic density changes with androgen supplementation [41].

Also of importance are data regarding other potential effects of testosterone therapy on other female reproductive cancers. No association was demonstrated between androgen levels and ovarian cancer risk in several prospective studies [42,43]. In addition, the European Prospective Investigation into Cancer and Nutrition concluded that there is a significantly increased risk of endometrial cancer in patients with high *endogenous* serum estrogen and free testosterone levels [44]. The association of elevated BMI and higher observed incidence of endometrial cancer is probably associated with higher androgen levels in this study population. Studies of androgen supplementation in menopausal women from Australia have not reported this association of increased endometrial hyperplasia or neoplasia [6].

Cardiovascular and metabolic concerns and androgen therapy

Carotid artery intima–media thickness estimated via carotid ultrasonography is a frequent marker used to investigate the progression of atherosclerosis [45,46]. Several studies have shown an inverse correlation of circulating androgen levels in postmenopausal women to intima–media thickness, providing some preliminary evidence suggesting a protective effect of androgens on the development of atherosclerosis [47]. There are many potential benefits of testosterone on

Table 4.3 Randomized controlled studies on testosterone treatment in postmenopausal women

Study (year)	Patients	Design	Duration	Testosterone administration	Effect of testosterone	Reference
Barrett-Connor et al. (1999)	n = 311; surgical menopause	Double-blind	2 years	CEE, 0.625 mg/day CEE, 1.25 mg/day CEE, 0.625 mg, + MT, 1.25 mg/day CEE, 1.25, + MT 2.5 mg/day	↓ menopausal symptoms ↑ HDL and TG in CEE alone ↑ BMD in CEE + MT	[54]
Braunstein et al. (2005)	n = 447; surgical postmenopausal women on ERT; 24–70 years	Double-blind, placebo-controlled, parallel-group	24 weeks	Placebo 150 μg/day TTP 300 μg/day TTP 450 μg/day TTP	↑ sexual desire and satisfying sexual activity in 300 and 450 μg/day group	[55]
Basaria et al. (2002)	n = 40; > 1 year menopause; natural and surgical; > 3 m ERT; > 21 years	Double-blind, parallel-group	16 weeks	1.25 mg CEE 1.25 mg CEE + 2.5 mg MT ↓ TG	↓ plasma viscosity ↑ fibrogen	[56]
Buster et al. (2005)	n = 533; HSDD, surgical menopause, on RT	Double-blind, placebo-controlled	24 weeks	Placebo TTP 300 μg/day	↑ satisfying sexual activity ↑ sexual desire ↓ personal distress	[57]
Chiuve et al. (2004)	n = 40; surgical menopause on ERT	Double-blind, parallel-group	10 weeks	1.25 mg CEE 1.25 mg CEE + 2.5 mg MT	↓ apoC I, II, III ↓ apoE ↓ TG, ↓ HDL	[58]
Davis et al. (1995)	n = 34; menopause natural/surgical	Single-blind	2 years	E 50 mg/3M implant E 50 mg + T 50 mg/3M implant	More ↑ in BMD, in all sites of the body More ↑ sexual energy ↓ LDL in both groups ↓ total body fat-free mass	[59]
Davis et al. (2000)	n = 34; menopause natural/surgical	Single-blind	2 years	E 50 mg/3M implant E 50 mg + T 50 mg/3M implant	↓ total chol ↓ LDL ↓ total body fat-free mass ↓ fat-free mass (FM:FFM) ratio	[60]
Davis et al. (2006)	n = 61; HSDD, surgical menopause on ERT	Double-blind, placebo-controlled	24 weeks	Placebo TTP 300 μg/day	↑ sexual desire ↓ personal distress	[61]
Davis et al. (2008)	n = 814; surgical and natural menopause 20–70 years	Double-blind, placebo-controlled	52 weeks	150 μg/day TTP 300 μg/day TTP Placebo	↑ sexual desire, higher in 300 μg group	[62]
Davis et al. (2009)	n = 279; surgical and natural menopause	Double-blind, placebo-controlled, parallel-group	52 weeks	150 μg/day TTP 300 μg/day TTP Placebo	No difference from placebo for total dense or nondense area on digital mammograms from baseline to week 52	[40]

Study	Population	Design	Duration	Intervention	Results	Ref
De Paula et al. (2007)	n = 85; HSDD, postmenopausal, on HRT	Double-blind, placebo-controlled, crossover	16 weeks	Placebo 16 week; MT 2.5 mg 16 week; Placebo 8 week + MT 2.5 mg 8 week; MT 2.5 mg 8 week + placebo 8 week	No change in liver enzymes and lipids; ↑ sexual satisfaction; ↑ sexual desire	[63]
Dobs et al. (2002)	n = 36; natural menopause	Double-blind, placebo-controlled, parallel	16 weeks	EE 1.25 mg/day; EE + MT 2.5 mg/day	↑ sexual activity; ↑ lean body mass; ↓ % body fat; ↑ body weight	[64]
El-Hage et al. (2007)	n = 36; HSDD, surgical menopause	Double-blind, placebo-controlled, crossover	12 weeks	T cream 10 mg/day	↑ sexual desire; ↑ frequency of sex; No change in lipids, blood pressure, or weight	[65]
Flöter et al. (2005)	n = 50; surgical menopause, on ERT, 45–60 years	Double-blind, placebo-controlled, crossover	24 weeks	(1) estradiol 2 mg + TU 40 mg; (2) estradiol 2 mg + placebo	↓ IGF-1; ↓ propeptide of type I procollagen; ↑ total lean body mass; No change in fat mass, BMI, BMD, and blood pressure	[66]
Nathorst-Böös et al. (2005)	n = 53; HSDD, postmenopausal, on ERT	Double blind, crossover	24 weeks	(1) ERT + T gel 10 mg/day; (2) ERT + placebo	↑ sex quality parameters; No change in liver enzymes, lipids, endometrium, or blood velocity	[67]
Hofling M et al. (2007)	n = 99; postmenopausal, on ERT (2 mg estradiol and 1 mg norethisterone acetate)	Prospective, double-blind, placebo-controlled	24 weeks	(1) ERT + TTP 300 µg/day; (2) ERT + placebo	↑ mammographic density in 18–30% of women, no differences between the two groups	[68]
Hofling M et al. (2007)	n = 99; postmenopausal, on ERT (2 mg estradiol and 1 mg norethisterone	Prospective, double-blind, placebo-controlled	24 weeks	(1) ERT + TTP 300 µg/day; (2) ERT + placebo	Placebo group: fivefold ↑ ($P < 0.001$) in breast cell proliferation (fine needle biopsy); T group: no ↑	[69]
Panay et al. (2010)	n = 272; HSDD, natural menopause	Double-blind, placebo-controlled	24 weeks	(1) TTP 300 µg/day; (2) placebo	↑ satisfying sexual episodes; ↑ desire ↓ distress	[31]
Penotti et al. (2001)	n = 40; postmenopausal, on HRT since 1–5 years (E_2 50 µg/day + MPA 10 mg/day for 12 days every other month)	Double-blind, placebo-controlled	32 weeks	(1) HRT + TU 40 mg/day; (2) HRT + placebo	↑ pulsatility index of Doppler cerebral artery; ↓ HDL; ↑ sexual desire	[70]

Table 4.3 (cont.)

Study (year)	Patients	Design	Duration	Testosterone administration	Effect of testosterone	Reference
Penteado et al. (2008)	n = 60; HSDD, postmenopausal, on HRT (CEE 0.625 mg + MPA 2.5 mg)	Double-blind, placebo-controlled	52 weeks	(1) HRT + MT 2.0 mg/day (2) HRT + placebo	↑ sexual energy No change in orgasmic capacity	[71]
Raisz et al. (1996)	n = 28; natural, menopause	Double-blind, placebo-controlled, parallel	9 weeks	(1) CEE 1.25 mg (2) CEE + MT 2.5 mg/day	↑ bone formation ↓ HDL, ↓ TG	[72]
Sherwin and Gelfand et al. (1985)	n = 53, surgical menopause	Double-blind, placebo-controlled, crossover	12 weeks	(1) EE (2) TE 150 mg (3) EE + TE 150 mg (4) placebo	↑ sexual desire, arousal, fantasies	[73,74]
Shifren et al. (2000)	n = 75, surgical menopause, on ERT (CEE 0.625 mg), 31–56 years	Double-blind, placebo-controlled	12 weeks	(1) ERT + TTP 150 µg/day (2) ERT + TTP 300 µg/day (3) ERT + placebo	↑ frequency of sexual activity, pleasure-orgasm, well-being at 300 µg/day	[29]
Shifren et al. (2006)	n = 549; HSDD, natural menopause, on ERT	Double-blind, placebo-controlled, parallel group	24 weeks	(1) 0.625 mg CEE (2) 0.625 mg CEE + TTP 300 µg/day	↑ total satisfying sexual episodes ↑ sexual desire ↓ personal distress	[75]
Simon et al. (2005)	n = 562; HSDD, surgical menopause, on ERT, 26–70 years	Double-blind, placebo-controlled	24 weeks	(1) ERT + TTP 300 µg/day (2) ERT + placebo	↑ total satisfying sexual activity ↑ sexual desire ↓ distress	[33]
Warnock et al (2005)	n = 102; HSDD, surgical menopause, on ERT (CEE 1.25 mg), 33–62 years	Double-blind	8 weeks	(1) ERT + MT 2.5mg (2) ERT + placebo	No changes in sexual desire/ interest	[76]
Watts et al. (1995)	n = 66; surgical menopause	Double-blind, placebo-controlled, parallel	24 months	(1) CEE 0.625 mg/day (2) CEE + MT 2.5 mg/day	↑ BMD (lumbar spine) ↓ HDL ↓ TG	[77]
Zang et al. (2006)	n = 63; natural menopause	Open, parallel group	12 weeks	(1) 40 mg TU/2 days (2) 2 mg estradiol valerate/day (3) both	(1) and (3): ↓ insulin-induced glucose disposal, ↑ lean body mass, ↓ HDL	[78]

BMD, bone mineral density; CEE, combined esterified estrogens; Chol, cholesterol; ERT, estrogen replacement therapy; HDL, high-density lipoprotein; HRT, hormonal replacement therapy; HSDD, hypoactive sexual desire disorder; LDL, low-density lipoprotein; MPA, medroxyprogesterone acetate; MT, methyltestosterone; T, testosterone; TE, testosterone enanthate; TG, triglycerides; TTP, transdermal testosterone patch; TU, testosterone undecanoate.

cardiovascular fitness validated in a secondary prevention study. In this study testosterone supplementation improved functional capacity, insulin resistance, and muscle strength in women with advanced congestive heart failure (CHF), and appeared to be an effective and safe therapy for elderly women with CHF [48].

Studies of the lipid profile in females so far seem to be in agreement, since total serum testosterone levels and free androgen index, in particular, have been associated with higher LDL and lower HDL cholesterol levels in the presence of elevated BMI [49,50]. Apolipoprotein CIII (ApoCIII) levels decrease in women receiving androgen therapy, and this is somewhat promising, since ApoCIII is an independent risk for cardiovascular disease [51]. Of note, obesity increases androgen levels, as observed in the context of the polycystic ovary syndrome, and has had a confounding impact on many studies. Encouraging data revealed testosterone therapy provided benefits to cardiovascular fitness in women with chronic heart failure in a double-blinded, randomized, placebo-controlled trial [48].

The long-term impact of unfavorable changes in cardiovascular risk markers, such as the decrease in HDL cholesterol, cannot be properly judged based on currently available data due to the confounding effects of obesity. This would also have to be weighed against potential beneficial effects on body composition and insulin sensitivity and can be assessed on an individual basis in testosterone-treated patients with serial lipid measurements.

Contraindications to androgen therapy

Relative contraindications of testosterone replacement in pre- and postmenopausal women include androgenic alopecia, acne, hirsutism, hyperlipidemia, and liver dysfunction. Absolute contraindications are the presence or increased risk of invasive breast cancer, endometrial cancer, and venothrombotic episodes and diagnosed cardiovascular disease [52].

After prescribing testosterone therapy for a "reliable" patient, the practitioner needs to be highly vigilant in monitoring both the response to treatment and potential overuse/abuse. Overuse can be assessed in several ways. Obviously, the patient's side effects (seborrhea appearing first, then hirsutism) can and should be monitored, and serum total testosterone concentrations can be used to confirm one's clinical suspicion that the patient is receiving an excessive amount of medication. Finally, under conditions of overuse, the

patient will require a refill of her prescription too early. This issue is not to be underestimated and could be a cause for litigation over androgenic alopecia and voice changes.

Conclusion

Given the potential role of androgens in improving sexual function and mood in androgen-deficient women, HSDD and FSD patients deserve at the least an empiric trial with testosterone supplementation. Table 4.3 provides a summary of clinical trials with postmenopausal testosterone replacement [53]. Further randomized controlled trials with adequate long-term follow-up are essential to evaluate the true efficacy of testosterone replacement therapy, in the presence or absence of estrogen therapy in women. The regulation of female sexual function is complex, not solely androgen dependent, with strong evidence for potential benefit that exceeds harm from testosterone properly utilized.

It is the responsibility of the clinician to offer androgen support and the patient's decision to embark on a monitored trial period to assess her individual response to therapy. Periodicity of laboratory clinical assessment must be judged on an individual basis as no clear and simple monitoring algorithm exists. As with all hormone therapy, testosterone therapy should be individualized and requires that each woman weigh her own risks and benefits. The questionnaire below provides a

The Decreased Sexual Desire Screener (DSDS) used in the non-treatment validation study.

Patient, please answer each of the following questions (yes/no)

• In the past was your level of sexual desire or interest good and satisfying to you?

• Has there been a decrease in your level of sexual desire or interest?

• Are you bothered by your decreased level of sexual desire or interest?

• Would you like your level of sexual desire or interest to increase?

• Please check all the factors that you feel may be contributing to your current decrease in sexual desire or interest:

– An operation, depression, injuries or other medical condition

– Medication, drugs or alcohol you are currently taking

– Pregnancy, recent childbirth, menopausal symptoms

– Other sexual issues you may be having (pain, decreased arousal or orgasm)

– Your partner's sexual problems

– Dissatisfaction with your relationship or partner

– Stress or fatigue

good starting point to help assess and follow up each patient over time. Only long-term safety studies will clarify concerns regarding ongoing use in FSD, underscoring the importance of ongoing clinical assessment.

References

1. Kingsberg S, Althof SE. Evaluation and treatment of female sexual dysfunction. *Int Urogynecol J Pelvic Floor Dysfunct* 2009; 20 (Suppl 1): S33–43.

2. Laumann EO, Paik A, Rosen RC. Sexual dysfunction in the United States: prevalence and predictors. *JAMA* 1999; 281(6): 537–44.

3. Laumann EO, Nicolosi A, Glasser DB, et al. Sexual problems among women and men aged 40–80 y: prevalence and correlates identified in the Global Study of Sexual Attitudes and Behaviors. *Int J Impot Res* 2005; 17(1): 39–57.

4. Shifren JL, Monz BU, Russo PA, et al. Sexual problems and distress in United States women: prevalence and correlates. *Obstet Gynecol* 2008; 112(5): 970–8.

5. Pontiroli AE, Cortelazzi D, Morabito A. Female sexual dysfunction and diabetes: a systematic review and meta-analysis. *J Sex Med* 2013; 10(4): 1044–51.

6. Davis SR, Davison SL, Donath S, Bell RJ. Circulating androgen levels and self-reported sexual function in women. *JAMA* 2005; 294(1): 91–6.

7. Shufelt CL, Braunstein GD. Safety of testosterone use in women. *Maturitas* 2009; 63(1): 63–6.

8. Grazziottin A, Leiblum SR. Biological and psychosocial pathophysiology of female sexual dysfunction during the menopausal transition. *J Sex Med* 2005; 2 (Suppl 3): 133–45.

9. Palacios S. Hypoactive sexual desire disorder and current pharmacotherapeutic options in women. *Women's Health* 2011; 7(1): 95–107.

10. Basson R, Berman J, Burnett A, et al. Report of the international consensus development conference on female sexual dysfunction: definitions and classifications. *J Urol* 2000; 163(3) 888–93.

11. Geiss IM, Umek WH, Dungl A, et al. Prevalence of female sexual dysfunction in gynecologic and urogynecologic patients according to the international consensus classification. *Urology* 2003; 62(3): 514–18.

12. Miller KK, Lawson EA, MacArthur V, et al. Androgens in women with anorexia nervosa and normal weight women with hypothalamic amenorrhea. *J Clin Endocrinol Metab* 2007; 92(4): 1334–9.

13. Kalantaridou SN, Calis KA, Venderhoof VH, et al. Testosterone deficiency in young women with 46XX spontaneous ovarian failure. *Fertil Steril* 2006; 86(5): 1475–82.

14. Meston CM, Hull E, Levin RJ, Sipski M. Disorders of orgasm in women. *J Sex Med* 2004; 1(1): 66–8.

15. Raghunandan C, Agrawal S, Dubey P, et al. A comparative study of the effects of local estrogen with or without local testosterone on vulvovaginal and sexual dysfunction in postmenopausal women. *J Sex Med* 2010; 7(3): 1284–90.

16. Goodman MP. Female genital cosmetic and plastic surgery: a review. *J Sex Med* 2011; 8(6): 1813–25.

17. Levine KB, Williams RE, Hartman KE. Vulvovaginal atrophy is strongly associated with female sexual dysfunction among sexually active postmenopausal women. *Menopause* 2008; 15(4-Pt 1): 661–6.

18. Bachman GA, Leiblum SR. The impact of hormones on menopausal sexuality: a literature review. *Menopause* 2004; 11(1): 120–30.

19. Alexander JL, Kotz K, Dennerstein L, et al. The effects of postmenopausal hormone therapies on female sexual functioning: a review of double-blind, randomized controlled trials. *Menopause* 2004; 11(6 Pt 2): 749–65.

20. Lieblum S, Bachman GA, Kemmann E, et al. The importance of sexual activity and hormones. *JAMA* 1983; 249: 2195.

21. Witherby S, Johnson J, Demers L, et al. Topical testosterone for breast cancer patients with vaginal atrophy related to aromatase inhibitors: a phase I/II study. *Oncologist* 2011; 16(4): 424–31.

22. Davis SR, Papalia MA, Norman RJ, et al. Safety and efficacy of a testosterone metered dose transdermal spray for treatment of decreased sexual satisfaction in premenopausal women: a placebo-controlled randomized, dose ranging study. *Ann Inter Med* 2008; 148: 569–77.

23. Goldstat R, Briganti E, Tran J, et al. Transdermal testosterone therapy improves well-being, mood, and sexual function in premenopausal women. *Menopause* 2003; 10(5): 390–8.

24. Davison SL, Bell R, Donath S, et al. Androgen levels in adult females: changes with age, menopause, and oophorectomy. *J Clin Endocrinol Metab* 2005; 90(7): 3847.

25. Mushayandebvu T, Castracane VD, Gimpel T, et al. Evidence for diminished midcycle ovarian androgen production in older reproductive aged women. *Fertil Steril* 1996; 65(4): 721–3.

26. Zumoff B, Strain GW, Miller LK, Rosner W. Twenty-four hour mean plasma Testosterone concentrations declines with age in normal premenopausal women. *J Clin Endocrinol Metab* 1995; 80(4): 1429–30.

27. Longcope C. Adrenal and gonadal secretion in normal females. *J Clin Endocrinol Metab* 1986; 15: 213–28.

28. Bachman G, Bancroft J, Braunstein G, et al. Female androgen insufficiency: the Princeton consensus statement on definition, classification and assessment. *Fertil Steril* 2002; 77(4): 660–5.

29. Shifren J, Braunstein G, Simon J, et al. Transdermal testosterone treatment in women with impaired sexual function after oophorectomy. *N Engl J Med* 2000; 343(10): 682–8.

30. Erekson EA, Martin DK, Zhu K. Sexual function in older women after oophorectomy. *ObGyn* 2012; 120(4): 833–42.

31. Panay N, Al-Azzawi F, Bouchard C, et al. Testosterone treatment of HSDD in naturally menopausal women: the ADORE study. *Climacteric* 2010; 13(2): 121–31.

32. Nachtigall L, Casson P, Lucas J, et al. Safety and tolerability of testosterone patch therapy for up to 4 years in surgically menopausal women receiving oral or transdermal oestrogen. *Gynecol Endocrinol* 2011; 27(1): 39–48.

33. Simon J, Braunstein G, Nachtigall L, et al. Testosterone patch increases sexual activity and desire in surgically menopausal women with hypoactive sexual desire disorder. *J Clin Endocrinol Metab* 2000; 90(9): 5226–33.

34. Swerdloff RS, Wang C, Cunningham G, et al. Long-term pharmacokinetics of transdermal testosterone gel in hypogonadal men. *J Clin Endocrinol Metab* 2000; 85(12): 4500–10.

35. Singh AB, Lee ML, Sinha-Hikim I, et al. Pharmacokinetics of a testosterone gel in healthy postmenopausal women. *J Clin Endocrinol Metab* 2006; 91(1): 136–44.

36. Pharmacy Compounding Accreditation Board. www.pcab.org/prescribers, accessed May 28, 2013.

37. LaCroix AZ, Chlebowski RT, Manson JE, et al. Health outcomes after stopping conjugated equine estrogens among postmenopausal women with prior hysterectomy: a randomized controlled trial. *JAMA* 2011; 305(13): 1305–14.

38. Tamimi RM, Hankinson SE, Chen WY, et al. Combined estrogen and testosterone use and risk of breast cancer in postmenopausal women. *Arch Intern Med* 2006; 166(14): 1483–9.

39. Danforth KN, Eliassen AH, Tworoger SS, et al. The association of plasma androgen levels with breast, ovarian and endometrial cancer risk factors among postmenopausal women. *Int J Cancer* 2010; 126(1): 199–207.

40. Davis SR, Hirschberg AL, Wagner LK, et al. The effect of transdermal testosterone on mammographic density in postmenopausal women not receiving systemic estrogen therapy. *J Clin Endocrinol Metab* 2009; 94(12): 4907–13.

41. McCormack VA, Dowsett M, Folkerd E, et al. Sex steroids, growth factors and mammographic density: a cross-sectional study of UK postmenopausal Caucasian and Afro-Caribbean women. *Breast Cancer Res* 2009; 11(3): R38.

42. Tworoger SS, Lee IM, Buring JE, Hankinson SE. Plasma androgen concentrations and risk of incident ovarian cancer. *Am J Epidemiol* 2008; 167(2): 211–18.

43. Li AJ, Karlan BY. Androgens and epithelial ovarian cancer: what's the connection? *Cancer Biol Ther* 2008; 7(11):1712–16.

44. Jansen I, Powell LH, Crawford S, et al. Menopause and the metabolic syndrome. The Study of Women's Health Across the Nation. *Arch Intern Med* 2008; 168: 1568–75.

45. Montalcini T, Gorgone G, Gazzaruso C, et al. Role of endogenous androgens on carotid atherosclerosis in non-obese postmenopausal women. *Nutr Metab Cardiovasc Dis* 2007; 17(10): 705–11.

46. Karim R, Hodis HN, Stanczyk FZ, et al. Relationship between serum levels of sex hormones and progression of subclinical atherosclerosis in postmenopausal women. *J Clin Endocrinol Metab* 2008; 93(1): 131–8.

47. Shaw LJ, Bairey Merz CN, Azziz R, et al. Post-menopausal women with a history of irregular menses and elevated androgen measurements at high risk for worsening cardiovascular event-free survival: results from the National Institutes of Health – National Heart, Lung and Blood Institute sponsored women's ischemia syndrome evaluation. *J Clin Endocrinol Metab* 2008; 93: 1276–84.

48. Iellamo F, Volterrani M, Caminiti G, et al. Testosterone therapy in women with chronic heart failure: a pilot double-blind, randomized, placebo-controlled study. *J Am Coll Cardiol* 2010; 56(16): 1310–16.

49. Mudali S, Dobs AS, Ding J, et al. Endogenous postmenopausal hormones and serum lipids: the atherosclerosis risk in communities study. *J Clin Endocrinol Metab* 2005; 90(2): 1202–9.

50. Lambrinoudaki I, Christodoulakos G, Rizos D, et al. Endogenous sex hormones and risk factors for atherosclerosis in healthy Greek postmenopausal women. *Eur J Endocrinol* 2006; 154(6): 907–16.

51. Manolakou P, Angelopoulou R, Bakoyiannis C, Bastounis E. The effects of endogenous and exogenous androgens on cardiovascular disease risk factors and progression. *Reprod Biol Endocrinol* 2009; 7: 44.

52. Basson R, Wierman ME, van Lankveld J, Brotto L. Summary of the recommendations on sexual dysfunctions in women. *J Sex Med* 2010; 7(1 Pt 2): 314–26.

53. Lebbe M, Hughes D, Reisch N, Arlt W. Androgen replacement therapy in women. *Expert Rev Endocrinol Metab* 2012; 7(5): 515–29.

54. Barrett-Connor E, Young R, Notelovitz M, et al. A two-year, double-blind comparison of estrogen-androgen and conjugated estrogens in surgically menopausal women. Effects on bone mineral density, symptoms and lipid profiles. *J Reprod Med* 1999; 44(12): 1012–20.

55. Braunstein GD, Sundwall DA, Katz M, et al. Safety and efficacy of a testosterone patch for the treatment of hypoactive sexual desire disorder in surgically menopausal women: a randomized, placebo-controlled trial. *Arch Intern Med* 2005; 165(14): 1582–9.

56. Basaria S, Nguyen T, Rosenson RS, Dobs AS. Effect of methyl testosterone administration on plasma viscosity in postmenopausal women. *Clin Endocrinol (Oxf)* 2002; 57(2): 209–14.

57. Buster JE, Kingsberg SA, Aguirre O, et al. Testosterone patch for low sexual desire in surgically menopausal women: a randomized trial. *Obstet Gynecol* 2005; 105(5 Pt 1): 944–52.

58. Chiuve SE, Martin LA, Campos H, Sacks FM. Effect of the combination of methyltestosterone and esterified estrogens compared with esterified estrogens alone on apolipoprotein CIII and other apolipoproteins in very low density, low density, and high density lipoproteins in surgically postmenopausal women. *J Clin Endocrinol Metab* 2004; 89(5): 2207–13.

59. Davis SR, McCloud P, Strauss BJ, Burger H. Testosterone enhances estradiol's effects on postmenopausal bone density and sexuality. *Maturitas* 1995; 21(3): 227–36.

60. Davis SR, Walker KZ, Strauss BJ. Effects of estradiol with and without testosterone on body composition and relationships with lipids in postmenopausal women. *Menopause* 2000; 7(6): 395–401.

61. Davis SR, van der Mooren MJ, van Lunsen RH, et al. Efficacy and safety of a testosterone patch for the treatment of hypoactive sexual desire disorder in surgically menopausal women: a randomized, placebo-controlled trial. *Menopause* 2006; 13(3): 387–96.

62. Davis SR, Moreau M, Kroll R, et al. APHRODITE Study Team. Testosterone for low libido in postmenopausal women not taking estrogen. *N Engl J Med* 2008; 359(19): 2005–17.

63. de Paula FJ, Soares JM Jr, Haidar MA, et al. The benefits of androgens combined with hormone replacement therapy regarding to patients with postmenopausal sexual symptoms. *Maturitas* 2007; 56(1): 69–77.

64. Dobs AS, Nguyen T, Pace C, Roberts CP. Differential effects of oral estrogen versus oral estrogen-androgen replacement therapy on body composition in postmenopausal women. *J Clin Endocrinol Metab* 2002; 87(4): 1509–16.

65. El-Hage G, Eden JA, Manga RZ. A double-blind, randomized, placebo-controlled trial of the effect of testosterone cream on the sexual motivation of menopausal hysterectomized women with hypoactive sexual desire disorder. *Climacteric* 2007; 10(4): 335–43.

66. Flöter A, Nathorst-Böös J, Carlström K, et al. Effects of combined estrogen/testosterone therapy on bone and body composition in oophorectomized women. *Gynecol Endocrinol* 2005; 20(3): 155–60.

67. Nathorst-Böös J, Jarkander-Rolff M, Carlström K, et al. Percutaneous administration of testosterone gel in postmenopausal women – a pharmacological study. *Gynecol Endocrinol* 2005; 20(5): 243–48.

68. Hofling M, Lundström E, Azavedo E, et al. Testosterone addition during menopausal hormone therapy: effects on mammographic breast density. *Climacteric* 2007; 10(2): 155–63.

69. Hofling M, Hirschberg AL, Skoog L, et al. Testosterone inhibits estrogen/progestogen-induced breast cell proliferation in postmenopausal women. *Menopause* 2007; 14(2): 183–90.

70. Penotti M, Sironi L, Cannata L, et al. Effects of androgen supplementation of hormone replacement therapy on the vascular reactivity of cerebral arteries. *Fertil. Steril* 2001; 76(2): 235–40.

71. Penteado SR, Fonseca AM, Bagnoli VR, et al. Effects of the addition of methyltestosterone to combined hormone therapy with estrogens and progestogens on sexual energy and on orgasm in postmenopausal women. *Climacteric* 2008; 11(1): 17–25.

72. Raisz LG, Wiita B, Artis A, et al. Comparison of the effects of estrogen alone and estrogen plus androgen on biochemical markers of bone formation and resorption in postmenopausal women. *J Clin Endocrinol Metab* 1996; 81(1): 37–43.

73. Sherwin BB, Gelfand MM, Brender W. Androgen enhances sexual motivation in females: a prospective, crossover study of sex steroid administration in the surgical menopause. *Psychosom Med* 1985; 47(4): 339–51.

74. Sherwin BB, Gelfand MM. Sex steroids and affect in the surgical menopause: a double-blind, cross-over study. *Psychoneuroendocrinology* 1985; 10(3): 325–35.

75. Shifren JL, Davis SR, Moreau M et al. Testosterone patch for the treatment of hypoactive sexual desire disorder in naturally menopausal women: results from the INTIMATE NM1 Study. *Menopause* 2006; 13(5): 770–9.

76. Warnock JK, Swanson SG, Borel RW, Zipfel LM, Brennan JJ; ESTRATEST Clinical Study Group.

Combined esterified estrogens and methyltestosterone versus esterified estrogens alone in the treatment of loss of sexual interest in surgically menopausal women. *Menopause* 2005; 12(4): 374–84.

77. Watts NB, Notelovitz M, Timmons MC, et al. Comparison of oral estrogens and estrogens plus androgen on bone mineral density, menopausal symptoms, and lipid-lipoprotein profiles in surgical menopause. *Obstet Gynecol* 1995; 85(4): 529–37.

78. Zang H, Carlström K, Arner P, Hirschberg AL. Effects of treatment with testosterone alone or in combination with estrogen on insulin sensitivity in postmenopausal women. *Fertil Steril* 2006; 86(1): 136–44.

The androgen receptor: basic structure and function

Leo Plouffe Jr.

The androgen receptor is central to the overall understanding of the effects of androgens throughout the body. In this chapter, we will review some of the key elements around the structure of the androgen receptor, the relationship between these structural parameters, and the function of the androgen receptor system. This knowledge enables the clinician to better understand and modulate the impact of androgens in health and in disease states.

Historical context

Understanding the role of androgens in gynecology was for many years driven by a simple approach to androgens and the androgen receptor complex. There were two primary constructs relative to androgens and their action:

- One construct highlights the broad range of activity of testosterone and its androgenic precursors. It states that these compounds act both directly as androgens, and as estrogens. The latter effect occurs via the conversion of testosterone to estradiol by the aromatase enzymatic system. This construct has fueled much scientific exploration, as well as heated debates, around which effects of testosterone truly reflect androgen-mediated effects, as opposed to estrogenic effects achieved through aromatization. In this model, much of the so-called androgenic effects are purely an extension of those ascribed to estrogens.
- The other construct emphasizes the activity of non-aromatizable androgens. These include dihydrotestosterone, as an endogenously produced androgen, and methyltestosterone, as a pharmacologic agent. The historical perspective is that these compounds can only act via the androgen receptor and, therefore, display uniquely androgenic effects.

In this chapter, we will review the current and evolving science of the androgen receptor, its complex dynamics, and how the traditional constructs must be revised in light of the evolving evidence.

High-level overview of the androgen receptor structure

Advances in molecular biology over the past decades have established the androgen receptor as a cornerstone of health and many disease states [1,2]. The androgen receptor is located on the X chromosome [3,4]. Most mutations detected so far in the androgen receptor are associated with disease.

The androgen receptor is similar in structure and function to other steroid hormone receptors, such as the estrogen and the progesterone receptors [5]. The last 20 years have contributed much to the overall understanding of the physiology of these receptors. The classic model for years was one of "lock and key," where a given hormone would bind to its receptor and trigger a specific activity. Competitive inhibitors, so-called antagonists, would have the ability to bind the receptor and block any further activity. It is now obvious that the reality is far more complex and dynamic [5,6].

All of the steroid receptors appear to function in a similar manner [6]. For purposes of the current description, we will refer to the androgen receptor, though much of the understanding has been derived initially from the study of the estrogen and glucocorticoid receptors [7,8].

The free form of the androgen receptor is predominantly found in the cytosol, where it is essentially inactive [9]. This neutral state is the result of a highly

Androgens in Gynecological Practice, ed. Leo Plouffe and Botros Rizk. Published by Cambridge University Press. © Cambridge University Press 2015.

dynamic set of interactions with an array of intracellular factors, referred to as chaperones [9].

The cytoplasm and nucleus are also rich in regulatory factors [10]. Some of these can amplify the activity of the hormone-bound receptor; they are known as co-activators [11]. Other regulatory factors cause downregulation; they are known as co-repressors [11]. These regulatory factors bind to the receptor, even in the absence of the primary androgen substrate. An important concept is that the amount and type of regulatory factors varies by cell type and the prevailing hormone environment within the cell [11].

The general mechanism of hormone action is now well described, though much remains to be further elucidated [5,6,12]. When the androgen reaches the cytoplasm, through diffusion, it binds to the receptor. Generally, two hormone-bound receptors pair up, thus forming a dimer. The hormone-bound dimer reaches the nucleus, where it binds to the target gene and triggers the hormone-linked effect. The binding of the receptor dimer on the DNA occurs in a region referred to as the hormone response element (HRE). While the accompanying regulatory factors are likely to bind the androgen receptor at any time, the most critical time for the modulatory effect is estimated to be at the time of binding of the receptor with DNA.

It is important to understand the broad range of effects that can be seen throughout the body. For the same androgen agonist, the effect varies depending on the target cell, and the magnitude of the effect is modulated based on the balance of co-activators and co-repressors [11]. Some substrates consistently bind to the androgen receptor and block any further activity, so-called pure antagonists [5]. Many more substrates have the ability to act both as agonists and antagonists, depending on the target cell, and on the prevailing balance of co-activators and co-repressors within that cell at that particular point in time [5,6].

One final key concept is that there is a lot of interaction between the various steroid receptors [13]. It has long been recognized that androgens have the ability to bind the estrogen and progesterone receptors [12]. Conversely, estrogens and progestins can bind the androgen receptor [12]. However, recent data have revealed that there are direct interactions between the steroid receptors, which creates even more opportunities for a wide range of activity from any given steroid, including all androgens [13].

Ultimately, the overall system anchored by the androgen receptor is a highly dynamic and complex one. It is very far from the simple and early models that fueled our clinical approach for much of the twentieth century. What is even more important to acknowledge is that we are likely to be still in the early stages of our understanding of the androgen and steroid receptor systems.

Detailed structure of the androgen receptor

The androgen receptor belongs to what is described as a superfamily of nuclear receptors [6]. All of these act as transcription factors. The tally, at the writing of this chapter, enumerates 48 of these in humans, including the receptors for the respective families of steroid hormones, for thyroid hormone, and for vitamin D.

For research purposes, the receptors have been grouped into subfamilies. One grouping (subfamily 3) includes the androgen, progesterone, glucocorticoid, and mineralocorticoid receptors [6].

The androgen receptor shares a common overall structure with other steroid receptors [5,6]. There are many ways of describing the structure of these receptors. A big picture approach describes three components: a region that allows for the binding of regulatory factors (Region A/B), a region that binds to DNA (Region C), and a region that binds the hormone ligand (Region E).

On a more detailed, "anatomical" level, this group of receptors consists of five distinct areas that comprise a total of six regions, as follows [5,6]:

1. The first area encompasses Regions A and B, generally written as Region A/B. This area shows the most variation from one type of steroid receptor to the other. This is the region where the co-regulatory factors described earlier bind to the receptor complex. Functionally, it is referred to as the hormone-independent transactivation function (activation function 1, AF-1) area.

2. The second area is known as Region C. This is the area of greatest similarity between steroid hormone receptors. This area binds directly to DNA, referred to as the DNA-binding domain. It also specifically recognizes target gene promoters (zinc fingers).

3. Next is Region D. It is referred to as the "hinge domain" and is responsible for modulating binding to DNA.

4. The fourth area is Region E. It is the area where the androgen binds to the receptor, or ligand-binding domain (LBD). Functionally, this area acts as a molecular "switch" that stops the receptor from being active in the absence of ligand (or when treated with an antagonist) and that converts it

into an active transcriptional factor in the presence of the hormone. This area is also the target for the binding of specific regulatory factors, dependent on the ligand binding to Region E. This is referred to as the transactivation factor (AF-2) area.

5. The fifth and final area is referred to as Region F. It is not present in all steroid receptors and the role of this region has yet to be elucidated. The androgen receptor does not possess this area.

Most steroid receptors have two forms: estrogen receptor: α and β, progesterone receptor: A and B, and for the androgen receptor: isoforms A (AR-A) and B (AR-B) [12,14]. The B form is the full-length androgen receptor, with a full length of 110 kDa [14].

Function of the androgen receptor

Androgen receptor in its inactive state

A number of receptors from the superfamily have the ability to bind to DNA and cause gene activation in the absence of their primary ligand [6]. This finding has led to the understanding that the neutral state for steroid receptors is very much an active process, linked to proteins found within the cytoplasm that bind the receptor and keep it inactive, so-called chaperone proteins, as already stated [9]. An updated understanding has expanded the number and type of these factors, which are now referred to as the foldosome [15]. Data are evolving that aberrations in the dynamics associated with the foldosome can either trigger or increase the susceptibility to specific disease states [15]. This will be discussed later.

The androgen receptor, in its inactive form, is predominantly found in the cytoplasm and bound by the foldosome, including chaperone proteins. Another set of factors is labeled co-chaperone proteins. Additionally, a number of heat shock proteins further contribute to the stabilization and neutralization process, such as HSP90 and HSP70. Other proteins are also involved in the process, such as p23.

This process of inactivation appears to serve at least two purposes. One is to keep the unbound receptor physiologically inactive. The other is to keep the receptor at the ready to bind its ligands, when they reach the cytoplasm.

The foldosome is best thought of as a dynamic functional unit within the cell. It now appears that the respective components are gathered in a sequential manner. Certain heat shock and chaperone proteins appear to initiate the process and recruit other components to produce the stable androgen receptor [16,17]. Early participants in the foldosome include HSP90, HSP70, HIP, HOP, the co-chaperone p23, the immunophilins cyclosporin A-binding protein (Cyp40), and FK506-binding proteins 51 and 52 (FKBP51 and FKBP52) [15]. The current understanding is that foldosome assembly consists of three individual stages, all of which coexist at any given time within a cell [15]. These stages represent orderly cycles of association and dissociation from the androgen receptor [15].

One stage, thought of as the early stage in the process, is related to the association of HSP70 and HSP40 with the androgen receptor [18]. The next stage, considered the intermediate stage, involves binding of proteins such as HOP to HSP70 and further recruitment of other heat shock proteins, such as HSP90, which are thought to bind directly to the ligand-binding area (Region E) of the androgen receptor [17]. The third stage, referred to as the late or final stage, consists of dissociation from the receptor of HSP70, HSP40, and HOP. This dissociation step allows proteins such as HSP90 to interact with other key components of the foldosome, such as p23 and immunophilins listed above, to allow the receptor complex to achieve full ligand-binding activity [19]. At all times, the process remains in constant flux and each of these stages is unstable until ligand binding occurs [16]. Ligand binding triggers a chain of dissociation of chaperone proteins and the activation of the receptor complex [15].

Ligand binding, receptor activation, and transcriptional activation

The primary ligands for the androgen receptor are so-called androgenic compounds, including agonists, antagonists, and selective agonist-antagonist compounds (or selective androgen receptor modulators – SARMs) [5]. In addition, estrogenic, progestogenic, and other compounds can bind the androgen receptor [12]. For the purpose of this section, we will purely use the term "androgen" or "ligands" to represent any compound that has the ability to bind the androgen receptor.

Androgens do not need any cell surface receptor and are thought to diffuse freely into the cytosol [6]. The androgen binds to the receptor at the ligand-binding region (Region E). This binding process triggers a chain of physical changes to the receptor complex, which leads to dissociation of several components of the foldosome [15]. These changes free up areas in the hinge region of the receptor (Region D), which facilitates (or may be entirely responsible) for the transfer of the receptor complex to the nucleus [20,21]. The areas in the hinge region are called the nuclear localization sequence (NLS), since they allow the re-localization of the androgen receptor complex from the cytosol to the nucleus.

For many years, it was believed that steroid receptors were predominantly found in the nucleus at all times, and that this is where steroid binding occurred. This has now been disproven, with binding primarily occurring in the cytosol [6]. In a similar manner, the process by which the receptor-bound complex reaches the nucleus has evolved from the notion of a passive transfer to the nucleus to a complex and active one [6,22].

The foldosome appears to play a critical role in the transfer of the bound receptor complex to the nucleus, and continues to play a role when the receptor reaches the nucleus and binds to DNA [23]. Transport of the receptor complex to the nucleus appears to be mediated, or at the very least modulated, by elements of the foldosome such as HSP90, FKBP51, and FKBP52 [15,23,24]. An attractive model to explain these complex interactions is that transport of the receptor complex to the nucleus is linked to tubulin proteins, which essentially serve as a highway within the cell to transport bound receptors to the nucleus [22]. The foldosome components would then play a key role in association of the receptor complex to the tubulin proteins. This construct opens up a number of new therapeutic opportunities targeting the foldosome complex, as well as tubulin proteins. This will be discussed later in the chapter.

Once in the nucleus, a number of additional transformational steps occur, many of which involve further changes to the interaction between the remaining foldosome components and the bound androgen receptor.

At some point in the transition from the cytosol to the nucleus, the receptor complex binds to another activated receptor complex, a process referred to as dimerization. Since there are two types of androgen receptors, as stated above, dimers may consist of coupled AR-A or AR-B, so-called homodimers, or coupling of an AR-A and an AR-B, referred to as heterodimers [6]. The dimerization process is linked to physico-chemical changes that occur within the DNA-binding area (Region C) [6].

At or near the time of binding of the androgen receptor to the DNA androgen response element, much dynamic activity occurs. This is the predominant time when co-activators or co-repressors bind to the androgen receptor complex [6]. These interactions occur in the activation factor 1 area (Region A/B) as well as in the activation factor 2 area (Region E – which is also the ligand-binding region). There appears to be much competition among these regulatory proteins for their binding to the receptor complex, and it is clear that the ligand itself modulates which regulatory proteins are more likely to bind and upregulate or downregulate transcriptional activity. It also appears that the process is not a static one, where one set of factors associates with the receptor dimer complex and results in a set transcriptional activity, but much more of a dynamic state where regulatory factors associate and dissociate in a cyclic fashion [15]. Last but not least, components of the foldosome also appear to play a key modulatory role at this stage of the process [15].

As already stated, the androgen receptor dimer binds DNA at areas known as androgen response elements (AREs). Over 1500 AREs have been identified in the human genome. This clearly highlights the wide array of complex actions that androgens may exert. In addition, more than half of these ARE sites can bind the androgen receptor, but are about half the size of regular response elements, suggesting the potential for even more complex interactions occurring at these so-called half sites [25].

In addition to binding to DNA at AREs, the androgen receptor can bind to glucocorticoid response elements (GREs), just like the progesterone and mineralocorticoid receptors [26].

Co-activators and co-repressors

As already stated, a number of proteins modulate the ultimate transcriptional activity of any compound that binds the androgen receptor. These are divided into two groups, co-activators, which upregulate activity, and co-repressors, which downregulate activity (5,6). As is the case for all other components of the androgen receptor system, these display a high level of complexity.

So far, many more co-activators have been found compared with co-repressors [11]. Classic co-activators include families of proteins described as SRC1, SRC2, and SRC3 [11]. A large number of additional co-activators continue to be identified, now numbering in the hundreds [11].

The mechanisms by which co-regulators exert their action continue to be the focus of much research [11]. They appear to regulate transcription through a wide variety of mechanisms, primarily targeting the configuration of DNA during the transcription process. The activity includes modulation of such key DNA-linked enzymes as histone acetyltransferase activity, histone deacetylase activity, arginine methyltransferase activity, ubiquitin ligase activity, and ATP-dependent chromatin-remodeling activity.

It is essential to grasp the concept that the ultimate effect of any given androgen depends on the complex interaction between the receptor, the components of the foldosome, and the balance of co-activators and co-repressors present in any particular cell at the time. Modulation of these co-activators and co-repressors, either from the ligand itself (e.g., agonist vs. SARM) or via the prevalent intracellular homeostasis, is critical to the ultimate physiologic effect seen. This system also lends itself to multi-site targeting for therapeutic purposes.

Androgen receptor linked molecular disorders

The androgen receptor gene is located on the long arm of the X chromosome, in region Xq11-q12 [3,4]. Structurally, the gene is more than 90 kB long. One exon (exon 1) encodes for Region A/B, and consists of 1586 bp. The DNA-binding domain (Region C) is encoded by two distinct exons (exon 2 and 3), respectively 152 and 117 bp in length. The androgen-binding domain (Region E) consists of five exons, with lengths from 131 to 288 bp.

Well over 600 mutations have been reported in the androgen receptor gene [27]. These have been associated with a wide variety of conditions relevant to the gynecologist, such as complete androgen insensitivity syndrome, male infertility, and malignancies of the breast, endometrium, and colorectum. It is important to remember that the associations do not necessarily imply a causal relationship,

Androgen insensitivity conditions

Androgen insensitivity syndrome presents as a continuum of disorders, from a fully female external phenotype (historically referred to as testicular feminization syndrome) to milder forms of androgen insufficiency states in otherwise phenotypic males [12]. It has classically been described as an X-linked disorder, where XY individuals inherit an abnormal androgen receptor on the maternal X chromosome. This is an oversimplification.

Molecular studies in individuals with complete androgen insensitivity syndrome (CAIS) have revealed a number of alterations within the androgen receptor gene. Nearly 50 mutations have been identified in individuals with CAIS [28]. Some individuals with CAIS are afflicted with defects in the ligand-binding region, while others show mutations in the DNA-binding domain [29,30]. The responsible mutations in many cases introduce a premature termination codon or produce aberrant mRNA splicing [31,32]. The clinical phenotype, thus, appears to vary depending on the extent to which the critical binding regions of the androgen receptor become truncated; most CAIS cases are associated with more encompassing stop deletions. Rarely, subjects with CAIS suffer from a complete deletion of the androgen receptor gene [33].

Mutations in the androgen receptor gene may be inherited or result from somatic cell mutations, including wild-type and mutant mosaicism for the receptor [34–36]. In these instances, the phenotype can range from classic CAIS to various forms of under-masculinization. It has been estimated that about 70% of androgen insensitivity syndromes are inherited in a classic X-linked recessive manner; the remaining 30% of cases represent somatic cell mutation, with a high potential for cellular mosaicism in such cases [37]. It is these latter situations that primarily lead to the broad range of presentations of androgen insensitivity states [29,37]. However, even familial clusters of androgen insensitivity can present with a wide phenotypic range, with some XY individuals presenting with a phenotype very close to CAIS, with the exception of some degree of pubic hair development, while siblings have a male phenotype with hypospadias and cryptorchidism [38].

In summary, androgen insensitivity syndrome, ranging from CAIS to slight under-masculinization, provides a window into the complexity of the androgen receptor system. A variety of mutations in different regions of the androgen receptor are implicated. The mutations may be transmitted predictably as an X-linked recessive disorder through large pedigrees across many generations, or result from germ cell mosaicism in a mother, with significant discordance across siblings. De novo somatic cell mutations with diffuse cellular mosaicism can also be seen in other individuals, in the absence of any family history. Last but not least, even among siblings carrying the same mutated androgen receptor, there appears to be considerable variation in the clinical presentation of androgen insensitivity demonstrated. Research is ongoing in this area, and the knowledge can be applicable to our overall understanding of the androgen receptor system.

Kennedy disease (spinal and bulbar muscular atrophy)

Kennedy disease, also known as spinal and bulbar muscular atrophy, is a rare X-linked recessive form of spinal muscular atrophy [39]. While this is an uncommon genetic neurological condition, research on this condition has expanded greatly our understanding of the androgen receptor and the broad range of consequences associated with its disorders. The disease is a slowly progressive condition characterized by a progressive limb and bulbar muscle weakness, usually with onset in the third to fifth decade of life. The neurological conditions progress to muscle weakness accompanied by muscle wasting, tremor, muscular fasciculations, especially perioral, and dysphagia. Gynecomastia, as well as infertility, has also been associated with the condition [40]. Hyperlipoproteinemia has also been described with these patients.

Among patients affected with infertility, testicular biopsy in some of the patients identified pronounced involutional changes in Leydig cells [41]. Circulating testosterone levels were also found to be decreased in these individuals.

The molecular defect associated with Kennedy disease is located on Xq21.3-q22 [42]. It consists of a trinucleotide CAG repeat expansion in exon 1 of the androgen receptor gene. The normal wild-type gene includes 10–36 such repeats in exon 1, in contrast to 38–62 repeats in patients with Kennedy disease. Extensive studies on a histological and molecular level have revealed a number of findings, all of which point to the ubiquitous nature of androgen receptor function. In addition to the impact at the level of the breast tissue (gynecomastia) and the testes, Kennedy disease highlights the contribution of the androgen receptor system to the integrity of the neuromuscular unit, with a direct impact on both sensory and motor neurons. Many of these findings have been directly linked to a degree of androgen resistance (e.g., gynecomastia, testicular findings).

In summary, CAIS and Kennedy disease, while rare disorders, highlight the far-ranging impact of the androgen receptor system, beyond what are considered its traditional targets. This encompasses the genital tract, systemic targets such as muscle mass, and even less obvious targets such as the sensory-neural system [43].

The androgen receptor and breast cancer

The role of androgens in managing breast cancer provides an interesting example of successive empiric approaches to medical therapy, reflecting the evolution of medical knowledge. The basic epidemiology of breast cancer, predominantly affecting women, led to the conclusion that breast cancer reflected a relative imbalance in the effect of estrogens and androgens [44]. The logical conclusion was that androgens would be an effective form of treatment for breast cancer [44]. With the realization that many androgenic compounds are readily metabolized to estrogen, the place of androgen therapy was reassessed, with the demonstration that androgens still yielded a positive therapeutic effect [45]. This leads to the interesting situation that androgens may exert a positive effect in treating breast cancer [45], while data suggest that androgens may have a very significant role in modulating, potentially increasing, the risk of breast cancer [46].

Extensive research is ongoing in this area. It has now become obvious that a number of the preclinical, in vitro models used show a great heterogeneity in the response to androgens, even when focusing on androgens that cannot be aromatized to estradiol [44]. Based on the factors discussed previously, much of this heterogeneity in the response of different breast cancer cell lines to a range of androgens may reflect marked differences in the co-activator, co-repressor, and chaperone environment seen across different experimental systems. The overall conclusion, however, is that androgens exert an important role in the modulation of the growth of breast cancer cells, and are therefore a promising target for breast cancer treatment [44].

The molecular basis for therapeutic interventions

Based on the understanding of the androgen receptor system described above, a number of opportunities for therapeutic intervention become obvious. From a therapeutic perspective, one can consider the following classification of compounds that offer therapeutic potential [5]:

- Full agonists: these compounds trigger all activation steps, and function essentially in the same manner as endogenous hormones. Physiologically, these molecules appear to bind the androgen receptor in a manner analogous to the endogenous androgens, and trigger a sequence of events fully aligned with those that occur in the presence of the endogenous ligand.
  - A variety of testosterone derivatives fall in this category, such as testosterone undecanoate and methyltestosterone [47,48].
- SARMs: these compounds trigger a complex range of predictable responses. In some cells they act as strong agonists while in others they elicit activity that is weaker than endogenous hormones. This is often referred to as a partial agonist or partial antagonist activity, since in a competitive binding state, the overall consequence is to attenuate the impact of endogenous androgens. Importantly, SARMs also hold the potential to completely block androgen activity in certain cells. It must be remembered that the activity of a SARM is dependent not only on the target cell, but also on the intracellular environment at the time, including regulatory proteins and components of the foldosome, discussed earlier.
  - Much research has been conducted on how SARMs exert their selective and differentiated mode of action. Most of the data point to a set of complex interactions at the time of binding to the hormone response element. SARMs appear to either allow full agonist expression or block, by structural re-alignment of the DNA receptor–dimer complex, the recruitment of regulatory proteins and other key factors required for transcription.
  - SARMs are not yet part of the clinical armamentarium, but they are the focus of intensive research. Of particular interest is their anabolic effects, especially on the musculoskeletal system under conditions of extreme stress, such as cancer [49].
- Full antagonists: these compounds completely block the activity of the androgen receptor. The mechanism of action is similar to that described above for SARMs, except that these compounds universally and predictably squelch any androgenic activity when they bind to the androgen receptor; and generally do so with great affinity, thus largely displacing endogenous androgens at the same time.
  - Some compounds of this group have been widely used in the clinical setting, including flutamide and bicalutamide [5]. While spironolactone has a wide range of other effects (e.g., diuresis), it falls in this general category relative to androgen antagonism.
- Destabilizers: these ligands impact the androgen receptor pathway by interfering with the receptor in the cytosol or during transport to the nucleus. The ultimate result is a reduction in the number of activated receptors, and therefore downregulation of androgenic activity. Compounds in this group are still in the research phase [5].
- Endogenous-like analogs: as has been discussed previously, steroid hormones can bind multiple hormone receptors. Progestins and estrogens, as well as many of their analogs, have the ability to bind the androgen receptor [5,12]. In most instances, the physiologic effect is less than what is seen with endogenous androgens. This partial antagonistic activity has long been leveraged in treating androgen excess disorders with progestins, such as cyproterone acetate or drospirenone [12].

Conclusion

Our understanding of the androgen receptor system has evolved markedly over the past two decades. The system is marked by a high level of complexity, shared with other steroid receptors. The range of receptor-linked defects is broad and is marked with a wide range of phenotypic variations for seemingly similar defects. Last but not least, it is now clearly established that the androgen receptor plays a key role well beyond the homeostasis of traditional androgen-target organs. This awareness opens up a range of diagnostic and therapeutic opportunities in the field of gynecology.

References

1. Kelly DM, Jones TH. Testosterone: a metabolic hormone in health and disease. *J Endocrinol* 2013;217(3):R25–45.
2. Kelly DM, Jones TH. Testosterone: a vascular hormone in health 1 and disease. *J Endocrinol* 2013;217(3):R47–71.

3. Lubahnn DB, Joseph DR, Sullivan PM, et al. Cloning of human androgen receptor complementary DNA and localization to the X chromosome. *Science* 1998;240(4850):327–330.

4. Brown CJ, Goss SJ, Lubahn DB, et al. Androgen receptor locus on the human X chromosome: regional localization to Xq11-12 and description of a DNA polymorphism. *Am J Hum Genet* 1989;44(2):264–269.

5. Cleve A, Fritzemeier KH, Haendler B, et al. Pharmacology and clinical use of sex steroid hormone receptor modulators. *Handb Exp Pharmacol* 2012;214:543–587.

6. Burris TP, Solt LA, Wang Y, et al. Nuclear receptors and their selective pharmacologic modulators. *Pharmacol Rev* 2013;65(2):710–778.

7. Toft D, Gorski J. A receptor molecule for estrogens: isolation from the rat uterus and preliminary characterization. *Proc Natl Acad Sci USA* 1966;55(6):1574–1581.

8. Hollenberg SM, Weinberger C, Ong ES, et al. Primary structure and expression of a functional human glucocorticoid receptor cDNA. *Nature* 1985;318(6047):635–641.

9. Pratt WB, Galigniana MD, Morishima Y, Murphy PJ. Role of molecular chaperones in steroid receptor action. *Essays Biochem* 2004;40:41–58.

10. Meyer ME, Gronemeyer H, Turcotte B, et al. Steroid hormone receptors compete for factors that mediate their enhance function. *Cell* 1989;57(3):433–442.

11. Dasgupta S, Lonard DM, O'Malley BW. Nuclear receptor coactivators: master regulators of human health and disease. *Annu Rev Med* 2014;65:279–292.

12. Fritz MA, Speroff L. *Clinical Gynecologic Endocrinology and Infertility* (8th Ed.). (2011) Wolters Kluwer, Philadelphia.

13. Grubisha MJ, DeFranco DB. Local endocrine, paracrine and redox signaling networks impact estrogen and androgen crosstalk in the prostate cancer microenvironment. *Steroid* 2013;78(6):538–541.

14. Wilson CM, McPhaul MJ. A and B forms of the androgen receptor are present in human genital skin fibroblasts. *Proc Natl Acad Sci USA* 1994;91(4):1234–1238.

15. Cano LQ, Lavery DN, Bevan CL. Mini-review: foldosome regulation of androgen receptor action in prostate cancer. *Mol Cell Endocrinol* 2013;369(1–2):52–62.

16. Heinlein CA, Chang C. Role of chaperones in nuclear translocation and transactivation of steroid receptors. *Endocrine* 2001;14(2):143–149.

17. Pratt WB, Toft DO. Steroid receptors interactions with heat shock protein and immunophilin chaperones. *Endocr Rev* 1997;18(3):306–360.

18. Dittmar KD, Banach M, Galigniana MD, Pratt WB. The role of DNA-like proteins in glucocorticoid receptor. hsp90 heterocomplex assembly by the reconstituted hsp90.p60.hsp70 foldosome complex. *J Biol Chem* 1998; 273(13):7358–7366.

19. Grenert JP, Sullivan WP, Fadden P, et al. The amino-terminal domain of heat shock protein 90 (hsp90) that binds geldanamycin is an ATP/ADP switch domain that regulates hsp90 conformation. *J Biol Chem* 1997;272(38):23843–23850.

20. Black BE, Paschal BM. Intranuclear organization and function of the androgen receptor. *Trends Endocrinol Metab* 2004;15(9):411–417.

21. Cutress ML, Whitaker HC, Mills LG, et al. Structural basis for the nuclear import of the human androgen receptor. *J Cell Sci* 2008;121(Pt 7):957–968.

22. Thadani-Mulero M, Nanus DM, Giannakakou P. Androgen receptor on the move: boarding the microtubule expressway to the nucleus. *Cancer Res* 2012;72(18):4611–4615.

23. Echeverria PC, Picard D. Molecular chaperones, essential partners of steroid hormone receptors for activity and mobility. *Biochim Biophys Acta* 2010;1803(6):641–649.

24. Vandevyver S, Dejager L, Libert C. On the trail of the glucocorticoid receptor: into the nucleus and back. *Traffic* 2012;13(3):364–374.

25. Massie CE, Adryan B, Barbosa-Morais NL, et al. New androgen receptor genomic targets show an interaction with the ETS1 transcription factor. *EMBO Rep* 2007;8(9):871–878.

26. Claessens F, Gewirth DT. DNA recognition by nuclear receptors. *Essays Biochem* 2004;40:59–72.

27. Gottlieb B, Beitel LK, Wu JH, Trifiro M. The androgen receptor gene mutations database (ARDB): 2004 update. *Hum Mutat* 2004;23(6): 527–533. Note: Erratum: *Hum Mutat* 24: 102 only, 2004.

28. Sultan C, Lumbroso S, Poujol N, et al. Mutations of androgen receptor gene in androgen insensitivity syndromes. *J Steroid Biochem Mol Biol* 1993;46(5): 519–530

29. McPhaul MJ, Marcelli M, Zoppi S, et al. Mutations in the ligand-binding domain of the androgen receptor gene cluster in two regions of the gene. *J Clin Invest* 1992;90(5):2097–2101.

30. McPhaul MJ, Marcelli M, Zoppi S, et al. Genetic basis of endocrine disease 4: the spectrum of mutations in the androgen receptor gene that causes androgen resistance. *J Clin Endocrinol Metab* 1993;76(1):17–23.

31. Ris-Stalpers C, Kuiper GGJM, Faber PW, et al. Aberrant splicing of androgen receptor mRNA results in synthesis of a nonfunctional receptor protein in a

patient with androgen insensitivity. *Proc Natl Acad Sci USA* 1990;87(20): 7866–7870.

32. Holterhus P-M, Bruggenwirth HT, Hiort O, et al. Mosaicism due to a somatic mutation of the androgen receptor gene determines phenotype in androgen insensitivity syndrome. *J Clin Endocrinol Metab* 1997;82(11):3584–3589.

33. Quigley CA, Friedman KJ, Johnson A, Complete deletion of the androgen receptor gene: definition of the null phenotype of the androgen insensitivity syndrome and determination of carrier status. *J Clin Endocrinol Metab* 1992;74(4):927–933.

34. Hiort O, Sinnecker GHG, Holterhus P-M, et al. Inherited and de novo androgen receptor gene mutations: investigation of single-case families. *J Pediatr* 1998;132(6):939–943.

35. Gottlieb B, Beitel LK, Trifiro MA. Variable expressivity and mutation databases: the androgen receptor gene mutations database. *Hum Mutat* 2001;17(5):382–388.

36. Holterhus P-M, Wiebel J, Sinnecker GHG, et al. Clinical and molecular spectrum of somatic mosaicism in androgen insensitivity syndrome. *Pediatr Res* 1999;46(6):684–690.

37. Kohler B, Lumbroso S, Leger J, et al. Androgen insensitivity syndrome: somatic mosaicism of the androgen receptor in seven families and consequences for sex assignment and genetic counseling. *J Clin Endocrinol Metab* 2005;90(1):106–111.

38. Rodien P, Mebarki F, Mowszowicz I, et al. Different phenotypes in a family with androgen insensitivity caused by the same M780I point mutation in the androgen receptor gene. *J Clin Endocrinol Metab* 1996;81(8):2994–2998.

39. Kennedy WR, Alter M, Sung JH. Progressive proximal spinal and bulbar muscular atrophy of late onset: a sex-linked recessive trait. *Neurology* 1968;18(7):671–680.

40. Harding AE, Thomas PK, Baraitser M, et al. X-linked recessive bulbospinal neuronopathy: a report of ten cases. *J Neurol Neurosurg Psychiatry* 1982;45(11):1012–1019.

41. Hausmanowa-Petrusewicz I, Borkowska J, Janczewski Z. X-linked adult form of spinal muscular atrophy. *J Neurol* 1983;229(3):175–188.

42. Fischbeck KH, Ionasescu V, Ritter AW, et al. Localization of the gene for X-linked spinal muscular atrophy. *Neurology* 1986;36(12):1595–1598.

43. Palazzolo I, Gliozzi A, Rusmini P, et al. The role of the polyglutamine tract in androgen receptor. *J Steroid Biochem Mol Biol* 2008;108(3–5):245–253.

44. Garay JP, Park BH. Androgen receptor as a targeted therapy for breast cancer. *Am J Cancer Res* 2012;2(4):434–445.

45. Council on Drugs, Subcommittee on Breast and Genital Cancer, Committee on Research, A.M.A. Androgens and estrogens in the treatment of disseminated mammary carcinoma – Retrospective study of 944 patients. *JAMA* 1960;172(12):1271–1283.

46. Campagnoli C, Pasanisi P, Castellano I, et al. Postmenopausal breast cancer, androgens, and aromatase inhibitors. *Breast Cancer Res Treat* 2013;139(1):1–11.

47. Edelstein D, Basaria S. Testosterone undecanoate in the treatment of male hypogonadism. *Expert Opin Pharmacother* 2010;11(12):2095–2106.

48. Thevis M, Schanzer W. Synthetic anabolic agents: steroids and nonsteroidal selective androgen receptor modulators. *Handb Exp Pharmacol* 2010; 195:99–126.

49. Dalton JT, Taylor RP, Mohler ML, Steiner MS. Selective androgen receptor modulators for the prevention and treatment of muscle wasting associated with cancer. *Curr Opin Support Palliat Care* 2013;7(4):345–351.

Chapter

6

Androgens in postmenopausal women: their practically exclusive intracrine formation and inactivation in peripheral tissues

Fernand Labrie

Summary

The specificity, reliability, and precision of mass spectrometry assays for steroids combined with the molecular biology-based identification of the enzymes responsible for sex steroid formation have permitted important progress in our understanding of androgen physiology in women. Women are protected at menopause from the stimulatory effect of estrogens on the endometrium by the arrest of ovarian estrogen secretion. The continuing need of sex steroids for the normal functioning of most tissues is provided by the cell-specific local formation of androgens and estrogens from dehydroepiandrosterone (DHEA) achieved according to the mechanisms of intracrinology, thus avoiding the potential systemic effects of sex steroids. Consequently, the low serum testosterone in women does not adequately represent the total pool of androgens, which are essentially made intracellularly from the precursor DHEA secreted mainly (about 80%) by the adrenals, while a significant amount (approximately 20%) is secreted by the postmenopausal ovary that does not secrete significant amounts of testosterone in the circulation. The estimate of total androgens in women can best be obtained by the serum level of the glucuronide and sulfate derivatives of the androgen metabolites, thus replacing the measurement of serum testosterone, which only corresponds to a variable and low leakage from peripheral tissues.

Introduction

The introduction of mass spectrometry (MS)based assays for testosterone has facilitated major progress in our understanding of androgen formation and metabolism in women. Most importantly, it became apparent that the concentration of serum testosterone is not a valid parameter of total androgen activity in women

[1], since only a small and variable fraction of testosterone made intracellularly from dehydroepiandrosterone (DHEA) diffuses into and can be measured in the circulation according to the mechanisms of intracrinology [2]. This conclusion is in agreement with the observation that the serum levels of all sex steroids are similar in postmenopausal women and castrated men of similar age [3], thus suggesting a negligible role of the ovary for the direct secretion of testosterone in the circulation. In both cases, the exclusive source of sex steroids is the adrenals. In fact, further studies have indicated that in postmenopausal women, the best estimate is that about 80% of circulating DHEA is of adrenal origin, while approximately 20% is from an ovarian source [4].

Well-recognized importance of mass spectrometry for reliable assays of testosterone

The uncertain reliability of the serum testosterone assays in women has been summarized in the following words from the position statement of the North American Menopause Society: "especially given the relative unreliability of most clinically available testosterone assays for women …" [5]. In fact, as well indicated by successive and recent position statements of the United States Endocrine Society [6,7], the ability to measure testosterone and estradiol (E₂) with sufficient reliability, sensitivity, and specificity in postmenopausal women is a continuing and serious problem. "None of the immunoassays tested was sufficiently reliable for the investigation of sera from children and women in whom very low and low serum testosterone concentrations are expected." [8,9] The basis of the problem is that immuno-based assays for testosterone lack specificity [10], thus leading to the now well-recognized major importance of using

Androgens in Gynecological Practice, ed. Leo Plouffe and Botros Rizk. Published by Cambridge University Press. © Cambridge University Press 2015.

MS-based assays [1,11–15]. The greater specificity of the MS-based assays explains the lower levels of serum steroid levels measured by gas chromatography (GC)-MS/MS and liquid chromatography (LC)-MS/MS compared with immuno-based assays [16].

Crucial role of intracrinology or extraovarian formation of androgens

It is of particular interest to summarize the mechanisms explaining the low levels of serum testosterone observed in women. It is also important to mention that the low circulating concentrations of testosterone in women do not mean that the tissue concentrations of testosterone are as low as the blood levels might suggest. In fact, the explanation for the low circulating levels of testosterone in women is that androgens are practically all made, act, and are metabolized intracellularly in the peripheral target tissues according to the

mechanisms of intracrinology [2,17]. In fact, a particularly remarkable and highly sophisticated achievement of evolution relates to the mechanisms that permit cell-specific production of sex steroids for a strictly local action without biologically significant release of active sex steroids in the circulation, thus avoiding systemic effects [18]. This mechanism has been named intracrinology [2,19], following observations made in the early 1980s in men castrated for prostate cancer [20] showing that an important proportion of the androgens present and acting in the prostate are made from DHEA in the prostate itself. The finding of a dual source of androgens in men has been the basis for the development of the first treatment shown to prolong life in prostate cancer, namely combined androgen blockade [20–22]. With combined androgen blockade, testicular androgens are eliminated by surgical or medical castration [23], while the action of the androgens made locally in the prostate is neutralized by a pure

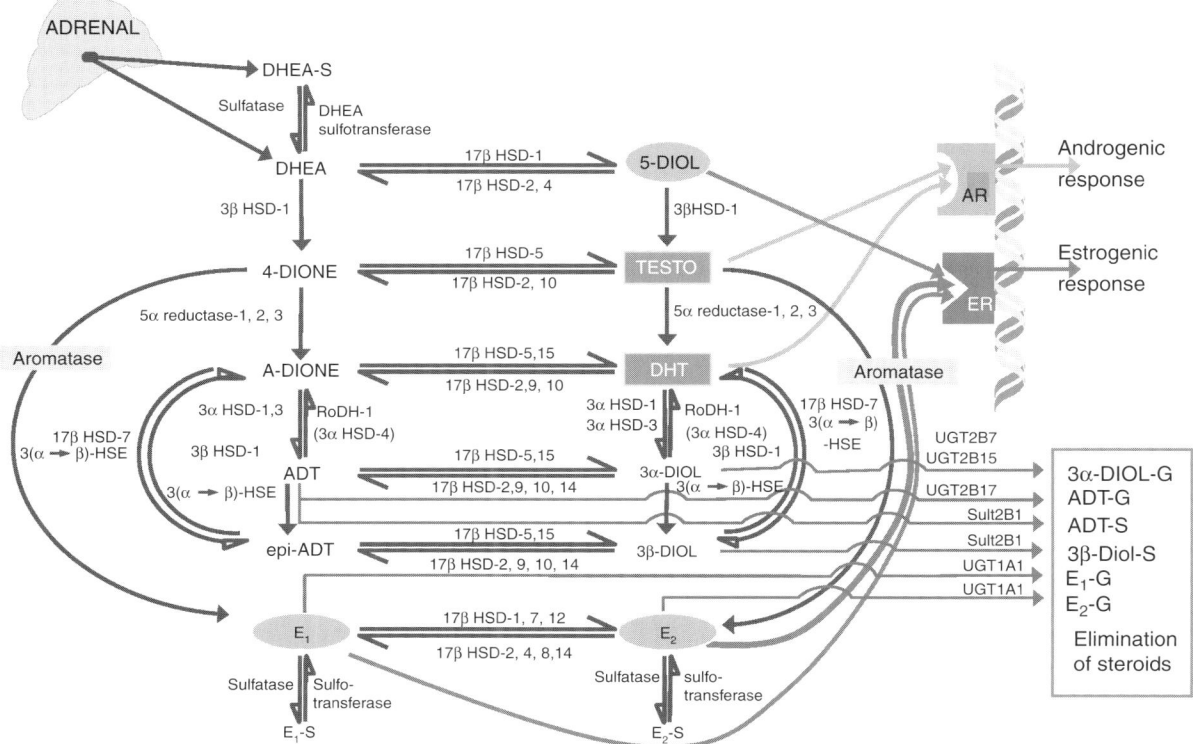

Figure 6.1 Human steroidogenic and steroid-inactivating enzymes in peripheral intracrine tissues. 4-dione, androstenedione; A-dione, 5α-androstane-3,17-dione; ADT, androsterone; epi-ADT, epiandrosterone; E_1, estrone; E_1-S, estrone sulfate; E_2, 17β-estradiol; E_2-S, estradiol sulfate; 5-diol, androst-5-ene-3α, 17β-diol; HSD, hydroxysteroid dehydrogenase; testo, testosterone; RoDH-1, Ro dehydrogenase 1; ER, estrogen receptor; AR, androgen receptor; UGT2B28 family (including UGT2B7, UGT2B15, and UGT2B17), uridine glucuronosyl transferase 2B28; Sult2B1, sulfotransferase 2B1; UGT1A1, uridine glucuronosyl transferase 1A1. A black and white version of this figure will appear in some formats. For the color version, please refer to the plate section.

antiandrogen, which blocks the access of the locally made androgens to the androgen receptor [20].

Through an estimated 500 million years [24], evolution has progressively inserted in the peripheral tissues all the enzymes required to locally make and inactivate sex steroids from DHEA (Figure 6.1). On the other hand, it is about 50 million years ago that the adrenals of primates gained the ability to secrete large amounts of the precursor DHEA, which has recently been demonstrated to be transformed into androgens and/or estrogens by the mechanisms of intracrinology, using the tissue-specific steroidogenic enzymes already in place [2,13,15,16,25,26] (Figure 6.1). Major and essential progress in this area has been made by elucidation of the structure of practically all the tissue-specific genes that encode the human steroidogenic enzymes responsible for the transformation of DHEA into androgens and/or estrogens in the peripheral target tissues [26–32] (Figure 6.1).

DHEA becomes the practically exclusive source of sex steroids (both androgens and estrogens) after menopause

At menopause, the ovary becomes completely depleted of estrogen-producing follicles and the secretion of E_2 by the ovary into the circulation practically ceases. The consequence is that serum E_2 decreases from values of at least 80 pg/mL in premenopausal women to an average of 4.2 pg/mL after menopause, with 95% of women having serum E_2 concentrations below 9.2 pg/mL [4]. These low biologically inactive E_2 concentrations avoid stimulation of the endometrium, which shows atrophy in all women after menopause. This extremely positive aspect of menopause, namely the protection from rapid appearance of endometrial cancer, has provided a decisive factor for evolutionary forces to choose the lineage of women having menopause and non-estrogen-secreting ovaries after the reproductive years. For serum testosterone, on the other hand, no significant change [33,34] or a small 15% decline [35] has been reported between pre- and postmenopause. In fact, as mentioned above, the serum levels of testosterone in postmenopausal women are similar to those of castrated men [3], while in intact men, serum testosterone is about 25-fold higher due to the direct secretion of testosterone in the bloodstream by the testicles.

While the cessation of reproduction and the arrest of ovarian estrogen secretion long before the end of life are essential characteristics of menopause, it is equally important to recognize that in the hypothetically complete absence of sex steroids, the life of women after menopause would have been of poor quality and likely to be seriously shortened, since practically all bodily functions are modulated, to various degrees, by estrogens and/or androgens [13,36]. In fact, in the absence of the estrogens and androgens made specifically in each cell type of each tissue from circulating DHEA by intracrine mechanisms, the problems presently affecting women at menopause, especially osteoporosis and fractures, hot flushes, muscle loss, type 2 diabetes, vulvovaginal atrophy, sexual dysfunction, memory loss, cognition loss, and possibly Alzheimer's disease, would be much more serious than presently observed with a likely significant reduction in quality of life and even in lifespan. In other words, while serum E_2 must remain at sub-threshold or biologically inactive concentrations in the bloodstream after menopause in order to avoid endometrial stimulation and the risk of endometrial cancer, the normal functioning of most of the other peripheral tissues requires intracellular biologically active concentrations of estrogens and/or androgens.

Medical research, however, has concentrated almost exclusively on the arrest of E_2 and progesterone secretion by the ovary and how to replace estrogens. One practically never envisaged that the arrest of secretion of E_2 by the ovary into the circulation at menopause could be a positive factor resulting from elimination of the risk of endometrial hyperplasia and carcinoma, instead of being exclusively a negative phenomenon requiring estrogen replacement. As mentioned above, in the presence of the steroid-forming and steroid-inactivating enzymes specifically expressed in each cell type of each human peripheral tissue (Figure 6.1), coupled with the availability of the precursor DHEA (Figure 6.2), all the elements were in place for delivery of the required amounts of sex steroids to each cell in each tissue, and thus minimize the negative impact of the arrest of E_2 secretion by the ovary at menopause. It does not mean that the arrest of E_2 secretion by the ovary has no causative role in the menopausal symptoms, but one has to consider that evolution/nature has designed an alternative to E_2 replacement, namely DHEA, a precursor inactive by itself, in order to distribute the sex steroids needed in a tissue-specific manner in the whole organism without biologically significant release of the active sex steroids in the circulation.

After menopause

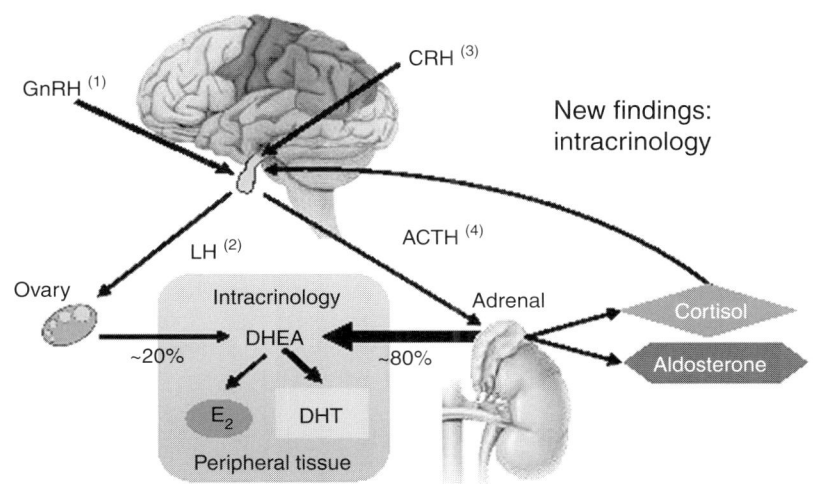

Figure 6.2 Schematic representation of dehydroepiandrosterone (DHEA) acting as the unique source of sex steroids in women after menopause. Approximately 80% of circulating DHEA is of adrenal origin while about 20% is released from the ovary [4]. Accordingly, after menopause, all androgens and all estrogens are made locally from DHEA in peripheral target tissues. The amount of sex steroids made depends upon the level of the steroid-forming enzymes specifically expressed in each cell in each tissue. GnRH, gonadotropin-releasing hormone; LH, luteinizing hormone; CRH, corticotropin releasing hormone; ACTH, adrenocorticotropic hormone; E_2, 17β-estradiol; DHT, dihydrotestosterone. From Labrie et al. [4]. A black and white version of this figure will appear in some formats. For the color version, please refer to the plate section.

(1) Gonadotropin-releasing hormone.
(2) Luteinizing hormone.
(3) Corticotropin-releasing hormone.
(4) Adrenocorticotropic hormone.

The secretion of DHEA, however, markedly decreases with age, with an average 60% loss already observed at the time of menopause [1,21,37,38]. The marked reduction in the secretion of DHEA with age [38] results in a parallel fall in the formation and availability of androgens and estrogens in peripheral target tissues, a situation believed to be associated with the series of medical problems of menopause mentioned above [32].

It is very important to mention that an essential aspect of intracrinology is the fact that the active sex steroids are not only made locally, but that they are also inactivated locally in the same cells where synthesis takes place. In fact, the sex steroids made from DHEA in peripheral tissues are essentially released outside the cells as inactive compounds. As illustrated in Figure 6.3, DHEA of either adrenal, ovarian, or exogenous (for example, tablet or cream) origin is distributed by the general circulation to all tissues indiscriminately. The transformation of DHEA into estrogens/androgens, however, is tissue specific, ranging from none in the endometrium to various levels in the other tissues of the human body. Most importantly, approximately 95% of the active estrogens and androgens made are inactivated locally before being released in the blood as inactive metabolites, thus avoiding inappropriate exposure of the other tissues [13] (Figure 6.3).

Variable/rate-limiting circulating levels of DHEA and indirect role of the ovary in androgen formation

Because, as indicated above, after menopause, circulating DHEA is the only source of sex steroids made locally in peripheral tissues by the process of intracrinology [2,15,25,39], it seems important to examine the interindividual variability in the serum levels of DHEA and its metabolites. It is also of interest to investigate the potential role of the postmenopausal ovary in the direct secretion of sex steroids by comparing the steroid levels in intact and ovariectomized (OVX) women using MS-based assays [4, 12–15]. The data obtained show a high interindividual variability in DHEA and all its metabolites, with no evidence of a direct secretion of estrogens or androgens by the postmenopausal ovary, which does seem, however, to contribute approximately 20% of total circulating DHEA, and consequently, around 20% of total sex steroids after menopause in this age group[4].

The serum DHEA levels in intact 46- to 74-year-old postmenopausal women were measured at 2.03 ± 1.33 ng/mL (mean ± SD) (5th–95th centiles, 0.55–4.34 ng/mL) with a median value of 1.73 ng/mL, whereas an 18.2% lower mean value (1.66±1.04 ng/mL) was found in OVX women of similar age (42–74 years

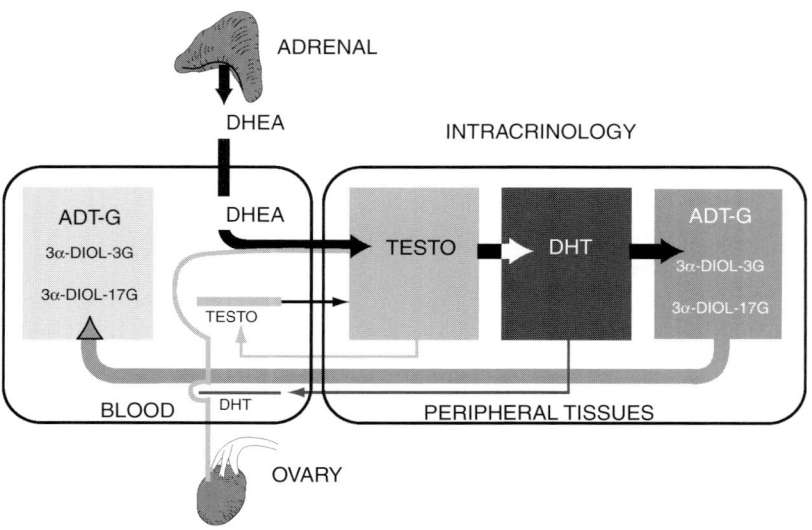

Figure 6.3 Schematic representation of the role of the precursor DHEA of both adrenal (~80%) and ovarian (~20%) origins in total androgenic activity in postmenopausal women. A very small proportion of the active androgens testosterone and DHT made intracellularly by the steroidogenic enzymes of the intracrine pathway diffuse into the circulation, thus avoiding systemic effect. The height of the bars is proportional to the concentration of each steroid. For androgens, the mechanisms of intracrinology [2] make it possible to keep intracellularly the formation and inactivation of androgens with only some leakage at very low levels in the circulation. In the circulation, the biologically significant and representative parameters are the precursor DHEA and the glucuronidated metabolites of androgens (ADT-G and 3α-diol-3G and 3α-diol-17G) as well as sulfate derivatives. A black and white version of this figure will appear in some formats. For the color version, please refer to the plate section.

old) (P = 0.0102). Compared with 30- to 39-year-old premenopausal women having DHEA measured at 4.47 ± 2.19 ng/mL, the mean serum DHEA concentration was 55% lower in intact postmenopausal women and 63% lower in OVX postmenopausal women, thus indicating the contribution of the ovary [4].

Similar observations were made for serum dehydroepiandrosterone sulfate (DHEA-S). The mean serum DHEA-S levels were 19.0% lower in OVX women compared with intact postmenopausal women, with values of 0.63 ± 0.41 μg/mL (5th–95th centiles = 0.15–1.38 μg/mL) in intact and 0.51 ± 0.39 μg/mL (5th–95th centiles = 0.13–1.00 μg/mL) in OVX postmenopausal women (P = 0.0181). Serum testosterone, on the other hand, was 21.4% lower in OVX compared with intact women. In fact, serum testosterone decreased from a mean value of 0.14 ± 0.08 ng/mL in intact women to 0.11 ± 0.05 ng/mL in OVX subjects (P <0.0001) [4].

When calculations are made to compensate for the higher serum DHEA levels observed in intact women compared with OVX women, no significant contribution of androgens by the postmenopausal ovary independently from DHEA could be found. The 18.2% lower serum DHEA observed in OVX women compared with intact women can well explain the 21.4% lower serum testosterone and the 16.6% lower serum

E_2 in OVX women compared with intact postmenopausal women, thus leaving no significant role for a direct contribution of the postmenopausal ovary to circulating testosterone and E_2 [4].

Because the lower serum DHEA levels found in OVX women can explain the lower serum concentrations of all measured sex steroids and metabolites, it can be logically concluded that the postmenopausal ovary does not secrete significant amounts of estrogens or androgens directly into the circulation. Sex steroid secretion by the postmenopausal ovary is a controversial area, especially in regard to immuno-based assays of testosterone. Most studies, however, have reported that the contribution of the postmenopausal ovary to circulating active androgens is not significant [40–45].

The data summarized above indicate that the persistence of low testosterone concentrations in the blood of postmenopausal women is not only "largely due," [33,46,47] but is rather entirely due to the peripheral conversion of DHEA into testosterone, which leaks in small amounts into the extracellular space and then into the general circulation. Such a possibility is supported by the finding that the postmenopausal ovary has persistent, although much reduced as compared with the corpus luteum of the premenopausal ovary, transcript levels of the cholesterol transport protein steroidogenic acute regulatory cholesterol side chain cleavage enzyme (CYP11A) and 17α-hydroxylase-17,20 lyase enzyme (CYP17), with extremely low levels of type 2 3βhydroxysteroid dehydrogenase and aromatase (CYP19) [45]. This possibility is further supported by the presence of CYP17 in the stroma of the postmenopausal ovary, whereas the type 2 hydroxysteroid dehydrogenase protein is absent, thus favoring the formation of DHEA [45]. From the serum levels of testosterone and the intraprostatic concentrations of dihydrotestosterone (DHT) observed in men before and after castration [3], it could be estimated that only about 5% of the intracellular testosterone diffuses into the general circulation, a situation which is likely to be similar in the peripheral tissues of women.

The 18%, but parallel lower serum levels of DHEA and all its metabolites, found in ovariectomized women, including E_2 and testosterone, strongly suggest that the postmenopausal ovary secretes approximately 20% of total DHEA in the 42- to 72-year-old age group of intact postmenopausal women, with no significant E_2 or testosterone direct secretion by the ovary into the general circulation. There is no reason to believe that the DHEA of ovarian origin is not submitted to the same intracrine mechanisms as the DHEA of adrenal origin.

The serum levels of androgen metabolites, but not of testosterone, as markers of androgen exposure in women

It is a well-recognized issue that the long series of case-control and prospective cohort studies, which analyzed a potential correlation between serum testosterone and the incidence of obesity, insulin resistance, sexual dysfunction, or other clinical problems in women, generally yielded contradictory results [37,48–56]. This lack of consistency between the serum levels of testosterone and the observed effect of exogenous androgens has raised serious doubts about the validity of measurements of total as well as free serum testosterone as markers of androgenic activity in women.

As described above, the lack of correlation between serum testosterone and clinical parameters believed to be under androgen control can be best explained by the above-summarized data showing that practically all androgens in women are made locally in peripheral target tissues from the inactive precursor DHEA of adrenal and ovarian origins [2,34,57]. Since

the androgens made locally do not originate from the very low circulating testosterone, one could reasonably conclude that measurement of serum testosterone is of highly questionable clinical significance. In fact, as indicated above, the androgens testosterone and DHT made in peripheral tissues from circulating DHEA exert their action locally in the same cells where synthesis takes place, with only minimal and variable release as active androgens in the circulation. Testosterone and DHT are then inactivated in the same cells into water-soluble glucuronide and sulfate derivatives, which diffuse quantitatively into the general circulation, where they can be measured before their elimination by the kidneys (Figure 6.3) [13,15].

Considering the major importance for clinicians and women to have access to a valid and reliable marker of androgenic activity in order to assess with confidence the role of androgens in a series of problems particularly frequent after menopause, namely type 2 diabetes, obesity and arteriosclerosis (metabolic syndrome), sexual dysfunction, osteoporosis, breast cancer, skin atrophy as well as loss of muscular strength, physical fitness, and well-being, we have used LCMS/MS and GC-MS to measure nine androgens, their precursors, and metabolites in 377 postmenopausal women in good health aged 55–65 years in order to analyze the correlation between serum testosterone and the true markers of the total pool of androgens, namely the glucuronide derivatives of androsterone (ADT) and androstane-3α, 17β-diol (3α-diol), the obligatory route of elimination of androgens. We have also compared the results with data obtained in 47 30- to 35-year-old normally cycling women (Table 6.1).

No convincing correlation was found between serum testosterone and ADTG (r = 0.37), this metabolite accounting by itself for about 93% of the metabolites of androgen glucuronide elimination [1]. While one would ideally like to know the level of androgenic activity in each specific tissue, such a direct measurement of the intratissular concentration of the active androgens is not possible in the human, except under exceptional circumstances such as in samples of tissue obtained at surgery [20,58,59]. However, while not permitting the assessment of androgenic activity in specific tissues, measurement of the glucuronide derivatives of ADT and 3α-diol by validated MS-based technology permits a precise measure of the total pool of androgens in the whole organism. In fact, it is now well established that uridine glucuronosyl transferase 2 B7 (UGT 2 B7), UGT 2 B15, and UGT 2 B17 are

Table 6.1 Average serum levels of DHEA and 11 of its metabolites in normal postmenopausal and premenopausal women

Group	Value	DHEA (ng/mL)	5-Diol (ng/mL)	Testo (ng/mL)	DHT (ng/mL)	E₁ (pg/mL)	E₂ (pg/mL)
55- to 65-year-old postmenopausal women (n = 377)	Mean	1.95	0.27	0.14	0.04	17.78	4.17
	SD	1.18	0.15	0.07	0.03	10.04	3.29
	Median	1.72	0.25	0.13	0.03	15.58	3.44
	5th–95th centiles	0.56–3.99	0.1–0.54	0.06–0.26	0.01–0.07	7.57–34.8	1.0–9.27
	(Min–max)	(0.1–11.2)	(0.1–0.85)	(0.03–0.57)	(0.01–0.29)	(4.0–103)	(1.0–30.0)
30- to 35-year-old premenopausal women (n = 47)	Mean	4.47	0.49	0.18	0.07	54.0	82.0
	SD	2.19	0.20	0.07	0.03	23.3	42.2
	Median	4.14	0.44	0.17	0.07	49.5	71.4
	5th–95th centiles	1.53–9.14	0.25–0.84	0.06–0.31	0.03–0.14	23.7–87.5	22.0–160
	(Min–max)	(1.41–10.4)	(0.25–0.96)	(0.05–0.32)	(0.03–0.17)	(18.3–123)	(17.7–181)

Group	Value	E₁-S (ng/mL)	DHEA-S (µg/mL)	4-Dione (ng/mL)	ADT-G (ng/mL)	3α-Diol-3G (ng/mL)	3α-Diol-17G (ng/mL)
55- to 65-year-old postmenopausal women (n = 377)	Mean	0.22	0.59	0.40	15.8	0.64	0.57
	SD	0.21	0.36	0.18	12.5	0.52	0.47
	Median	0.17	0.55	0.37	13.1	0.55	0.25
	5th–95th centiles	0.04–0.59	0.15–1.24	0.17–0.71	3.27–41.7	0.25–1.69	0.25–1.54
	(Min–max)	(0.04–2.00)	(0.04–2.44)	(0.10–1.37)	(1.00–79.4)	(0.25–3.48)	(0.25–3.56)
30- to 35-year-old premenopausal women (n = 47)	Mean	1.19	1.27	0.96	40.2	1.21	1.43
	SD	0.93	0.62	0.35	29.3	0.83	0.93
	Median	0.87	1.04	0.92	31.6	1.06	1.35
	5th–95th centiles	0.31–3.50	0.56–2.65	0.45–1.64	12.2–118	0.25–2.78	0.25–2.56
	(Min–max)	(0.21–4.40)	(0.45–2.71)	(0.31–1.77)	(6.86–133)	(0.25–4.33)	(0.25–5.71)

Serum concentrations measured in 30- to 35-year-old premenopausal (n = 47) and 55- to 65-year-old postmenopausal (n = 377) women are indicated (Labrie, Bélanger et al. 2006 [1]).

the three enzymes responsible for the glucuronidation of all androgens and their metabolites in the human [60]. Recent completion of the identification and characterization of all the human UDP glucuronosyl transferases makes possible the use of the glucuronide derivatives of androgens as markers of the total pool of androgens in both women and men. ADTS is also a metabolite excreted in large amounts, but this steroid is exclusively of adrenal origin and apparently does not reflect androgenic activity in peripheral tissues [61].

Relatively high level of exposure to androgens in women

The availability of MS-based assays of androgens has permitted the recognition that postmenopausal women synthesize approximately 50% as much androgens as men of the same age [3]. This estimate is very different from the comparison of serum testosterone concentrations, which are at least 20 times higher in men than in women. This is due to the fact that in men, approximately 50% of the total androgens come from the direct secretion of testosterone by the testicles into the circulation, thus explaining the high circulating levels of testosterone, while, in women, practically all androgens are made locally in peripheral tissues and only a very small percentage of the testosterone made intracellularly (estimated at about 5%) diffuses into the blood. Measurements of serum testosterone concentrations in women are thus highly misleading and dramatically underestimate the true exposure of women to androgens. As mentioned above, the problem with DHEA, as a source of androgens, however, is that it decreases on average by 60% between the age of 30 years and menopause, and continues to decline thereafter. Moreover, since there is no feedback mechanism to stimulate DHEA secretion in women having low DHEA, women suffer from progressively low DHEA and menopausal symptoms of increasing severity due to a progressive deficiency in estrogens and androgens, although approximately 20% of women have sufficient DHEA to avoid the symptoms of menopause [17,18].

Conclusion

The classical concept of androgen and estrogen secretion in women assumed that all sex steroids, following their secretion by the ovaries, had to be transported by the general circulation before reaching the target tissues. According to this traditional concept, it was thought that the active steroids could be measured directly in the blood, thus providing an easily accessible estimate of the general exposure to sex steroids. In fact, this concept is valid only for the animal species lower than primates, but it does not apply to the human, especially in postmenopausal women where all estrogens and all androgens are made locally from DHEA in the peripheral tissues, which possess the enzymes required to synthesize the physiologically required active sex steroids. Such a local intracrine biosynthesis and action of androgens in target tissues eliminates the exposure of the other tissues to androgens, and thus minimizes the risks of undesirable masculinizing or other androgen-related side effects.

The present review indicates that the most practical and valid means of assessing androgenic activity in women is to measure ADTG, the metabolite that accounts for about 93% of the total androgen glucuronide derivatives, using validated LC/MS-MS assays. This replaces the measurement of serum testosterone, which does not provide a reliable measure of androgenic exposure in women. This strategy should help to identify the cases of true androgen deficiency, thus offering the possibility of an appropriate androgen therapy or prevention regimen. It is likely, however, that measurement of the sulfate metabolites will become important for the estimation of total androgen metabolites [13,14].

References

1. Labrie F, Bélanger A, Bélanger P, et al. Androgen glucuronides, instead of testosterone, as the new markers of androgenic activity in women. *J Steroid Biochem Mol Biol* 2006;99(4–5):182–8.

2. Labrie F. Intracrinology. *Mol Cell Endocrinol* 1991;78(3):C113–C8.

3. Labrie F, Cusan L, Gomez JL, et al. Comparable amounts of sex steroids are made outside the gonads in men and women: strong lesson for hormone therapy of prostate and breast cancer. *J Steroid Biochem Mol Biol* 2009;113(1–2):52–6.

4. Labrie F, Martel C, Balser J. Wide distribution of the serum dehydroepiandrosterone and sex steroid levels in postmenopausal women: role of the ovary? *Menopause* 2011;18(1):30–43.

5. NAMS – The North American Menopause Society. The role of testosterone therapy in postmenopausal women: position statement of The North American Menopause Society. *Menopause* 2005;12(5):496–511; quiz 649.

6. Rosner W, Auchus RJ, Azziz R, et al. Position statement: utility, limitations, and pitfalls in measuring

testosterone: an Endocrine Society position statement. *J Clin Endocrinol Metab* 2007;92(2):405–13.

7. Rosner W, Hankinson SE, Sluss PM, et al. Challenges to the measurement of estradiol: an endocrine society position statement. *J Clin Endocrinol Metab* 2013;98(4):1376–87.

8. Taieb J, Mathian B, Millot F, et al. Testosterone measured by 10 immunoassays and by isotope-dilution gas chromatography-mass spectrometry in sera from 116 men, women, and children. *Clin Chem* 2003;49(8):1381–95.

9. Moal V, Mathieu E, Reynier P, et al. Low serum testosterone assayed by liquid chromatography-tandem mass spectrometry. Comparison with five immunoassay techniques. *Clin Chim Acta* 2007;386(1–2):12–19.

10. Sikaris K, McLachlan RI, Kazlauskas R, et al. Reproductive hormone reference intervals for healthy fertile young men: evaluation of automated platform assays. *J Clin Endocrinol Metab* 2005;90(11):5928–36.

11. Shackleton C. Clinical steroid mass spectrometry: a 45-year history culminating in HPLC-MS/MS becoming an essential tool for patient diagnosis. *J Steroid Biochem Mol Biol* 2010;121(3–5):481–90.

12. Ke, Y., J. Bertin, et al. (2014). "A Sensitive, Simple and Robust LC-MS/MS Method for the Simultaneous Quantification of Seven Androgen- and Estrogen-Related Steroids in Postmenopausal Serum." *J Steroid Biochem Mol Biol* 144: 523–534.

13. Ke, Y., R. Gonthier, et al. (2015). "A rapid and sensitive UPLC-MS/MS method for the simultaneous quantification of serum androsterone glucuronide, etiocholanolone glucuronide, and androstan-3alpha, 17beta diol 17-glucuronide in postmenopausal women." *JSBMB*: In press.

14. Dury, A. Y., Y. Ke, et al. (2015). "Validated LC-MS/MS simultaneous assay of five sex steroid/neurosteroid-related sulfates in human serum." *JSBMB* 149C: 1–10.

15. Labrie, F., Y. Ke, et al. (2015). "Why both LC-MS/MS and FDA-compliant validation are essential for accurate estrogen assays?" *J Ster Biochem Mol Biol*. 149C: 89–91.

16. Hsing AW, Stanczyk FZ, Belanger A, et al. Reproducibility of serum sex steroid assays in men by RIA and mass spectrometry. *Cancer Epidemiol Biomarkers Prev* 2007;16(5):1004–8.

17. Labrie F. DHEA after menopause – sole source of sex steroids and potential sex steroid deficiency treatment. *Menopause Management* 2010;19:14–24.

18. Labrie F, Labrie C. DHEA and intracrinology at menopause, a positive choice for evolution of the human species. *Climacteric*. 2013;16(2):205–13.

19. Labrie C, Bélanger A, Labrie F. Androgenic activity of dehydroepiandrosterone and androstenedione in the rat ventral prostate. *Endocrinology* 1988;123(3):1412–17.

20. Labrie F, Dupont A, Bélanger A. Complete androgen blockade for the treatment of prostate cancer. In: de Vita VT, Hellman S, Rosenberg SA, editors. *Important Advances in Oncology*. Philadelphia: J.B. Lippincott; 1985. pp. 193–217.

21. Labrie F, Dupont A, Bélanger A, et al. New hormonal therapy in prostatic carcinoma: combined treatment with an LHRH agonist and an antiandrogen. *Clin Invest Med* 1982;5(4):267–75.

22. Crawford ED, Eisenberger MA, McLeod DG, et al. A controlled trial of leuprolide with and without flutamide in prostatic carcinoma. *N Engl J Med* 1989;321(7):419–24.

23. Labrie F, Bélanger A, Cusan L, et al. Antifertility effects of LHRH agonists in the male. *J Androl* 1980;1:209–28.

24. Baker ME. Co-evolution of steroidogenic and steroid-inactivating enzymes and adrenal and sex steroid receptors. *Mol Cell Endocrinol* 2004;215(1–2):55–62.

25. Labrie F, Luu-The V, Bélanger A, et al. Is DHEA a hormone? Starling Review. *J Endocrinol* 2005;187(2):169–96.

26. Luu-The V. Assessment of steroidogenic pathways that do not require testosterone as intermediate. *Horm Mol Biol Clin Invest* 2011;5:161–5.

27. Labrie F, Simard J, Luu-The V, et al. Structure, function and tissue-specific gene expression of 3b-hydroxysteroid dehydrogenase/5-ene-4-ene isomerase enzymes in classical and peripheral intracrine steroidogenic tissues. *J Steroid Biochem Mol Biol* 1992;43(8):805–26.

28. Labrie F, Sugimoto Y, Luu-The V, et al. Structure of human type II 5α-reductase gene. *Endocrinology* 1992;131(3):1571–3.

29. Labrie Y, Durocher F, Lachance Y, et al. The human type II 17β-hydroxysteroid dehydrogenase gene encodes two alternatively-spliced messenger RNA species. *DNA Cell Biol* 1995;14(10):849–61.

30. Luu-The V, Zhang Y, Poirier D, et al. Characteristics of human types 1, 2 and 3 17b-hydroxysteroid dehydrogenase activities: oxidation-reduction and inhibition. *J Steroid Biochem Mol Biol* 1995;55(5–6):581–7.

31. Labrie F, Simard J, Luu-The V, et al. The 3β-hydroxysteroid dehydrogenase/isomerase gene family: lessons from type II 3β-HSD congenital deficiency. *Signal Transduction in Testicular Cells, Ernst Schering Research Foundation Workshop* 1996;Suppl. 2:185–218.

32. Labrie F, Luu-The V, Lin SX, et al. The key role of 17b-HSDs in sex steroid biology. *Steroids* 1997;62(1):148–58.

33. Longcope C, Franz C, Morello C, et al. Steroid and gonadotropin levels in women during the peri-menopausal years. *Maturitas* 1986;8(3):189–96.

34. Burger HG, Dudley EC, Cui J, et al. A prospective longitudinal study of serum testosterone dehydroepiandrosterone sulfate and sex hormone binding globulin levels through the menopause transition. *J Clin Endocrinol Metab* 2000;85(8):2832–8.

35. Rannevik G, Jeppsson S, Johnell O, et al. A longitudinal study of the perimenopausal transition: altered profiles of steroid and pituitary hormones, SHBG and bone mineral density. *Maturitas* 1995;21(2):103–13.

36. Labrie F. Drug Insight: breast cancer prevention and tissue-targeted hormone replacement therapy. *Nat Clin Pract Endocrinol Metab* 2007;3(8):584–93.

37. Labrie F, Luu-The V, Labrie C, et al. Endocrine and intracrine sources of androgens in women: inhibition of breast cancer and other roles of androgens and their precursor dehydroepiandrosterone. *Endocr Rev* 2003;24(2):152–82.

38. Labrie F, Bélanger A, Cusan L, et al. Marked decline in serum concentrations of adrenal C19 sex steroid precursors and conjugated androgen metabolites during aging. *J Clin Endocrinol Metab* 1997;82(8):2396–402.

39. Luu-The V, Labrie F. The intracrine sex steroid biosynthesis pathways. In: Martini L, Chrousos GP, Labrie F, et al., editors. *Neuroendocrinology, Pathological Situations and Diseases, Progress in Brain Research*: Elsevier; 2010. pp. 177–92.

40. Couzinet B, Meduri G, Lecce MG, et al. The postmenopausal ovary is not a major androgen-producing gland. *J Clin Endocrinol Metab* 2001;86(10):5060–6.

41. Abraham GE, Lobotsky J, Lloyd CW. Metabolism of testosterone and androstenedione in normal and ovariectomized women. *J Clin Invest*; 1969;48(4):696–703.

42. Jabara S, Christenson LK, Wang CY, et al. Stromal cells of the human postmenopausal ovary display a distinctive biochemical and molecular phenotype. *J Clin Endocrinol Metab* 2003;88(1):484–92.

43. Cauley JA, Gutai JP, Kuller LH, et al. The epidemiology of serum sex hormones in postmenopausal women. *Am J Epidemiol* 1989;129(6):1120–31.

44. Nagamani M, Urban RJ. Expression of messenger ribonucleic acid encoding steroidogenic enzymes in postmenopausal ovaries. *J Soc Gynecol Investig* 2003;10(1):37–40.

45. Havelock JC, Rainey WE, Bradshaw KD, et al. The post-menopausal ovary displays a unique pattern of steroidogenic enzyme expression. *Hum Reprod* 2006;21(1):309–17.

46. Spark RF. Dehydroepiandrosterone: a springboard hormone for female sexuality. *Fertil Steril* 2002;77 Suppl 4:S19–25.

47. Zumoff B, Strain GW, Miller LK, et al. Twenty-four-hour mean plasma testosterone concentration declines with age in normal premenopausal women. *J Clin Endocrinol Metab* 1995;80(4):1429–30.

48. Basson R. A new model of female sexual desire. *Endocrine News*. 2004;29:22.

49. Tchernof A, Labrie F. Dehydroepiandrosterone, obesity and cardiovascular disease risk. A review of human studies. *Eur J Endocrinol* 2004;151(1):1–14.

50. Shifren JL, Braunstein GD, Simon JA, et al. Transdermal testosterone treatment in women with impaired sexual function after oophorectomy. *N Engl J Med* 2000;343(10):682–8.

51. Sherwin BB, Gelfand MM. The role of androgen in the maintenance of sexual functioning in oophorectomized women. *Psychosom Med* 1987;49(4):397–409.

52. Cameron DR, Braunstein GD. Androgen replacement therapy in women. *Fertil Steril* 2004;82(2):273–89.

53. Garland CF, Friedlander NJ, Barrett-Connor E, et al. Sex hormones and postmenopausal breast cancer: a prospective study in an adult community. *Am J Epidemiol* 1992;135(11):1220–30.

54. Leiblum S, Bachmann G, Kemmann E, et al. Vaginal atrophy in the postmenopausal women. The importance of sexual activity and hormones. *JAMA* 1983;249(16):2195–8.

55. Lipworth L, Adami HO, Trichopoulos D, et al. Serum steroid hormone levels, sex hormone-binding globulin, and body mass index in the etiology of postmenopausal breast cancer. *Epidemiology* 1996;7(1):96–100.

56. Davis SR, McCloud P, Strauss BJ, et al. Testosterone enhances estradiol's effects on postmenopausal bone density and sexuality. *Maturitas* 1995;21(3):227–36.

57. Labrie F, Bélanger A, Cusan L, et al. Physiological changes in dehydroepiandrosterone are not reflected by serum levels of active androgens and estrogens but of their metabolites: intracrinology. *J Clin Endocrinol Metab* 1997;82(8):2403–9.

58. Poortman J, Thijssen JH, von Landeghem AA, et al. Subcellular distribution of androgens and oestrogens in target tissue. *J Steroid Biochem* 1983;19(1C):939–45.

59. Bélanger B, Bélanger A, Labrie F, et al. Comparison of residual C-19 steroids in plasma and prostatic tissue of human, rat and guinea pig after castration: unique importance of extratesticular androgens in men. *J Steroid Biochem* 1989;32(5):695–8.

60. Bélanger A, Pelletier G, Labrie F, et al. Inactivation of androgens by UDP-glucuronosyltransferase enzymes in humans. *Trends Endocrinol Metab* 2003;14(10):473–9.

61. Zwicker H, Rittmaster RS. Androsterone sulfate: physiology and clinical significance in hirsute women. *J Clin Endocrinol Metab* 1993;76(1):112–16.

Diagnostic criteria of polycystic ovary syndrome

Heather R. Burks and Robert A. Wild

Diagnostic criteria for polycystic ovary syndrome

The polycystic ovary syndrome (PCOS) was first coined as a term in 1935 by Stein and Leventhal, who identified a series of patients with the combination of oligo-ovulation and hyperandrogenism [1]. Since that time, diagnosis of this syndrome has undergone multiple iterations of application of diagnostic criteria. Abnormal uterine bleeding was the most common symptom associated with the condition in the late 1800s, but over time, new and better evidence has become available to inform the diagnosis. As recognition of this entity as a complex constellation of symptoms and metabolic derangements expanded, multiple professional groups have attempted to better and more precisely characterize this common syndrome.

In the past 25 years, three major sets of diagnostic criteria have emerged to define PCOS (Table 7.1) [2–4]. The first set of criteria, outlined at the National Institutes of Health (NIH) in Bethesda, Maryland, in 1990, is the most stringent. It has largely been replaced in clinical practice by the more inclusive Rotterdam criteria. In Rotterdam, The Netherlands, in 2003, a task force sponsored by the European Society of Human Reproduction and Embryology (ESHRE) and the American Society for Reproductive Medicine (ASRM) met to review the available data and to propose a revision of the 1990 NIH diagnostic paradigm. More recently, in 2009, the Androgen Excess Society (AES) outlined its own set of criteria for PCOS. Due to the subtle heterogeneities between the various diagnostic criteria, the reported prevalence of PCOS can vary depending on which definition is used by investigators. The discriminating reader should take note of the specific criteria utilized in the reporting.

The NIH meeting in 1990 was the first international conference on PCOS. The information available to attendees at the time was based largely on expert opinion [2]. The criteria set forth included (1) chronic anovulation and (2) clinical or biochemical signs of hyperandrogenism. Both criteria must be present under NIH guidelines, and other diagnoses must be excluded. This initial step to clearly define the syndrome provided a backbone upon which the ensuing body of new analytic research could be built. A large volume of new studies thus emerged, providing additional information subsequently evaluated by the Rotterdam ESHRE/ASRM-sponsored PCOS Consensus Workshop Group to revise the original NIH proposed set of diagnostic criteria.

The Rotterdam consensus includes three diagnostic criteria, and states that any two of the three must be present in order to make the diagnosis [3]. The revised criteria included (1) oligo- or anovulation, (2) clinical or biochemical signs of hyperandrogenism, and (3) polycystic-appearing ovaries (PCO) on imaging. The diagnosis depends upon the exclusion of other causes of hirsutism and anovulatory bleeding, including nonclassic congenital adrenal hyperplasia (NC-CAH), Cushing's syndrome, and androgen-secreting tumors, as well as some other more common entities such as thyroid dysfunction and hyperprolactinemia. By including morphological appearance of polycystic ovaries, the Rotterdam consensus added two additional phenotypes not previously included in the diagnosis: (1) women with ovulatory dysfunction and polycystic ovaries but without hyperandrogenism and (2) ovulatory women with hyperandrogenism and polycystic ovaries. By including these phenotypes, PCOS was canonized as a spectrum, with associated long-term health risk implications, such as type 2 diabetes mellitus and cardiovascular disease

Androgens in Gynecological Practice, ed. Leo Plouffe and Botros Rizk. Published by Cambridge University Press. © Cambridge University Press 2015.

Table 7.1 Diagnostic criteria and their associated phenotypes

	Potential phenotypes															
	A	B	C	D	E	F	G	H	I	J	K	L	M	N	O	P
Hyperandrogenemia	+	+	+	+	-	-	+	-	+	-	+	-	-	-	+	-
Hirsutism	+	+	-	-	+	+	+	+	-	-	+	-	-	+	-	-
Oligo-anovulation	+	+	+	+	+	+	-	-	-	-	-	-	-	-	-	-
Polycystic ovaries	+	-	+	-	+	-	+	+	+	+	-	+	-	-	-	-
NIH 1990	√	√	√	√	√	√										
Rotterdam 2003	√	√	√	√	√	√	√	√	√	√						
AE-PCOS 2006	√	√	√	√	√	√	√	√	√							

Adapted from Azziz et al. 2009 [4].

consequences. The Rotterdam consensus statement advocated widening the inclusion criteria to avoid missing patients with the potential for these increased health risks. Deeper explorations reveal that these subcategories within PCOS, identified based on the Rotterdam diagnostic criteria, manifest as subtle but distinct hormonal and metabolic milieus when compared with cases of PCOS identified based on the more stringent NIH criteria.

The most recent set of diagnostic criteria released was from the AES in 2009 [4]. This consensus group reexamined the key features of PCOS, including menstrual dysfunction, hyperandrogenemia, clinical signs of hyperandrogenism, and polycystic ovarian morphology. A thorough review of existing literature was used to examine each feature for its appropriateness for inclusion as a defining criterion. A slightly modified version of the criteria for the diagnosis of PCOS emerged in this process: (1) hyperandrogenism, including hirsutism and/or hyperandrogenemia, (2) ovarian dysfunction, including oligo-anovulation and/or polycystic appearing ovaries, and (3) exclusion of other androgen excess or related disorders. Under the AES's PCOS criteria, related disorders of hyperandrogenism or anovulation also must be excluded, such as androgen-secreting neoplasms, Cushing's syndrome, congenital adrenal hyperplasia, thyroid disorders, hyperprolactinemia, and premature ovarian failure. The clinician may take into account the prevalence of these differential diagnoses when deciding what test to order. Similar to the NIH criteria, androgen excess is an essential component of the diagnosis by AES criteria. Therefore, the phenotype of ovulatory dysfunction and polycystic ovaries (PCO) alone – permissible under Rotterdam – does not qualify for a diagnosis of the syndrome by AES criteria. The combination of menstrual dysfunction and polycystic ovaries, in the absence of features of hyperandrogenism or evidence of hyperandrogenemia has, in fact, been shown to have the most similar anthropometric, hormonal, and metabolic risk factors to control subjects without PCOS matched for age and body mass index. The AES consensus criteria for defining PCOS are thus more inclusive than the NIH version, but less inclusive than the Rotterdam criteria.

Anti-müllerian hormone (AMH) has recently been proposed as a biochemical marker to replace ultrasonographic assessment of PCO morphology. Specificity and sensitivity of 97.1% and 94.6% when using the Rotterdam criteria, or 97.2% and 95.5% using the NIH criteria, have been reported [5]. AMH levels correlate independently with both PCO morphology and androgenic profile [6]. Ovarian stromal volume, measured as a ratio of the stromal area to total area of the ovary (S/A ratio), has been proposed as another ultrasonographic parameter of the ovarian morphology highly associated with hyperandrogenemia. Although this S/A ratio performs well when discriminating between women with and without PCOS, and correlates well with androgen levels, it has not been adopted as part of any of the existing diagnostic criteria [7,8].

Both the clinical and biochemical determination of hyperandrogenism in females can be challenging. Laboratory assays for androgens were initially designed for detection in males, and have been calibrated accordingly. For example, total testosterone assays are typically calibrated for normal male levels, the lower end of which is 250 ng/dL. The upper end of normal female total testosterone ranges between 55 ng/dL and 80 ng/dL (inter-laboratory differences exist and clinicians should familiarize themselves with the assay range for the laboratories serving their patient population). The female normal range occurs well below the fifth percentile for the assay detection range commonly utilized in men. At these lower levels, results may be less reliable; notably in most commercial assays. An additional diagnostic dilemma is that the reporting of clinical hyperandrogenism is examiner-dependent and can be quite subjective. While a standard tool such as the Ferriman–Gallwey score is frequently applied in an attempt to create an objective evaluation, this method has been shown to have better intra-observer reliability than inter-observer reliability [9]. Furthermore, a universal application of such tools across all ethnic groups may discount the normal ethnic variation in the appearance of body hair.

As evidenced by the variation among diagnostic criteria, inclusion of ultrasonographic evidence of PCO morphology into the definition of PCOS is controversial. The NIH criteria do not address ovarian morphology. The Rotterdam criteria in 2003 included polycystic ovaries as a phenomenon distinct from menstrual irregularities. The AES groups ovarian morphology into an "ovarian dysfunction" category along with oligo-anovulation, and they require only one or the other to suffice as a diagnostic criterion. It is important to appreciate that PCO morphology is not specific to PCOS and can be found in 20–30% of the general population of women 20–25 years of age. Isolated PCO, therefore, should not be considered an indication of the syndrome in the absence of menstrual irregularities, infertility, or complaints of hirsutism [10].

In some ways, efforts to agree on diagnostic criteria are artificial. There continues to be controversy and lack of complete agreement for what elements constitute optimal criteria for PCOS diagnosis, in part because of the natural clinical desire to move to discrete categorical criteria for the ease of diagnosis. In truth, there is a continuum of presentation from those persons minimally affected, with regular menses and only mild excess of androgens, to those who have a unilateral PCO, to those who manifest more severe grades of androgen excess. Efforts to include hyperandrogenemia as diagnostic criteria will remain inadequate until the sensitivity of androgen assays is better refined because of our current inability to accurately quantify circulating androgens in women. It is useful to remember that androgen production is affected by metabolic clearance and concentrations, and often people with normal androgen levels in the circulation have increased androgen production. It is simply not practical to measure androgen production, so we are limited to imperfect representation of hyperandrogenemia using current outpatient widely available laboratory measurements in the blood.

Prevalence of polycystic ovary syndrome

The prevalence of PCOS varies by region and ethnicity, and within any given population, as noted above, it depends upon the diagnostic criteria used. Most reports on the prevalence of PCOS range between 2% and 20%, with NIH criteria being the most restrictive and Rotterdam being the most inclusive. Several studies have compared the use of different diagnostic criteria within the same population (Table 7.2). A retrospective birth cohort in Australia found a prevalence of 8.7% using NIH criteria, 17.8% using Rotterdam criteria, and 12.0% using AES criteria [11]. A similar prevalence pattern was found in Turkey, where 6.1% met NIH criteria, 19.9% met Rotterdam criteria, and 15.3% met AES criteria [12]. In Iran, the estimated prevalence of PCOS was 7% based on the NIH criteria, 15.2% using Rotterdam criteria, and 7.92% using AES criteria [13].

In North America, estimates of the prevalence in the United States range from 4% to 8% in the literature, although most of this information comes from an unselected population of white and black women in the southeast region [14,15]. Mexican-American women have a higher prevalence than either of those groups, reportedly as high as 13% [16]. In contrast, the estimated prevalence of PCOS among women within Mexico itself is much lower at 6% [17]. These discrepancies in reporting seem to highlight that lifestyle and ethnic background interact in the occurrence of PCOS.

Prevalence varies on the Eurasian continent as well. In India, PCOS is reported as 9% of adolescents [18]. Among Indian women 15–35 years of age, evaluated at a rural gynecology clinic, 13% presented with menstrual irregularities, half of whom were found to have PCOS, estimating the prevalence to be around 6.5% [19]. In Sri Lanka, a similar prevalence of 6.3% was noted among women aged 15–39 [20]. In Iran, the prevalence of PCOS is reported as 8.5% of a sample of reproductive age women selected for participation in the Tehran Lipid and Glucose Study [21]. A Greek study on the island of Lesbos found a prevalence of 6.8% [22]. The overall prevalence of PCOS among a population of urban indigenous Australian women was particularly high using NIH criteria, at 15.3% [23]. A study in the United Kingdom found the prevalence

Table 7.2 Relative prevalence of PCOS categorized by diagnostic criteria

	Diagnostic criteria		
	NIH[a]	Rotterdam[b]	AES[c]
March et al. 2010 [11]	8.7	17.8	12.0
Yildiz et al. 2012 [12]	6.1	19.9	15.3
Mehrabian et al. 2011 [13]	7.0	15.2	7.9

[a] National Institutes of Health international conference, 1990.
[b] Task force sponsored by the European Society of Human Reproduction and Embryology (ESHRE) and the American Society for Reproductive Medicine (ASRM), 2003.
[c] Androgen Excess Society diagnostic criteria, 2009.

to be 8% using stricter NIH criteria, while 26% of their population met Rotterdam criteria. This again illustrates the variation that occurs when using different diagnostic criteria. In Spain, while screening a population of Caucasian women presenting spontaneously for blood donation, investigators found a prevalence of 6.5% [24]. By any measure, PCOS is one of the most prevalent endocrine disorders worldwide for young reproductive age women, and there is obvious regional and ethnic variation.

Unwanted excess facial and body hair and intractable acne are common presenting complaints that eventually lead to a diagnosis of PCOS. Rates of hirsutism vary among ethnic groups not necessarily commensurate with the diagnosis of PCOS. In the United States, the rates are similar in black and white women (around 5%) [25], but in Kashmir, India, the prevalence is reported as much higher at 10.5% [26]. Among women with hirsutism, up to one-third have an underlying diagnosis of PCOS. Around 27% of women presenting with acne were found in one study to have undiagnosed PCOS, compared with 8% of controls [27]. Patients presenting with acne resistant to standard treatment have an even higher prevalence rate, near 50% [28]. Among adolescents with irregular menses, after a 6-year follow-up period, 62% continued to have irregular menses, 59% of whom were diagnosed with PCOS. In other words, approximately one-third of the original adolescent population with irregular menses was diagnosed with PCOS within the study period [29].

Summary

PCOS is one of the most common chronic endocrine disorders among women. There are currently three subtly divergent sets of diagnostic criteria based on various combinations of hyperandrogenism, menstrual irregularities, and polycystic ovarian morphology. The prevalence of PCOS in a given population varies by region and ethnicity, as well as the specific diagnostic criteria used. Women often seek care initially for troublesome acne or hirsutism; women with these signs should be screened for menstrual irregularities. Early identification of PCOS is important due to the high occurrence of medical comorbidities that may affect a woman's cardiovascular risk profile (such as metabolic syndrome, impaired glucose tolerance, type 2 diabetes mellitus, dyslipidemia, and/or depression or anxiety), allowing for early intervention. Lifestyle modification is recommended for all patients with PCOS and obesity, with the goal of minimizing the overall health risk in this population.

References

1. Stein I, Leventhal M. Amenorrhea associated with bilateral polycystic ovaries. *Am J Obstet Gynecol* 1935;29:181–185.

2. Zawadski JK, Dunaif A. Diagnostic criteria for polycystic ovary syndrome: towards a rational approach. In: Dunaif A, Givens JR, Haseltine FP, Merriam GR, eds. *Polycystic Ovary Syndrome*. Boston: Blackwell Scientific Publications; 1992:377–384.

3. Rotterdam ESHRE/ASRM-Sponsored PCOS Consensus Workshop Group. Revised 2003 consensus on diagnostic criteria and long-term health risks related to polycystic ovary syndrome. *Fertil Steril* 2004;81(1):19–25.

4. Azziz R, Carmina E, Dewailly D, et al. The Androgen Excess and PCOS Society criteria for the polycystic ovary syndrome: the complete task force report. *Fertil Steril* 2009;91(2):456–488.

5. Eilertsen TB, Vanky E, Carlsen SM. Anti-Mullerian hormone in the diagnosis of polycystic ovary syndrome: can morphologic description be replaced? *Hum Reprod* 2012;27(8):2494–2502.

6. Rosenfield RL, Wroblewski K, Padmanabhan V, et al. Antimüllerian hormone levels are independently related to ovarian hyperandrogenism and polycystic ovaries. *Fertil Steril* 2012;98(1):242–249.

7. Belosi C, Selvaggi L, Apa R, et al. Is the PCOS diagnosis solved by ESHRE/ASRM 2003 consensus or could it include ultrasound examination of the ovarian stroma? *Hum Reprod* 2006;21(12):3108–3115.

8. Fulghesu AM, Ciampelli M, Belosi C, et al. A new ultrasound criterion for the diagnosis of polycystic ovary syndrome: the ovarian stroma/total area ratio. *Fertil Steril* 2001;76(2):326–331.

9. Wild RA, Vesely S, Beebe L, et al. Ferriman Gallwey Self-Scoring I: performance assessment in women with polycystic ovary syndrome. *J Clin Endocrinol Metab* 2005;90(7):4112–4114.

10. Michelmore KF, Balen AH, Dunger DB, Vessey MP. Polycystic ovaries and associated clinical and biochemical features in young women. *Clin Endocrinol* 1999;51:779–786.

11. March WA, Moore VM, Willson KJ, et al. The prevalence of polycystic ovary syndrome in a community sample assessed under contrasting diagnostic criteria. *Hum Reprod* 2010;25(2):544–551.

12. Yildiz BO, Bozdag G, Yapici Z, et al. Prevalence, phenotype and cardiometabolic risk of polycystic ovary syndrome under different diagnostic criteria. *Hum Reprod* 2012;27(10):3067–3073.

13. Mehrabian F, Khani B, Kelishadi R, Ghanbari E. The prevalence of polycystic ovary syndrome in Iranian women based on different diagnostic criteria. *Pol J Endocrinol* 2011;62(3):238–242.

14. Knochenhauer ES, Key TJ, Kahsar-Miller M, et al. Prevalence of the polycystic ovary syndrome in unselected black and white women of the southeastern United States: a prospective study. *J Clin Endocrinol Metab* 1998;83:3078–3082.

15. Azziz R, Woods KS, Reyna R, et al. The prevalence and features of the polycystic ovary syndrome in an unselected population. *J Clin Endocrinol Metab* 2004;89:2745–2749.

16. Goodarzi MO, Quiñones MJ, Azziz R, et al. Polycystic ovary syndrome in Mexican-Americans: prevalence and association with the severity of insulin resistance. *Fertil Steril* 2005;84(3):766–769.

17. Moran C, Tena G, Moran S, et al. Prevalence of polycystic ovary syndrome and related disorders in Mexican women. *Gynecol Obstet Invest* 2010;69:274–280.

18. Nidhi R, Padmalatha V, Nagarathna R, Amritanshu R. Prevalence of polycystic ovarian syndrome in Indian adolescents. *J Pediatr Adolesc Gynecol* 2011;24:223–227.

19. Chhabra S, Venkatraman S. Menstrual dysfunction in rural young women and the presence of polycystic ovarian syndrome. *J Obstet Gynaecol* 2010;30(1):41–5.

20. Kumarapeli V, Seneviratne RA, Wijeyaratne CN, et al. A simple screening approach for assessing community prevalence and phenotype of polycystic ovary syndrome in a semiurban population in Sri Lanka. *Am J Epidemiol* 2008;168:321–328.

21. Tehrani FR, Rashidi H, Azizi F. The prevalence of idiopathic hirsutism and polycystic ovary syndrome in the Tehran Lipid and Glucose Study. *Reprod Biol Endocrinol* 2011;9:144.

22. Diamanti-Kandarakis E, Kouli CR, Bergiele AT, et al. A survey of the polycystic ovary syndrome in the Greek island of Lesbos: hormonal and metabolic profile. *J Clin Endocrinol Metab* 1999;84:4006–4011.

23. Boyle JA, Cunningham J, O'Dea K, et al. Prevalence of polycystic ovary syndrome in a sample of Indigenous women in Darwin, Australia. *Med J Aust* 2012;196(1):62–66.

24. Asunción M, Calvo RM, San Millán JL, et al. A prospective study of the prevalence of the polycystic ovary syndrome in unselected caucasian women from Spain. *J Clin Endocrinol Metab* 2000;85:2434–2438.

25. DeUgarte CM, Woods KS, Bartolucci AA, Azziz R. Degree of facial and body terminal hair growth in unselected black and white women: toward a populational definition of hirsutism. *J Clin Endocrinol Metab* 2006;91(4):1345–1350.

26. Zargar AH, Wani AI, Masoodi SR, et al. Epidemiologic and etiologic aspects of hirsutism in Kashmiri women in the Indian subcontinent. *Fertil Steril* 2002;77(4):674–678.

27. Kelekci KH, Kelekci S, Incki K, et al. Ovarian morphology and prevalence of polycystic ovary syndrome in reproductive aged women with or without mild acne. *Int J Dermatol* 2010;49:775–779.

28. Maluki AH. The frequency of polycystic ovary syndrome in females with resistant acne vulgaris. *J Cosmet Dermatol* 2010;9:142–148.

29. Wiksten-Almströmer M, Hirschberg AL, Hagenfeldt K. Prospective follow-up of menstrual disorders in adolescence and prognostic factors. *Acta Obstet Gynecol Scand* 2008;87:1162–1168.

Chapter

Androgen effects on the skin

Grace E. Kim and Alexa B. Kimball

Introduction

Androgens play an important role in the normal development and functioning of human skin and its accompanying structures. They are necessary for a wide variety of cutaneous processes, ranging from sweat production to hair growth. However, when their careful regulation goes awry, androgens can contribute to the pathogenesis of multiple dermatoses. A deeper understanding of androgens and their impact on human skin allows us to work towards more effective therapeutic options for these skin disorders.

Role of androgens in normal human skin

Skin-related androgens

The adrenal cortex is largely responsible for synthesizing the circulating hormones dehydroepiandrosterone (DHEA) and DHEA sulfate (DHEA-S). Both the adrenal cortex and the ovaries contribute equally to the production of androstenedione in women [1]. DHEA, DHEA-S, and androstenedione are weak prohormones, which exert their effects after their conversion to their more potent counterparts, testosterone and 5α-dihydrotestosterone (DHT). In females of reproductive age, androstenedione is secreted by the ovary and adrenal cortex and converted into testosterone in the peripheral organs, including the skin [2] (Figure 8.1). Because of its concentration and potency, testosterone is the main circulating androgen [3]. In both genders, peripheral organs are the main synthesizers of DHT, the most potent tissue androgen produced from the metabolism of testosterone by 5α-reductases (Figures 8.1 and 8.2) [4].

Although the skin is not the major site of androgen production, the circulating prohormones DHEA and androstenedione can be converted into testosterone and DHT in sebocytes, sweat glands, and dermal papilla cells [4]. Consequently, these potent androgens can impact dermal functioning. Sweat glands and sebaceous glands account for the vast majority of androgen metabolism in the skin [5]. The peripheral organs produce 100% of the active sex steroids in postmenopausal women [6] (Figure 8.1). However, it remains unclear what proportion of the androgens made in the skin can be attributed to de novo synthesis from epidermally formed cholesterol versus cutaneous conversion of circulating prohormones into potent androgens [7].

Androgen receptor

The effects of testosterone and DHT are mediated through a nuclear receptor called the androgen receptor (AR). Testosterone and DHT are both able to bind to AR, but DHT has a tenfold higher affinity for AR compared with testosterone [8]. Located in the cytoplasm, AR consists of multiple protein subunits, including the heat shock proteins HSP90, HSP70, and HSP56. Upon binding its androgen ligand, AR dissociates from the heat shock proteins, exposing a nuclear translocation signal that allows the androgen–AR complex to move from the cytoplasm into the nucleus. Here, the complex acts as a transcription factor, binding to the promoters of androgen-regulated genes and influencing their expression [4].

The *AR* gene is located on the X chromosome and expressed in epidermal and follicular keratinocytes, sebocytes, sweat gland cells, dermal papilla cells, dermal fibroblasts, endothelial cells, and genital melanocytes [9]. *AR* expression has been shown to be

Androgens in Gynecological Practice, ed. Leo Plouffe and Botros Rizk. Published by CAMBRIDGE UNIVERSITY PRESS. © Cambridge University Press 2015.

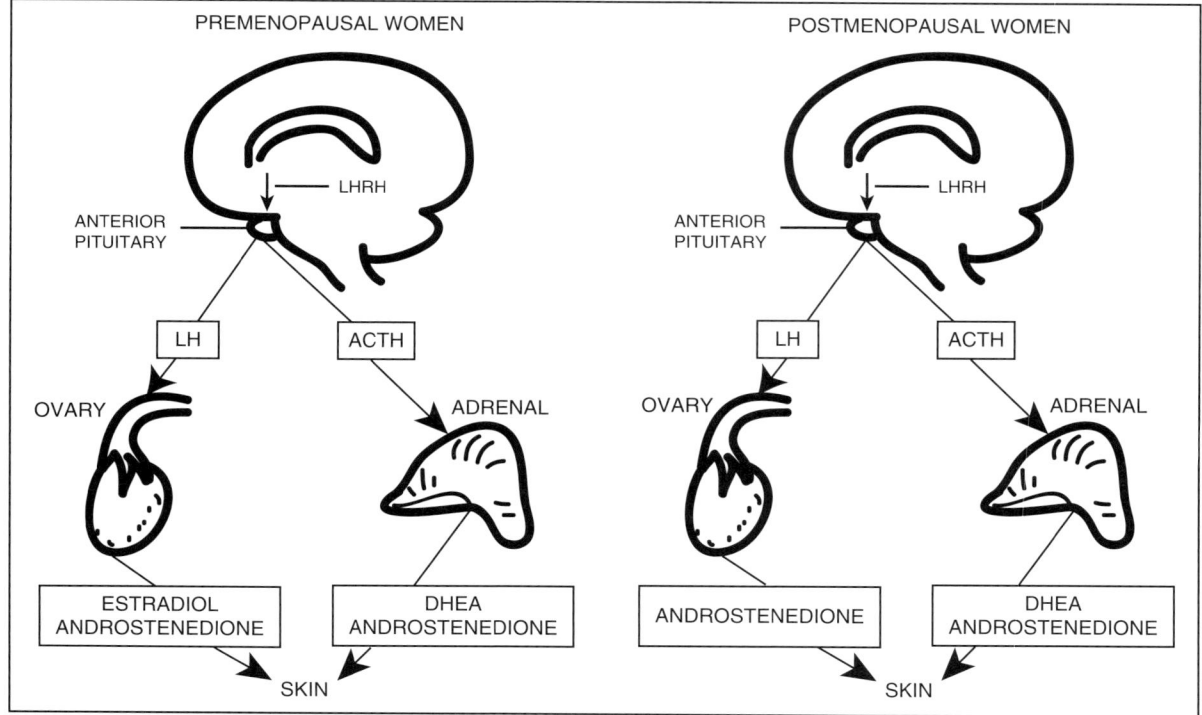

Figure 8.1 Representation of the role of ovarian and adrenal sources of sex steroids in women. After menopause, the secretion of estradiol by the ovaries ceases and almost all of the sex steroids are made locally in peripheral target intracrine tissues including the skin. LH, luteinizing hormone; LHRH, luteinizing hormone-releasing hormone; ACTH, adrenocorticotropic hormone. Adapted from "Intracrinology and the skin," by F. Labrie, V. Luu-The, C. Labrie, G. Pelletier, and M. El-Alfy, 2000, *Hormone Res*, 54, p. 219.

upregulated in genital skin fibroblasts and sebocytes [10]. In addition to its activation by androgen binding, AR function can be further regulated through binding of its N-terminal domain or ligand-binding domain by more than 200 identified co-regulators, including transcription factors, kinases, chaperones, cytoskeletal proteins, and histone modifiers [11]. The complexity of its regulation reflects the role of AR as an important synthesizer and convergence point of cellular signaling and function.

Androgen metabolism in the skin

In contrast to classic steroidogenic organs, such as the gonads and adrenal glands, the skin is considered a peripheral endocrine organ that both produces and is targeted by androgens [12]. The local concentration of each androgen depends on the expression level of each androgen-synthesizing enzyme in a particular cell type, especially the sebaceous glands and sweat glands [13] (Figure 8.2). The skin is capable of producing cholesterol, but expresses very little cytochrome P450c17,

the enzyme necessary for synthesizing DHEA and androstenedione. However, sebocytes, sweat glands, and dermal papilla cells in the skin do possess adequate amounts of the enzymes needed to convert DHEA and androstenedione (and possibly DHEA-S) into the more potent testosterone and DHT [12].

Five key enzymes play a role in the cutaneous metabolism of androgens. First, steroid sulfatase hydrolyzes DHEA-S to DHEA. Next, 3β-hydroxysteroid dehydrogenase (3β-HSD) converts DHEA to androstenedione in sebaceous glands. Androstenedione is then activated by converting it to testosterone via isoforms 3 and 5 of androgenic 17β-hydroxysteroid dehydrogenase (17β-HSD). 17β-HSD types 2 and 4 catalyze the reverse oxidation reaction that inactivates testosterone into androstenedione. 5α-Reductase irreversibly converts testosterone to DHT, particularly in sebaceous and sweat glands and to a lesser extent in epidermal cells and hair follicles in the skin [5]. There are two isoforms of 5α-reductase: type I dominates in the skin while type II occurs mainly in beard hair follicles [8]. Finally, 3α-hydroxysteroid dehydrogenase (3α-HSD) breaks

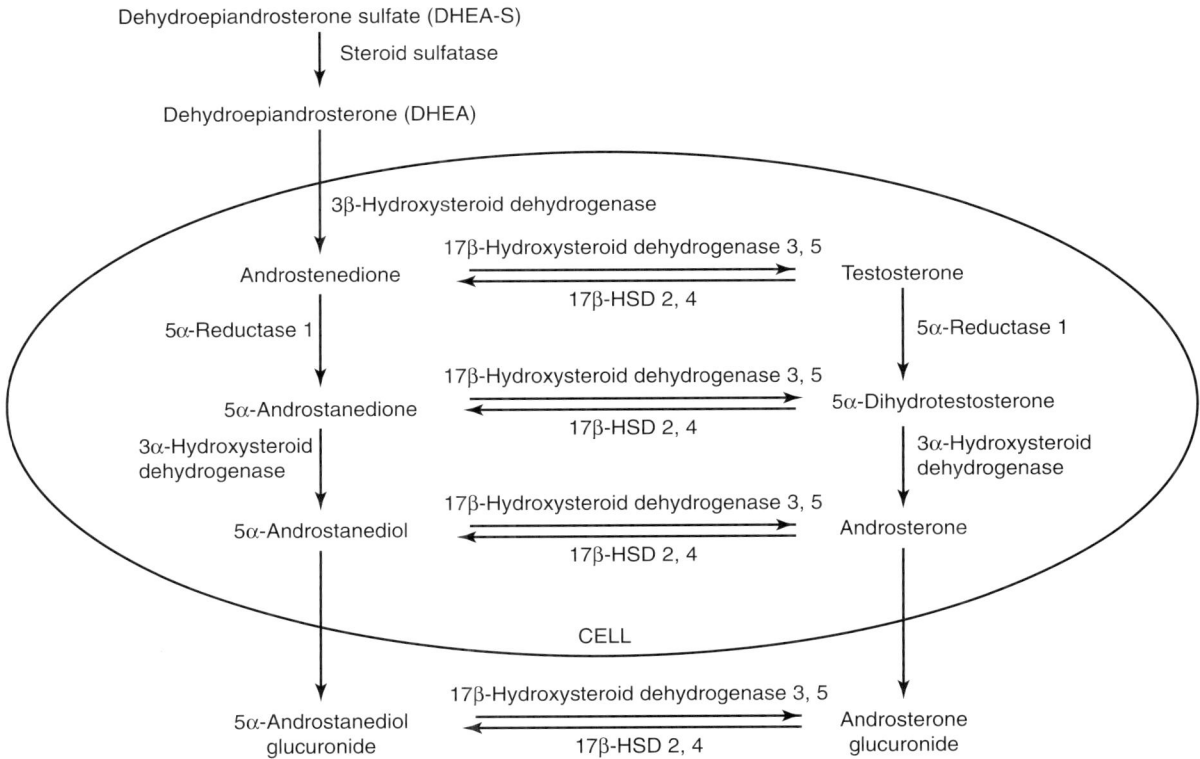

Dehydroepiandrosterone sulfate (DHEA-S)

Steroid sulfatase

Dehydroepiandrosterone (DHEA)

3β-Hydroxysteroid dehydrogenase

Androstenedione

17β-Hydroxysteroid dehydrogenase 3, 5

17β-HSD 2, 4

Testosterone

5α-Reductase 1

5α-Reductase 1

5α-Androstanedione

17β-Hydroxysteroid dehydrogenase 3, 5

17β-HSD 2, 4

5α-Dihydrotestosterone

3α-Hydroxysteroid dehydrogenase

3α-Hydroxysteroid dehydrogenase

5α-Androstanediol

17β-Hydroxysteroid dehydrogenase 3, 5

17β-HSD 2, 4

Androsterone

CELL

5α-Androstanediol glucuronide

17β-Hydroxysteroid dehydrogenase 3, 5

17β-HSD 2, 4

Androsterone glucuronide

Figure 8.2 Pathways of cutaneous androgen metabolism and the converting enzymes. Adapted from "Cutaneous androgen metabolism: basic research and clinical perspectives," by W. Chen, D. Thiboutot, and C. C. Zouboulis, 2002, *J Invest Dermatol*, 119, p. 993.

down active androgens into compounds that no longer bind AR [13]. Alternatively, aromatase can convert testosterone and androstenedione to estrogens in sebaceous glands, hair follicles, and dermal papilla cells [11].

Effects on the sebaceous gland

Sebaceous gland enlargement and sebum production both rely on androgens [5]. AR has been identified in basal and differentiating sebocytes, indicating that androgens contribute to cell proliferation and lipogenesis [14]. Androgens are known to stimulate sebocyte proliferation, especially in facial sebocytes. However, androgens alone are unable to modify sebocyte differentiation and require peroxisome proliferator-activated receptor (PPAR) ligands in order to stimulate sebocytes to differentiate [15].

Effects on the hair follicle

Androgens influence hair growth by acting through type 2 5α-reductase and AR in dermal papilla cells. By releasing growth factors that act in paracrine fashion

on other follicle cells, dermal papilla cells mediate the growth-stimulating effect of androgens [16]. Androgens promote the enlargement of hair follicles in some androgen-dependent areas, such as the axillary and pubic regions, but cause miniaturization and shortage of hair in other areas, such as the scalp in susceptible individuals [3]. The differential response of dermal papilla cells to androgens may depend on genetically determined differences. For example, AR mRNA has been shown to be expressed at high levels in dermal papilla cells found in axillary hair, but at low levels in dermal papilla from occipital scalp hair [16].

Effects on the sweat gland

Given that at puberty males sweat at a higher rate than females in similar situations, androgens are thought to stimulate perspiration [17]. Over half of the skin's 5α-reductase activity is present in sweat glands; moreover, sweat glands express AR and contain the enzymes required to convert DHEA to DHT. However, androgens are not thought to regulate the secretion rate of sweat glands directly, because androgen treatment has failed to stimulate sweat production in women and antiandrogen therapy has not decreased sweat production in men [2]. Rather than stimulating or maintaining the function of sweat glands, androgens likely exert their effect by initiating the factors required for differential sweat secretion between the sexes during puberty. Androgens are thought to influence the differentiation of apoeccrine sweat glands, which have a sevenfold higher secretory rate in response to similar innervation [18]. In addition, studies have reported that regardless of gender, apocrine glands of individuals with excessive or abnormal odor are typical target organs of androgens and contain predominantly type I 5α-reductase [19].

Human axillary odor results from a mixture of volatile organic compounds, including the steroids androstenol and androstenone, which arrive on the skin surface in apocrine secretions [20]. Variants in the *ABCC11* gene, which encodes an ATP-dependent efflux pump expressed in apocrine sweat glands, is an important determinant of axillary odor. Individuals who are homozygous for a single-nucleotide polymorphism (SNP) (538G>A) were found to have a significantly lower amount of axillary odorants than either those who were heterozygous for this SNP or those who had the wild-type gene. The 538G>A SNP is predominant in Asians and rare in Africans and Europeans [20].

Further effects on skin

The skin of adult males has been shown to be thicker and drier than that of adult females. This difference is due in part to the epidermal hyperplasia and suppressed epidermal barrier function that results from androgens [21]. Furthermore, androgens are thought to inhibit cutaneous wound healing through binding to AR and modulating inflammatory responses [4]. The effect of androgens on wound healing remains an active area of research.

Pathogenesis and hormonally mediated treatment of androgen-related skin disorders

Acne vulgaris

Pathogenesis

Acne vulgaris is a disease of pilosebaceous follicles that involves follicular hyperkeratinization, excessive sebum production, *Propionibacterium acnes* (*P. acnes*)

in the follicle, and excessive inflammation. Androgens contribute to the development of acne by promoting the growth and secretory function of sebaceous glands. The crucial role of androgens in the pathophysiology of acne (Figure 8.3) has long been confirmed. Clinical evidence includes the close correlation between the onset of acne in prepubertal children and the adrenarcheal rise in circulating levels of DHEA-S, acne formation in children with virilizing tumors or congenital adrenal hyperplasia, hyperandrogenism in women with sudden exacerbation of acne, acne induction by systemic or topical steroid use, and positive associations between serum androgen levels and acne lesion counts [22]. Furthermore, experimental studies have shown that sebaceous glands contain most of the steroidogenic enzymes needed for the conversion of DHEA and DHEA-S into testosterone and DHT [7], while immunohistochemistry and biochemical binding assays have shown that AR is present in the epithelial cells of sebaceous glands [23].

The biological mechanism by which androgens stimulate sebocyte activity in acne is not fully understood. One possibility is that AR might enhance the activities of fibroblast growth factor receptor 2 (FGFR2), which previous studies have shown to be important for the development and homeostasis of sebaceous glands [24]. Another plausible mechanism is AR might increase lipogenesis in sebocytes by upregulating the expression of sterol-regulatory element-binding proteins (SREBPs). Androgens may also influence the activity of insulin-like growth factor-1 (IGF-1), which has been found to be able to induce SREBP-1 expression and thereby stimulate lipogenesis in sebocytes [25]. Finally, other studies in animal models have suggested that the androgen–AR complex may enhance the inflammatory responses of neutrophils and macrophages, and as a result, androgens not only promote sebaceous gland activity, but also contribute to the inflammation that leads to acne formation and progression [26]. Further research is needed to determine the exact mechanism of androgen-induced sebocyte activity.

Diagnosis

Evaluation of acne vulgaris requires particular attention to endocrine function. In addition to a review of cosmetic use for comedogenic products and medication history for acne-inducing drugs, such as glucocorticoids, phenytoin, and epidermal growth factor receptor (EGFR) inhibitors, the skin should

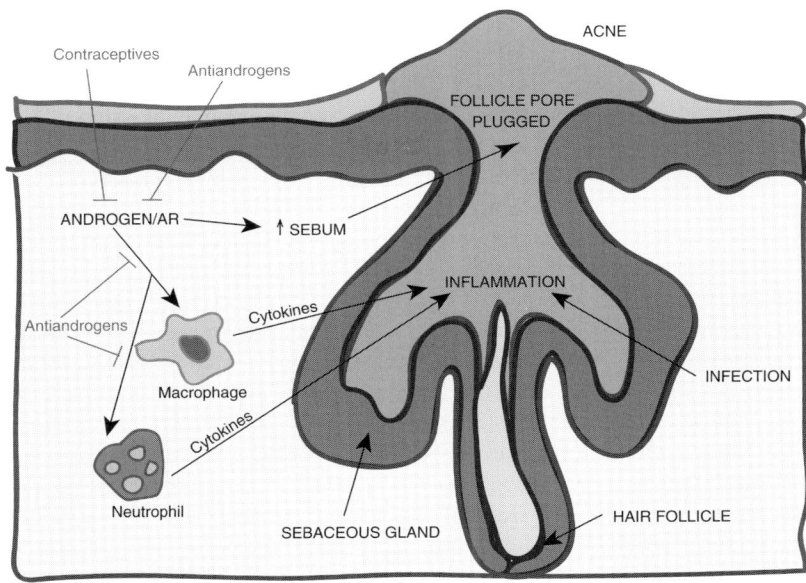

Figure 8.3 The role of androgens/AR in acne formation and progression. Acne formation results from excessive sebum formation accompanied by excessive inflammation and infections in the hair follicle. Macrophages and neutrophils are recruited to the inflamed follicles and secrete cytokines and other factors that promote inflammation and infection clearance. However, the inflammatory response also damages the normal tissues in the follicles. AR can promote the inflammatory response mediated by macrophages and neutrophils as well as directly promote sebum production that plugs the follicle pore. Current treatments for acne include antiandrogens and contraceptives, which reduce androgen levels, thereby reducing sebum production and suppressing the function of macrophages and neutrophils to dampen the inflammatory response. Adapted from "The role of androgen and androgen receptor in skin-related disorders," by J. J. Lai, P. Chang, K. P. Lai, L. Chen, and C. Chang, 2012, *Arch Dermatol Res*, 304, p. 504. A black and white version of this figure will appear in some formats. For the color version, please refer to the plate section.

be carefully examined for the type and location of lesions. Although acne is a common condition, a variety of disorders should be considered in the differential diagnosis. Rosacea, which is characterized by erythema, telangiectases, and papules or pustules on the central face, is distinguished from acne vulgaris, which is characterized by the presence of comedones and the absence of telangiectases [27]. Perioral dermatitis, which manifests as small, grouped, erythematous papules in a perioral (or occasionally perinasal or periorbital) distribution, has a rim of spared skin around the vermilion border of the lip. Sebaceous hyperplasia, or visible enlargement of sebaceous glands, commonly occurs in adults with a history of oily skin and appear as umbilicated yellowish papules usually on the forehead and cheeks. Staphylococcal, eosinophilic, or pseudomonal folliculitis may resemble inflammatory acne, but comedones are absent and lesions are monomorphous, unlike the polymorphous lesion in different developmental stages that are typical of acne [27]. Keratosis pilaris, which results from plugging of keratotic follicles, presents as small follicular papules on the face and extensor surfaces of the upper arms or thighs with or

without erythema. Favre–Racouchot syndrome caused by sun damage is seen in middle-aged or older adults, who have open and closed comedones in areas of photodamage, often the lateral upper cheeks.

Treatment

Hormonal therapy is not usually the first option for treatment of acne vulgaris. However, a significant proportion of acne patients, anywhere between 30% and 80%, have been shown to display varying degrees of hyperandrogenemia [28]. A positive correlation between acne severity and markers of androgenicity was not found, but it is still possible to treat hyperandrogenemic patients with therapies that lower serum androgen levels or inhibit the action of androgens on sebaceous glands [4]. In female acne patients, oral contraceptives decrease free testosterone levels by 40–50% by suppressing luteinizing hormone, reduce androgen bioavailability by increasing the level of androgen-binding protein (sex hormone-binding globulin), and prevent the conversion of free testosterone to DHT by blocking AR and inhibiting 5α-reductase activity [29]. Flutamide, a nonsteroidal antiandrogen that prevents binding of DHT to AR, has been studied mostly in patients with hirsutism and has been shown to produce satisfactory results in treating both acne and hirsutism in these patients [30]. Topical inocoterone, which reduced the size of sebaceous glands in animal experiments, did not significantly improve acne in clinical studies [31]; however, inocoterone does lower inflammation in acne lesions, most likely by suppressing AR function. In contrast, the antiandrogen cyproterone failed to suppress testosterone-induced proliferation of sebocytes and keratinocytes in vitro, but oral and topical cyproterone acetate treatments have been shown to improve acne lesions [4]. Other inhibitors of AR, such as spironolactone, achieve some level of therapeutic effect on acne, but are limited by their significant adverse effects on menstruation, liver function, and pregnancy. Isotretinoin, an effective treatment for acne that suppresses sebum production, has been shown to modify AR signaling [32].

Androgenetic alopecia

Pathogenesis

Androgenetic alopecia is the most common type of hair loss, affecting 30–40% of adult men and women [33]. The hallmark of androgenetic alopecia is follicular miniaturization, the process by which the hair growth phase (anagen) is shortened, resulting in the production of a shorter, thinner hair shaft. As follicles are miniaturized, new hair growth continues, but becomes shorter and thinner. Follicular miniaturization is a hormonally mediated process that occurs at the level of the follicle. Although androgens normally stimulate hair production in many sites of the body, they have an opposite effect in the scalp: androgens suppress hair growth in susceptible areas of the frontal and vertex scalp. In the hair follicle, testosterone is converted to DHT, which is found at increasing concentrations in the balding scalp. DHT binds to AR in susceptible follicles, and the hormone-receptor complex moves into the nucleus, where it activates genes responsible for the transformation of large, terminal follicles into miniaturized follicles. As evidence for the role of AR in androgenetic alopecia, individuals lacking functional ARs do not experience balding, and AR expression levels were found to be elevated in frontal and vertex regions, but normal in the parietal and occipital regions of the scalps of patients afflicted with androgenetic alopecia [34]. The AR gene is the only

risk gene for androgenetic alopecia confirmed to date, and AR gene variant(s) are generally accepted to be the primary culprit for the onset and development of androgenetic alopecia via abnormal expression of AR protein in the scalp follicle [4].

Within hair follicles, androgens act primarily on the dermal papilla cells, which contain AR with low capacity and high affinity. The molecular mechanism by which androgens exert their effect on the dermal papilla remains unclear. The current hypothesis is that the androgen–AR complex changes the production of autocrine and paracrine factors in the dermal papilla and thereby alters the growth of epithelial cells in the follicle [4]. A recent study found that androgens increase the synthesis and secretion of transforming growth factor beta (TGF-β) in the dermal papilla cells from the bald scalp and that TGF-β may be the androgen-induced growth inhibitor of co-cultured epithelial cells [35]. Wnt proteins are another possible regulatory factor that diffuses from the dermal papilla to the epithelium, triggering the androgenetic alopecia phenotype [36].

Diagnosis

Although the diagnosis of androgenetic alopecia is usually straightforward in men, diffuse frontal to vertex thinning may make the diagnosis more challenging in women. The scalp should be examined for other signs of hair disease, including scarring or follicular plugging. Other causes of hair loss, such as hyper- or hypothyroidism, iron deficiency, medications, and trauma, should be ruled out. Medications that can cause hair loss include anticoagulants, anticonvulsants, beta blockers, and antidepressants. A family history of similar hair loss supports androgenetic alopecia. Most women with androgenetic alopecia have normal hormone levels, but a careful history should be taken and signs of androgen excess, such as menstrual irregularities, acne, and hirsutism, should be noted. The differential diagnosis includes telogen effluvium, whose diffuse pattern of hair loss may be difficult to distinguish from extensive androgenetic alopecia. However, telogen effluvium is usually associated with an acute stressful event, such as pregnancy, severe weight loss, major illness or surgery, and traumatic psychological events, and the results of the hair pull are positive [37]. Diffuse alopecia areata is another consideration, but its onset is more acute than androgenetic alopecia, its distribution does not follow a classic distribution, and it is associated with hair loss at other body sites [38].

Treatment

Minoxidil, which is available as 2% and 5% solutions, is the only topical treatment for androgenetic alopecia approved by the United States Food and Drug Administration (FDA). Minoxidil, which acts as a peripheral vasodilator, promotes the growth of follicle keratinocytes and lengthens the anagen phase. Its suggested targets include potassium channels and/or prostaglandins, but the exact mechanism is unknown. Finasteride is a systemic treatment option for androgenetic alopecia that inhibits type II 5α-reductase, the enzyme responsible for converting testosterone into DHT in the scalp. Although finasteride stabilizes hair loss in most patients, only 37–61% of patients achieve mild to moderate hair growth. Moreover, improvement in hair density may require 6–12 months of treatment, and maximal effect requires 2 years of use [39]. Because of its known teratogenicity, finasteride is not officially approved for use in women, but it is being prescribed off-label in combination with an oral contraceptive. The largest controlled trial to date failed to demonstrate a significant improvement in hair loss for women on finasteride, but several uncontrolled studies and case reports suggest that finasteride may be effective in an as-yet unidentified subgroup of female patients [40]. Current treatment options for androgenetic alopecia are limited in that they require long-term use, vary greatly in their efficacy, and have sexual side effects. Recent research has focused on developing inhibitors of both type I and type II 5α-reductase as well as topical options to prevent the adverse effects of systemic treatment.

Hirsutism

Pathogenesis

Hirsutism, or excessive terminal hair growth in a male-like pattern, affects 5–10% percent of women of reproductive age and results from the overproduction of or increased sensitivity of hair follicles to androgens. Androgen-dependent areas that are commonly affected include the chin, upper lip, chest, breasts, abdomen, back, and anterior thighs. In these regions of the body, androgens increase the size and diameter of the hair fibers and prolong the anagen phase of the hair growth cycle. In non-androgen-dependent areas like the forehead and cheeks, androgens cause the enlargement of sebaceous glands, but the hair follicles remain vellus in nature. Either an exogenous or endogenous increase in circulating androgens or increased sensitivity of the hair follicle to normal serum androgen levels can lead to hirsutism. There exist several etiologies of hyperandrogenism that may cause elevated levels of one or more of the key androgens: testosterone from the ovary, DHEA-S from the adrenal gland, and androstenedione from the ovary or adrenal gland.

Polycystic ovary syndrome (PCOS) is the most common cause of hirsutism in women and is characterized by menstrual irregularity and hyperandrogenism, leading to hirsutism, acne, and/or androgenetic alopecia [41]. The major source of hyperandrogenemia in PCOS is gonadotropin-dependent functional ovarian hyperandrogenism, resulting in higher circulating levels of total testosterone, free testosterone, and androstenedione. Another cause of hirsutism is congenital adrenal hyperplasia, which usually results from 21-hydroxylase deficiency, leading to increased levels of its substrate 17-hydroxyprogesterone and androstenedione. In addition, ovarian tumors can cause elevated levels of testosterone, while rare adrenal tumors can lead to elevated levels of testosterone, DHEA, and/or DHEA-S. Women with Cushing's disease often have hirsutism as well from adrenal overactivity. Severe insulin resistance syndromes and the resulting hyperinsulinemia cause hirsutism from ovarian hyperandrogenism, which may be mediated by the theca cell receptors for insulin and insulin-like growth factor-1 (IGF-1) [42]. Finally, exogenous androgen therapies that administer testosterone or DHEA are also associated with hirsutism.

Diagnosis

Since many women with hirsutism have PCOS, the challenge is to identify the small subset of women with other causes for their hirsutism. The goals are to exclude serious, but rare, causes of hirsutism, such as ovarian and adrenal androgen-secreting tumors, and to document the degree of androgen excess. Findings that suggest a serious cause of hirsutism include abrupt onset, short duration (less than a year), progressive worsening, onset in the third decade of life or later, symptoms of virilization, and elevated serum androgen concentrations. Menstrual history, time course of symptoms, weight history, medication history, and family history should be elicited. The Endocrine Society Clinical Practice Guidelines recommend biochemical testing in women with hirsutism that is moderate or severe, rapid in onset or progression, or associated with irregular menses, obesity, or virilization [43].

Treatment

Hormonal treatment of hirsutism is achieved through antiandrogens or AR blockers. Although these therapies may have side effects in men, many of the adverse effects in women can be avoided. Spironolactone is an AR blocker that competes with androgens to bind to AR, effectively reducing hair growth. The detrimental side effects, including menstrual irregularities, breast tenderness, fatigue, gastritis, and headaches, can be evaded with the use of a contraceptive pill. Cyproterone acetate is another competitive inhibitor of androgens for binding to AR that also blocks androgen secretion. However, cyproterone acetate is not used in the United States because of the elevated risk of liver cancer. Flutamide is an AR blocker that leads to increasingly lower serum androgen levels over time and significantly retards hair growth. However, it too raises concerns of hepatotoxicity. Finasteride, which is currently used for male androgenetic alopecia, has been shown to be as effective as spironolactone or flutamide in treating hirsutism, most likely by preventing the conversion of testosterone to the much more potent DHT [44]. There is concern about finasteride causing birth defects when used during pregnancy. Eflornithine, a topical treatment, diminishes hair growth by irreversibly inhibiting ornithine decarboxylase, which is influenced by androgens and regulates cell proliferation in the hair follicle. However, its effect is only temporary and reverses with discontinued use. Finally, chlormadinone acetate is a steroidal progestin with antiandrogen and anti-gonadotropic effects that has been shown to reduce hirsutism in women.

Hidradenitis suppurativa

Pathogenesis

Hidradenitis suppurativa (HS) is a chronic follicular occlusive disease that involves intertriginous skin of the axillary, groin, perianal, perineal, and inframammary regions. The etiology and pathogenesis of HS are not fully understood, but most likely involve occlusion of the terminal parts of the follicular infundibulum due to a defect in terminal differentiation that impedes follicular epithelial shedding. Consequently, the follicular duct expands and ruptures, releasing numerous antigens that trigger inflammation. The combination of reactive inflammation and cellular proliferation may result in the cutaneous abscesses and sinus tracts that occur in HS.

The role of androgens in HS is elusive and controversial. They are thought to contribute to HS development by stimulating follicular occlusion [45]. Another hypothesis is that abnormal androgen levels or an increased sensitivity of the pilosebaceous unit to circulating androgens affect local gene expression [46]. Evidence for a link between HS and androgens includes the observation of premenstrual flare-ups, female predominance and improvement during pregnancy, and the positive effect of antiandrogen therapy on HS [47]. However, counterarguments include the observation that onset of HS usually occurs many years after puberty, the presence of normal serum levels of androgens in HS, and the lack of an association between HS and endocrinopathies [48]. Future research is needed to clarify the possible role of androgens in HS.

Diagnosis

Diagnosis of HS is made according to characteristic clinical manifestations: typical lesions are multiple deep-seated nodules, comedones, and/or fibrosis; typical locations are bilateral involvement of axillae, groin, and inframammary areas; and typical disease course shows relapses and chronicity. The first lesion is usually a single, painful, deep nodule in an intertriginous area that often progresses to form an abscess. Secondary lesions are characterized by recurrence at the same site, appearance of new adjacent lesions, coalescence of existing lesions, and fibrosis of surrounding skin [49]. Sinus tracts are also typical findings. Tertiary lesions present as unique hypertrophic fibrous scarring, leading to indurated plaques with active inflammatory nodules and sinuses. Open comedones may appear in long-lasting HS as multi-headed comedones found at the openings of burnt-out lesions or sinuses, forming the classic "tombstone comedones."

Only a few disorders cause recurrent abscesses and sinus tract formation in intertriginous skin. Follicular pyodermas, including folliculitis, furuncles, and carbuncles, are infections arising from hair follicles that are generally superficial, heal quickly, and do not produce the widespread scarring, pitting, and induration seen in HS. Granuloma inguinale presents as red ulcers with granulation tissue that bleed easily and have a foul odor, typically occurring on the vulva, penis, glans, inguinal folds, or perianal skin. The presence of an enlarging ulcer with an undermined border suggests granuloma inguinale, but diagnosis requires visualization of Donovan bodies on tissue crush preparation or biopsy. Differentiation of HS from Crohn's disease, which may cause perianal or vulvar abscesses, recto-perineal and rectovaginal fistulae, sinus tracts, fenestrations, scarring, and "knife-cut" ulcers, depends on a previous history of gastrointestinal involvement. Acne is distinguished from HS by its distribution on the face, upper chest, and back compared with the distribution of HS on the axillae, groin, buttocks, and inframammary fossae. In addition, acne is responsive to medical therapy, whereas HS is generally poorly responsive to medical or surgical therapy.

Treatment

Treatment of HS is notoriously challenging. Hormonally mediated treatment options rely on antiandrogens, such as ethinylestradiol and cyproterone acetate. Both of these therapeutic agents bind to estrogen receptors and inhibit gonadotropin secretion. Cyproterone acetate also acts as a partial agonist at androgen receptors in androgen-sensitive target tissue. Both treatments have been suggested to provide considerable improvement in HS severity, with some females responding better to antiandrogen therapy than to antibiotics [50]. The development of more effective treatment options for HS requires a better understanding of the hormonal influence on this debilitating disease.

References

1. Rosenfield RL. Role of androgens in growth and development of the fetus, child, and adolescent. *Adv Pediatr* 1972; 19: 171–213.

2. Zouboulis CC, Chen WC, Thornton MJ, et al. Sexual hormones in human skin. *Horm Metab Res* 2007; 39(2): 89–95.

3. Rosenfield RL. Hirsutism and the variable response of the pilosebaceous unit to androgen. *J Investig Dermatol Symp Proc* 2005; 10(3): 205–8.

4. Lai JJ, Chang P, Lai KP, et al. The role of androgen and androgen receptor in skin-related disorders. *Arch Dermatol Res* 2012; 304(7): 499–510.

5. Deplewski D, Rosenfield RL. Role of hormones in pilosebaceous unit development. *Endocr Rev* 2000; 21(4): 363–92.

6. Labrie F, Luu-The V, Labrie C, et al. Intracrinology and the skin. *Hormone Res* 2000; 54(5–6): 213–29.

7. Chen W, Thiboutot D, Zouboulis CC. Cutaneous androgen metabolism: basic research and clinical perspectives. *J Invest Dermatol* 2002; 119(5): 992–1007.

8. Chen W, Zouboulis CC, Fritsch M, et al. Evidence of heterogeneity and quantitative differences of the type 1 5alpha-reductase expression in cultured human skin cells – evidence of its presence in melanocytes. *J Invest Dermatol* 1998; 110(1): 84–9.

9. Zouboulis CC. The human skin as a hormone target and an endocrine gland. *Hormones* 2004; 3(1): 9–26.

10. Gad YZ, Berkovitz GD, Migeon CJ, et al. Studies of up-regulation of androgen receptors in genital skin fibroblasts. *Mol Cell Endocrinol* 1988; 57(3): 205–13.

11. Heemers HV, Tindall DJ. Androgen receptor (AR) coregulators: a diversity of functions converging on and regulating the AR transcriptional complex. *Endocr Rev* 2007; 28(7): 778–808.

12. Zouboulis CC. Human skin: an independent peripheral endocrine organ. *Horm Res* 2005; 54(5–6): 230–42.

13. Fritsch M, Orfanos CE, Zouboulis CC. Sebocytes are the key regulators of androgen homeostasis in human skin. *J Invest Dermatol* 2001; 116(5): 793–800.

14. Choudhry R, Hodgins MB, Van der Kwast TH, et al. Localization of androgen receptors in human skin by immunohistochemistry: implications for the hormonal regulation of hair growth, sebaceous glands and sweat glands. *J Endocrinol* 1992; 133(3): 467–75.

15. Chen W, Yang C-C, Sheu H-M, et al. Expression of peroxisome proliferator-activated receptor and CCAT/enhancer binding protein transcription factors in cultured human sebocytes. *J Invest Dermatol* 2003; 121(3): 441–7.

16. Ando Y, Yamaguchi Y, Hamada K, et al. Expression of mRNA for androgen receptor, 5alpha-reductase and 17beta-hydroxysteroid dehydrogenase in human dermal papilla cells. *Br J Dermatol* 1999; 141(5): 840–5.

17. Kim SS, Rosenfield RL. Hyperhydrosis as the only manifestation of hyperandrogenism in an adolescent girl. *Arch Dermatol* 2000; 136(3): 430–1.

18. Sato K, Sato F. Sweat secretion by human axillary apoeccrine sweat gland in vitro. *Am J Physiol* 1987; 252(1 Pt 2): R181–7.

19. Sato T, Sonoda T, Itami S, et al. Predominance of type I 5alpha-reductase in apocrine sweat glands of patients with excessive or abnormal odour derived from apocrine sweat (osmodrosis). *Br J Dermatol* 1998; 139(5): 806–10.

20. Preti G, Leyden JJ. Genetic influences on human body odor: from genes to the axillae. *J Invest Dermatol* 2010; 130(2): 344–6.

21. Kao JS, Garg A, Mao-Qiang M. Testosterone perturbs epidermal permeability barrier homeostasis. *J Invest Dermatol* 2001; 116(3): 443–51.

22. Cappel M, Mauger D, Thiboutot D. Correlation between serum levels of insulin-like growth factor 1, dehydroepiandrosterone sulfate, and dihydrotestosterone and acne lesion counts in adult women. *Arch Dermatol* 2005; 141(3): 333–8.

23. Choudhry R, Hodgins MB, Van der Kwast TH, et al. Localization of androgen receptors in human skin by immunohistochemistry: implications for the hormonal regulation of hair growth, sebaceous glands and sweat glands. *J Endocrinol* 1992; 133(3): 467–75.

24. Melnik BC. Role of FGFR2-signaling in the pathogenesis of acne. *Dermatoendocrinology* 2009; 1(3): 141–56.

25. Melnik BC. FoxO1 – the key for the pathogenesis and therapy of acne? *J Dtsch Dermatol Ges* 2010; 8(2): 105–14.

26. Lee WJ, Jung HD, Chi SG, et al. Effect of dihydrotestosterone on the regulation of inflammatory cytokines in cultured sebocytes. *Arch Dermatol Res* 2010; 302(6): 429–33.

27. Goodheart HP. *Goodheart's Photoguide of Common Skin Disorders*. 2nd ed. Philadelphia: Lippincott Williams & Wilkins, 2003.

28. Henze C, Hinney B, Wuttke W. Incidence of increased androgen levels in patients suffering from acne. *Dermatology* 1998; 196(1): 53–4.

29. Arowojolu AO, Gallo MF, Lopez LM, Grimes DA. Combined oral contraceptive pills for treatment of acne. *Cochrane Database Syst Rev* 2012; 7: CD004425.

30. Schmidt JB. Other antiandrogens. *Dermatology* 1998; 196(1): 153–7.

31. Lookingbill DP, Abrams BB, Ellis CN, et al. Inocoterone and acne. The effect of a topical antiandrogen: results of a multicenter clinical trial. *Arch Dermatol* 1992; 128(9): 1197–200.

32. Melnik BC. Isotretinoin and FoxO1: a scientific hypothesis. *Dermatoendocrinology* 2011; 3(3): 141–65.

33. Olsen EA. Androgenetic alopecia. In: *Disorders of Hair Growth: Diagnosis and Treatment*, Olsen EA (Ed.), McGraw-Hill, New York, 1994. Vol. 257.

34. Hibberts NA, Howell AE, Randall VA. Balding hair follicle dermal papilla cells contain higher levels of androgen receptors than those from non-balding scalp. *J Endocrinol* 1998; 156(1): 59–65.

35. Inui S, Itami S. Molecular basis of androgenetic alopecia: from androgen to paracrine mediators through dermal papilla. *J Dermatol Sci* 2011; 61(1): 1–6.

36. Kitagawa T, Matsuda K, Inui S, et al. Keratinocyte growth inhibition through the modification of Wnt signaling by androgen in balding dermal papilla cells. *J Clin Endocrinol Metab* 2009; 94(4): 1288–94.

37. Mounsey AL, Reed SW. Diagnosing and treating hair loss. *Am Fam Physician* 2009; 80(4): 356–62.

38. Gilhar A, Etzioni A, Paus R. Alopecia areata. *N Engl J Med* 2012; 366(16): 1515–25.

39. Otberg N, Finner AM, Shapiro J. Androgenetic alopecia. *Endocrinol Metab Clin North Am* 2007; 36(2): 379–98.

40. Stout SM, Stumpf JL. Finasteride treatment of hair loss in women. *Ann Pharmacother* 2010; 44(6): 1090–7.

41. Knochenhauer ES, Key TJ, Kahsar-Miller M, et al. Prevalence of the polycystic ovary syndrome in unselected black and white women of the southeastern United States: a prospective study. *J Clin Endocrinol Metab* 1998; 83(9): 3078–82.

42. Cebeci F, Onsun N, Mert M. Insulin resistance in women with hirsutism. *Arch Med Sci* 2012; 8(2): 342–6.

43. Miller KK, Rosner W, Lee H, et al. Measurement of free testosterone in normal women and women with androgen deficiency: comparison of methods. *J Clin Endocrinol Metab* 2004; 89(2): 525–33.

44. Moghetti P, Castello R, Negri C, et al. Flutamide in the treatment of hirsutism: long-term clinical effects, endocrine changes, and androgen receptor behavior. *Fertil Steril* 1995; 64(3): 511–17.

45. Stellon AJ, Wakeling M. Hidradenitis suppurativa associated with use of oral contraceptives. *BMJ* 1989; 298(6665): 28–9.

46. Harrison BJ, Read GF, Hughes LE. Endocrine basis for the clinical presentation of hidradenitis suppurativa. *BJS* 1988; 75(10): 972–5.

47. Barth JH, Layton AM, Cunliffe WJ. Endocrine factors in pre- and postmenopausal women with hidradenitis suppurativa. *Br J Dermatol* 1996; 134(6): 1057–9.

48. Mortimer PS, Dawber RPR, Gales MA, Moore RA. A double-blind controlled cross-over trial of cyproterone acetate in females with hidradenitis suppurativa. *Br J Dermatol* 1986; 115(3): 263–8.

49. Jemec GB. Clinical practice. Hidradenitis suppurativa. *N Engl J Med* 2012; 366: 158–64.

50. Kraft JN, Searles GE. Hidradenitis suppurativa in 64 female patients: Retrospective study comparing oral antibiotics and antiandrogen therapy. *J Cutan Med Surg* 2007; 11(4): 125–31.

Chapter

9

Polycystic ovary syndrome and cardiovascular risk

Heather R. Burks and Robert A. Wild

Women diagnosed with polycystic ovary syndrome (PCOS) have a higher prevalence of several well-known cardiovascular risk factors, which likely translates into a higher risk of cardiovascular events such as myocardial infarction, stroke, and sudden cardiac death. The most commonly implicated risk factors include disordered insulin regulation, including impaired glucose tolerance, insulin resistance, and diabetes, as well as dyslipidemia, metabolic syndrome, and nonalcoholic steatohepatitis. Each of these risk factors is influenced to varying degrees by the higher rate of obesity among women with PCOS. An increased rate of obstructive sleep apnea and depression/anxiety is also thought to contribute to an increase in cardiovascular risk in PCOS patients. Identifying PCOS women and screening for these disorders affords the clinician an opportunity to prevent or lessen the incidence of adverse outcomes in these women and potentially reduce the population burden of cardiovascular disease by early intervention.

Three major sets of diagnostic criteria currently define PCOS. The first set was outlined at the National Institutes of Health (NIH) in Bethesda, Maryland, in 1990 [1]. This has largely been replaced in clinical practice by the criteria defined in Rotterdam, The Netherlands, in 2003 by a task force sponsored by the European Society of Human Reproduction and Embryology (ESHRE) and the American Society for Reproductive Medicine (ASRM) [2]. More recently the Androgen Excess Society (AES) outlined its own set of criteria in 2009 [3,4]. All criteria stipulate that other diagnoses, such as nonclassic congenital adrenal hyperplasia (NC-CAH), Cushing's syndrome, androgen-secreting tumors, other causes of anovulatory cycles such as hyperprolactinemia or thyroid disease, and premature ovarian failure, be ruled out. The reported prevalence of PCOS in a given population depends largely on the diagnostic criteria utilized by the investigators.

Increased prevalence and risks for impaired glucose tolerance (IGT) and type 2 diabetes mellitus (T2DM) among women with PCOS are well described. It is estimated that 60–80% of women with PCOS have one of these diagnoses, and the prevalence increases to a remarkable 95% among obese PCOS patients [5]. Prevalence increases with age. A study of 27 adolescent girls with PCOS found that 8 (30%) had IGT and 1 (4%) already had developed T2DM [6]. A larger study of 254 women with PCOS with a wide age range (14–44) found that 31.1% had IGT and 7.5% had T2DM diagnosed by 2-hour 75 g oral glucose tolerance test. This was higher among the obese PCOS population and higher than age- and BMI-matched controls [7]. By their fourth decade of life, 40% of women with classic PCOS will have developed IGT or T2DM [8,9]. Even after adjusting for age and BMI, women with PCOS have a twofold increased risk of developing T2DM [10]. New diagnoses of T2DM occur at an estimated 5.7 cases per 1000 patient-years, compared with 1.7 per 1000 patient-years among reproductive age women without PCOS. This phenomenon is noted despite a similar fat distribution between PCOS and non-PCOS subjects [11]. Criteria for diagnosing diabetes have changed through the years. This of course will affect reported prevalence.

The prevalence of IGT depends upon the PCOS criteria utilized as well, with a higher prevalence found among women with classic PCOS using NIH criteria [12]. The severity of hyperinsulinemia correlates with the severity of clinical hyperandrogenism. There is also some ethnic variation in predisposition to insulin resistance among women with PCOS. For example, Mexican-American women have higher rates of insulin resistance than Caucasian American women. This

Androgens in Gynecological Practice, ed. Leo Plouffe and Botros Rizk. Published by Cambridge University Press. © Cambridge University Press 2015.

Table 9.1 Comparison of metabolic syndrome definitions

	Waist (cm)	Serum TG (mg/dL)	Serum HDL-C (mg/dL)	SBP or DBP (mmHg)	Glucose (mg/dL)
NCEP ATP III[a]	≥88	≥150	<50	≥130 or ≥85	≥100
IDF[b]	≥80	≥150	<50	≥130 or ≥85	≥100
AHA/NHLBI[c]	≥88	≥150	<50	≥130 or ≥85	≥100
Joint definition[d]	≥80	≥150	<50	≥130 or ≥85	≥100

TG, triglycerides; HDL-C, high-density lipoprotein cholesterol; SBP, systolic blood pressure; DBP, diastolic blood pressure.
[a] National Cholesterol Education Program Adult Treatment Panel III. Any three of the five criteria is diagnostic. This has since been revised in 2010 to include glucose <100 mg/dL.
[b] International Diabetes Federation. Requires waist circumference plus any two of the other criteria.
[c] American Heart Association/National Heart, Lung, and Blood Institute. Any three of the five criteria.
[d] Joint definition: any three of the five criteria.
Adapted from Grundy et al. [17].

finding prompted the authors to propose using different cut-off values for screening different ethnic groups [13]. The severity of insulin resistance (IR) has also been associated with the degree of menstrual derangement, with amenorrheic women displaying more pronounced IR than women with oligomenorrhea or polymenorrhea [14].

Although metabolic syndrome (MetS) can be defined using various criteria, it is very prevalent among populations of women with PCOS regardless of the criteria used. Definitions for having MetS have been proposed by the National Cholesterol Education Program Adult Treatment Panel III (NCEP ATP III) [15], International Diabetes Federation (IDF) [16], American Heart Association/National Heart, Lung and Blood Institute (AHA/NHLBI) [17], and the recent joint definition proposed by the IDF, NHLBI, AHA, World Heart Federation, International Atherosclerosis Society and International Association for the Study of Obesity (Joint definition) [18]. (Table 9.1). The most commonly used criteria in prevalence studies of women with PCOS are included in the NCEP ATP III definition.

The reported prevalence of MetS in the United States among patients with classic PCOS varies from 33% to 47%. Notably, this is two to three times higher than the prevalence in age-matched controls. It is related to obesity rates, particularly abdominal adiposity. Outside the United States, prevalence reports range from 8% to 30%. Among a population of Swedish women with PCOS, 23.8% met NCEP ATP III criteria for MetS, compared with 8.0% of controls [19]. A South Asian study found an even greater disparity, with 30.6% of PCOS patients and only 6.3% of controls

identified as having MetS [6–20]. Women in southwestern China with PCOS are five times more likely to have MetS than controls, with a prevalence of 25.6% [21]. A direct comparison of a Chinese PCOS population to a Dutch PCOS population showed that the Chinese women had greater rates of hyperandrogenism, increased waist circumference, and higher BMI. The authors concluded that Chinese women have an increased risk of metabolic complications, demonstrating ethnic and regional variations [22]. One additional example of this variation is a study in Iran showing no association between PCOS and the incidence of MetS, despite showing an association with IR [23].

There may be a stronger association between MetS and PCOS when stricter (NIH or AES) diagnostic criteria are used to define PCOS. A study in Turkey found a higher metabolic risk among women who met NIH criteria than among those who met only Rotterdam criteria [24]. A contrasting study in Turkey found that the risk of MetS is approximately double for PCOS, regardless of whether NIH, Rotterdam, or AES criteria are used to define the syndrome [25]. There is some evidence that this relationship may be entirely dependent upon obesity rates. One recent study found no independent association between PCOS and MetS, particularly when newer MetS criteria (IDF, AHA/NHLBI, or joint definition) were used rather than NCEP ATP III criteria. Obesity appeared to be the primary determinant in this population; while there were more obese patients in the PCOS group, multivariable logistic regression showed no relationship between PCOS and MetS when controlling for BMI [26]. Similarly, among adolescents, MetS is associated with visceral adiposity independent of PCOS [27].

Further contributing to a worsened cardiovascular risk profile is the increased rate of dyslipidemia. In the United States, about 70% of women with PCOS have abnormal lipids, compared with 52.9% in southwestern China [21]. In both cases, this is approximately double the rate among controls. The most common abnormality is low high-density lipoprotein cholesterol (HDL-C at 57.6%), followed by high triglycerides (28.3%) [28]. The former is dependent on glucose and insulin levels, while the latter varies with age. Elevated low-density lipoprotein cholesterol (LDL-C) is also associated with PCOS. This is most likely because of more circulating small atherogenic LDL particles and higher apolipoprotein CIII. IR impairs the ability of insulin to suppress lipolysis, increasing circulating free fatty acid release from adipose stores [5].

Women and adolescents with PCOS often display elevated blood pressure (BP) compared with age- and weight-matched controls, as well as a higher prevalence of hypertension [29,30]. One study of 34 adolescents showed that obese girls with PCOS had significantly higher 24-hour mean BP, daytime mean BP, and daytime diastolic BP, as well as a less pronounced nighttime diastolic dip in BP compared with obese controls [31]. Hypertension, however, is a late stage component of MetS and many studies do not find higher prevalence of hypertension.

Nonalcoholic fatty liver disease (NAFLD) is an important disorder, not only because of its potential for progression to more devastating liver pathology such as cirrhosis or hepatocellular carcinoma, but also because of its association with development of diabetes and cardiovascular disease. This condition is more prevalent in women with PCOS. A cohort of 31 premenopausal PCOS patients was found to have an overall rate of 62% [32]. Compared with age-matched obese controls, obese PCOS women had a significantly higher rate of NAFLD (73.3% compared with 46.7%) [33]. Among patients with PCOS, those with hyperandrogenemia have higher rates of NAFLD than those with normal androgens, even when controlling for overall and visceral fat stores and for IR [34]. IR has been shown to be an independent contributor to the development of NAFLD in PCOS [35]. Patients with PCOS also show a higher rate of disease progression to steatohepatitis and fibrosis compared with the overall population of patients with NAFLD [36,37].

Obstructive sleep apnea (OSA) is a disorder characterized by multiple episodes of apnea due to collapse of the upper airway during sleep, causing transient hypoxia and arousal throughout the night. This syndrome is associated with obesity and is characterized by an increased incidence of IR, diabetes, atherosclerosis, and cardiovascular disease [38]. Women with PCOS are at higher risk of OSA than reproductively normal controls [39], and the apnea–hypopnea index correlates positively with waist-to-hip ratios and serum total and free testosterone. A study investigating the prevalence of OSA among a cohort of premenopausal women with PCOS found 70% of women met diagnostic criteria, independent of obesity status [40]. However, a case-control study comparing obese and non-obese PCOS patients to controls found no cases of OSA in either non-obese group and concluded that the increased risk of OSA in PCOS is related primarily to the increased prevalence of obesity in this population [41]. Certainly the increased risk of hyperandrogenism and central obesity in this population should prompt the experienced clinician to screen for OSA once the diagnosis of PCOS has been established, as treatment can improve insulin sensitivity and possibly mediate cardiovascular risk [38].

It is estimated that as many as 40–85% of women with PCOS are overweight or obese, and even non-obese women or girls with PCOS are reported to have an elevated waist-to-hip ratio compared with non-PCOS BMI-matched controls. The various metabolic derangements of PCOS that contribute to a worsened cardiovascular risk profile are linked to obesity, in particular central obesity often manifested as an elevated waist-to-hip ratio. As reported above, obese patients with PCOS have higher rates of IR, diabetes, hypertension, dyslipidemia, NAFLD, and MetS. The degree to which these risk factors are due to obesity alone or obesity interacting with PCOS remains unclear. For example, age- and BMI-matched controls have lower rates of IR and hyperinsulinemia than their counterparts with PCOS among both lean and obese women, suggesting that this relationship is at least partially independent of obesity [38]. However, BMI has been shown to best predict IGT and MetS in PCOS [42].

Adipose tissue is an endocrine tissue, releasing multiple adipokines whose functions are increasingly becoming understood [43]. Leptin is one of the most well described, and levels may be increased in women with PCOS independent of obesity [44–46]. However, this is not replicated in all studies [47,48], and has not been linked to an increased risk of cardiovascular disease. Adiponectin, an adipokine with antidiabetic, anti-inflammatory, and antiatherosclerotic properties, may be lower in women with PCOS, but there is some disagreement between studies [49–51]. Visfatin, expressed preferentially in visceral adipose tissue such as that found in central obesity [52], is positively correlated with BMI, fasting insulin, IR, and BP, but has an unclear role in PCOS [53–55]. It has been shown to be elevated in lean, glucose-tolerant women with PCOS compared with controls [56]. The chemotactic properties of chemerin may contribute to the inflammatory component of obesity as a disease. It is involved in processes such as adipocyte differentiation, glucose homeostasis, and lipolysis [57,58], and recruits such inflammatory cells as lymphocytes and mononuclear cells. Given their known properties, adipokines may play a role in the pathogenesis of the metabolic derangements of PCOS, although these direct relationships have yet to be elucidated.

Various direct markers of cardiovascular risk are higher in PCOS patients, and available evidence suggests that this may translate to increased cardiovascular morbidity and mortality [5]. Women with PCOS have a lower event-free survival, and cardiovascular events increase as the number of diagnostic criteria for PCOS increases [59]. The extent of cardiovascular risk does vary by PCOS phenotype [60], with the phenotype of polycystic ovarian morphology with anovulation behaving most like control subjects [61]. A short-term prospective study with 4.7 years average of follow-up found no increased risk of adverse outcomes, including new diabetes, cancer, large vessel disease, or mortality, among PCOS women compared to controls [62], although the women in this study were relatively young. A Swedish cohort study with 21 years of follow-up, however, showed no difference in the incidence of myocardial infarction, stroke, cancer, or overall mortality between a group of PCOS patients and control subjects [63], despite an increased rate of hypertension and elevated triglycerides in the PCOS group. Variations among population studies reveal that more epidemiologic studies are required to precisely characterize the increased cardiovascular risk associated with PCOS.

Studies of carotid intima–media thickness (CIMT), a marker of atherosclerosis and future cardiovascular risk, show statistically significantly thicker measurements in PCOS compared with controls. Among premenopausal patients at least 40 years of age, CIMT was significantly thicker (0.68 mm versus 0.63 mm) in PCOS patients than in controls [64]. Similarly, women at least 45 years of age with PCOS had a thicker CIMT of 0.78 mm compared with 0.70 mm in those without, a statistically significant difference [65]. A meta-analysis of 19 studies found a 0.072 mm difference in CIMT between PCOS and age- and BMI-matched controls among high-quality studies [66].

Epicardial fat, or the layer of visceral adipose tissue deposited around the heart, can be measured using echocardiography. The epicardial fat layer correlates with visceral fat content, and has been developed as a marker of cardiovascular disease risk. The thickness of this layer has been positively correlated with the presence of coronary heart disease (CHD) and with the severity of coronary artery occlusion. Women with PCOS have an increased epicardial fat thickness compared with age- and BMI-matched controls, which has been shown to correlate with increased IR, total cholesterol, triglycerides, and androgen levels, as well as decreased adiponectin levels [67]. Epicardial fat thickness also correlates positively with BMI, waist-to-hip ratio, Ferriman–Gallwey score of hirsutism, fasting insulin, 17-OH progesterone, CIMT, and negatively with HDL cholesterol [68]. This measure appears to be emerging as an important indicator of cardiovascular risk and metabolic derangements in PCOS.

Women with PCOS within all weight categories have been shown to have a higher prevalence of cardiac dysfunction, including increased left ventricular mass [69,70], left ventricular stiffness [69,71], and degrees of diastolic dysfunction [69,71]. They have a higher incidence of coronary artery calcification independent of BMI and age [72–74].

Obesity, as previously stated, is an inflammatory condition with adipose tissue acting as an endocrine organ, supporting the release of multiple cytokines that have been linked to increased cardiovascular risk. Many of these markers have been shown to be significantly increased in PCOS, including C-reactive protein (CRP), homocysteine, plasminogen activator inhibitor-1 (both antigen and activity are increased), vascular endothelial growth factor (VEGF), asymmetric dimethylarginine (ADMA), advanced glycation end product (AGEs), and lipoprotein(a). Borderline significant elevations were also noted in tumor necrosis factor-alpha (TNF-α), endothelin-1, and fibrinogen [76].

Many features associated with PCOS have been shown to affect quality of life, particularly an increase in mood disorders. Up to 57% of women with PCOS have at least one psychiatric diagnosis [77]. Healthcare providers should be aware of the dramatically increased risk of depression and anxiety among women with PCOS, and should administer validated screening tools as part of their routine care [78]. Having PCOS confers a fourfold increased risk of depression [79], and nearly sevenfold increased risk of anxiety symptoms [80]. The overall prevalence of depression among the PCOS population has been reported at 26.4%. When NIH criteria are used to make the diagnosis, the prevalence is even higher at 35%, even adjusted for BMI, family history of depression, and infertility. Even so, obesity is known to interact with depression in the general population, and BMI, education, and parity have been identified as predictors of mild to moderate depressive symptoms [81]. Depression scores also correlate with levels of IR, lipid abnormalities, and components of the MetS, suggesting an association between depression and an increased cardiovascular risk [82]. Hyperandrogenism and hirsutism have not been shown to be independent risk factors for depression, but scores do correlate with patients' assessment of self-worth and evaluation to their own appearance [83].

Patients diagnosed with PCOS warrant a full evaluation for detection of cardiovascular risk, including measurement of height, weight, blood pressure, and waist-to-hip ratio. A complete lipid profile should be determined, and reassessment every 2 years if it is normal, as recommended by the AE-PCOS Society [5]. A 2-hour 75 g oral glucose challenge test should be performed if the BMI is greater than 30 kg/m^2, or in cases of age >40, and personal or family history of diabetes. In patients with normal testing, this may be repeated every 2 years or upon the appearance of additional risk factors. Strong consideration should be given to screening for depression and OSA as well, as these comorbidities may contribute to a woman's cardiovascular risk and overall quality of life.

References

1. Zawadski JK, Dunaif A. Diagnostic criteria for polycystic ovary syndrome: towards a rational approach. In: Dunaif A, Givens JR, Haseltine FP, Merriam GR, eds. *Polycystic Ovary Syndrome*. Boston, MA: Blackwell Scientific Publications, 1992:377–84.

2. Rotterdam ESHRE/ASRM-Sponsored PCOS Consensus Workshop Group. Revised 2003 consensus on diagnostic criteria and long-term health risks related to polycystic ovary syndrome. *Fertil Steril* 2004;81(1):19–25.

3. Azziz R, Carmina E, Dewailly D, et al. The Androgen Excess and PCOS Society criteria for the polycystic ovary syndrome: the complete task force report. *Fertil Steril* 2009;91(2):456–88.

4. Yilmaz M, Isaoglu U, Delibas IB, Kadanali S. Anthropometric, clinical and laboratory comparison of four phenotypes of polycystic ovary syndrome based on Rotterdam criteria. *J Obstet Gynaecol Res* 2011;37(8):1020–6.

5. Wild RA, Carmina E, Diamanti-Kandarakis E, et al. Assessment of Cardiovascular Risk and Prevention of Cardiovascular Disease in Women with the Polycystic Ovary Syndrome: A Consensus Statement by the Androgen Excess and Polycystic Ovary Syndrome (AE-PCOS) Society. *J Clin Endocrinol Metab* 2010;95(5):2038–49.

6. Palmert MR, Gordon CM, Kartashov AI, et al. Screening for abnormal glucose tolerance in adolescents with polycystic ovary syndrome. *J Clin Endocrinol Metab* 2002;87(3):1017–23.

7. Legro RS, Gnatuk CL, Kunselman AR, Dunaif A. Changes in glucose tolerance over time in women with polycystic ovary syndrome: a controlled study. *J Clin Endocrinol Metab*. 2005 Jun;90(6):3236–42.

8. Norman RJ, Masters L, Milner CR, et al. Relative risk of conversion from normoglycaemia to impaired glucose tolerance or non-insulin dependent diabetes mellitus in polycystic ovarian syndrome. *Hum Reprod* 2001;16(9):1995–8.

9. Boudreaux MY, Talbott EO, Kip KE, et al. Risk of T2DM and impaired fasting glucose among PCOS subjects: results of an 8-year follow-up. *Curr Diab Rep* 2006;6(1):77–83.

10. Barber TM, Golding SJ, Alvey C, et al. Global adiposity rather than abnormal regional fat distribution characterizes women with polycystic ovary syndrome. *J Clin Endocrinol Metab* 2008;93(3):999–1004.

11. Kumarapeli V, Seneviratne RA, Wijeyaratne CN, et al. A simple screening approach for assessing community prevalence and phenotype of polycystic ovary syndrome in a semiurban population in Sri Lanka. *Am J Epidemiol* 2008;168:321–8.

12. Tehrani FR, Rashidi H, Azizi F. The prevalence of idiopathic hirsutism and polycystic ovary syndrome in the Tehran Lipid and Glucose Study. *Reprod Biol Endocrinol* 2011;9:144.

13. Legro RS, Kunselman AR, Dodson WC, Dunaif A. Prevalence and predictors of risk for type 2 diabetes mellitus and impaired glucose tolerance in polycystic ovary syndrome: a prospective, controlled study in 254 affected women. *J Clin Endocrinol Metab* 1999;84(1):165–9.

14. Panidis D, Tziomalos K, Chatzis P, et al. Association between menstrual cycle irregularities and endocrine and metabolic characteristics of the polycystic ovary syndrome. *Eur J Endocrinol* 2013;168(2):145–52.

15. National Cholesterol Education Program (NCEP) Expert Panel on Detection, Evaluation, and Treatment of High Blood Cholesterol in Adults (Adult Treatment Panel III). Third Report of the National Cholesterol Education Program (NCEP) Expert Panel on Detection, Evaluation, and Treatment of High Blood Cholesterol in Adults (Adult Treatment Panel III) final report. *Circulation* 2002;106(25):3143–421.

16. Alberti KG, Zimmet P, Shaw J; IDF Epidemiology Task Force Consensus Group. The metabolic syndrome–a new worldwide definition. *Lancet* 2005;366(9261):1059–62.

17. Grundy SM, Cleeman JI, Daniels SR, et al.; American Heart Association; National Heart, Lung, and Blood Institute. Diagnosis and management of the metabolic syndrome: an American Heart Association/National Heart, Lung, and Blood Institute Scientific Statement. *Circulation* 2005;112(17):2735–52.

18. Alberti KG, Eckel RH, Grundy SM, et al.; International Diabetes Federation Task Force on Epidemiology and Prevention; National Heart, Lung, and Blood Institute; American Heart Association; World Heart Federation; International Atherosclerosis Society; International Association for the Study of Obesity. Harmonizing the metabolic syndrome: a joint interim statement of the International Diabetes Federation Task Force on Epidemiology and Prevention; National Heart, Lung, and Blood Institute; American Heart Association; World Heart Federation; International Atherosclerosis Society; and International Association for the Study of Obesity. *Circulation* 2009;120(16):1640–5.

19. Hudecova M, Holte J, Olovsson M, et al. Prevalence of the metabolic syndrome in women with a previous diagnosis of polycystic ovary syndrome: long-term follow-up. *Fertil Steril* 2011;96(5):1271–4.

20. Wijeyaratne CN, Seneviratne RA, Dahanayake S, et al. Phenotype and metabolic profile of South Asian women with polycystic ovary syndrome (PCOS): results of a large database from a specialist Endocrine Clinic. *Hum Reprod* 2011;26(1):202–13.

21. Zhang J, Fan P, Liu H, et al. Apolipoprotein A-I and B levels, dyslipidemia and metabolic syndrome in south-west Chinese women with PCOS. *Hum Reprod* 2010;27(8):2484–93.

22. Guo M, Chen ZJ, Eijkemans MJE, et al. Comparison of the phenotype of Chinese versus Dutch Caucasian women presenting with polycystic ovary syndrome and oligo/amenorrhoea. *Hum Reprod* 2012;27(5):1481–8.

23. Hosseinpanah F, Barzin M, Tehrani FR, Azizi F. The lack of association between polycystic ovary syndrome and metabolic syndrome: Iranian PCOS prevalence study. *Clin Endocrinol* 2011;75:692–7.

24. Anaforoglu I, Algun E, Incecayir O, Ersoy K. Higher metabolic risk with National Institutes of Health versus Rotterdam diagnostic criteria for polycystic ovarian syndrome in Turkish women. *Metab Syndr Relat Disord* 2011;9(5):375–80.

25. Yildiz BO, Bozdag G, Yapici Z, et al. Prevalence, phenotype and cardiometabolic risk of polycystic ovary syndrome under different diagnostic criteria. *Hum Reprod* 2012;27(10):3067–73.

26. Panidis D, Macut D, Tziomalos K, et al. Prevalence of metabolic syndrome in women with polycystic ovary syndrome. *Clin Endocrinol (Oxf)* 2013;78(4):586–92.

27. Rossi B, Sukalich S, Droz J, et al. Prevalence of metabolic syndrome and related characteristics in obese adolescents with and without polycystic ovary syndrome. *J Clin Endocrinol Metab* 2008;93(12):4780–6.

28. Rocha MP, Marcondes JAM, Barcellos CRG, et al. Dyslipidemia in women with polycystic ovary syndrome: incidence, pattern and predictors. *Gynecol Endocrinol* 2011;27(10):814–19.

29. Elting MW, Korsen TJ, Bezemer PD, Schoemaker J. Prevalence of diabetes mellitus, hypertension and cardiac complaints in a follow-up study of a Dutch PCOS population. *Hum Reprod* 2001;16(3):556–60.

30. Vrbíková J, Cífková R, Jirkovská A, et al. Cardiovascular risk factors in young Czech females with polycystic ovary syndrome. *Hum Reprod* 2003;18(5):980–4.

31. Zachurzok-Buczynska A, Szydlowski L, Gawlik A, et al. Blood pressure regulation and resting heart rate abnormalities in adolescent girls with polycystic ovary syndrome. *Fertil Steril* 2011;96(6):1519–25.

32. Gutierrez-Grobe Y, Ponciano-Rodríguez G, Ramos MH, et al. Prevalence of nonalcoholic fatty liver disease in premenopausal, postmenopausal, and polycystic ovary syndrome women. The role of estrogens. *Ann Hepatol* 2010;9(4):402–9.

33. Zueff LF, Martins WP, Vieira CS, Ferriani RA. Ultrasonographic and laboratory markers of metabolic and cardiovascular disease risk in obese women with polycystic ovary syndrome. *Ultrasound Obstet Gynecol* 2012;39(3):341–7.

34. Jones H, Sprung VS, Pugh CJA, et al. Polycystic ovary syndrome with hyperandrogenism is characterized by an increased risk of hepatic steatosis compared to nonhyperandrogenic PCOS phenotypes and healthy controls, independent of obesity and insulin resistance. *J Clin Endocrinol Metab* 2012;97:3709–16.

35. Cerda C, Pérez-Ayuso RM, Riquelme A, et al. Nonalcoholic fatty liver disease in women with polycystic ovary syndrome. *J Hepatol* 2007;47(3):412–17.

36. Brzozowska MM, Ostapowicz G, Weltman MD. An association between non-alcoholic fatty liver disease and polycystic ovarian syndrome. *J Gastroenterol Hepatol* 2009;24(2):243–7.

37. Setji TL, Holland ND, Sanders LL, et al. Nonalcoholic steatohepatitis and nonalcoholic fatty liver disease in young women with polycystic ovary syndrome. *J Clin Endocrinol Metab* 2006;91(5):1741–7.

38. Randeva HS, Tan BK, Weickert MO, et al. Cardiometabolic aspects of the polycystic ovary syndrome. *Endocr Rev* 2012;33(5):812–41.

39. Fogel RB, Malhotra A, Pillar G, et al. Increased prevalence of obstructive sleep apnea syndrome in obese women with polycystic ovary syndrome. *J Clin Endocrinol Metab* 2001;86(3):1175–80.

40. Gopal M, Duntley S, Uhles M, Attarian H. The role of obesity in the increased prevalence of obstructive sleep apnea syndrome in patients with polycystic ovarian syndrome. *Sleep Med* 2002;3(5):401–4.

41. Mokhlesi B, Scoccia B, Mazzone T, Sam S. Risk of obstructive sleep apnea in obese and nonobese women with polycystic ovary syndrome and healthy reproductively normal women. *Fertil Steril* 2012;97(3):786–91.

42. Liang SJ, Liou TH, Lin HW, et al. Obesity is the predominant predictor of impaired glucose tolerance and metabolic disturbance in polycystic ovary syndrome. *Acta Obstet Gynecol Scand* 2012; 91:1167–72.

43. Ahima RS, Flier JS. Adipose tissue as an endocrine organ. *Trends Endocrinol Metab* 2000;11(8):327–32.

44. Pusalkar M, Meherji P, Gokral J, et al. Obesity and polycystic ovary syndrome: association with androgens, leptin and its genotypes. *Gynecol Endocrinol* 2010;26(12):874–82.

45. Yildizhan R, Ilhan GA, Yildizhan B, et al. Serum retinol-binding protein 4, leptin, and plasma asymmetric dimethylarginine levels in obese and nonobese young women with polycystic ovary syndrome. *Fertil Steril* 2011;96(1):246–50.

46. Lecke SB, Morsch DM, Spritzer PM. Association between adipose tissue expression and serum levels of leptin and adiponectin in women with polycystic ovary syndrome. *Genet Mol Res* 2013;12(4):4292–6.

47. Pirwany IR, Fleming R, Sattar N, et al. Circulating leptin concentrations and ovarian function in polycystic ovary syndrome. *Eur J Endocrinol* 2001;145(3):289–94.

48. Telli MH, Yildirim M, Noyan V. Serum leptin levels in patients with polycystic ovary syndrome. *Fertil Steril* 2002;77(5):932–5.

49. Bobbert T, Rochlitz H, Wegewitz U, et al. Changes of adiponectin oligomer composition by moderate weight reduction. *Diabetes* 2005;54(9):2712–19.

50. Panidis D, Kourtis A, Farmakiotis D, et al. Serum adiponectin levels in women with polycystic ovary syndrome. *Hum Reprod* 2003;18(9):1790–6.

51. Toulis KA, Goulis DG, Farmakiotis D, et al. Adiponectin levels in women with polycystic ovary syndrome: a systematic review and a meta-analysis. *Hum Reprod Update* 2009;15(3):297–307.

52. Samal B, Sun Y, Stearns G, et al. Cloning and characterization of the cDNA encoding a novel human pre-B-cell colony-enhancing factor. *Mol Cell Biol* 1994;14(2):1431–7.

53. Chan TF, Chen YL, Chen HH, et al. Increased plasma visfatin concentrations in women with polycystic ovary syndrome. *Fertil Steril* 2007;88(2):401–5.

54. Jongwutiwes T, Lertvikool S, Leelaphiwat S, et al. Serum visfatin in Asian women with polycystic ovary syndrome. *Gynecol Endocrinol* 2009;25(8):536–42.

55. Tan BK, Chen J, Digby JE, et al. Increased visfatin messenger ribonucleic acid and protein levels in adipose tissue and adipocytes in women with polycystic ovary syndrome: parallel increase in plasma visfatin. *J Clin Endocrinol Metab* 2006;91(12):5022–8.

56. Yildiz BO, Bozdag G, Otegen U, et al. Visfatin and retinol-binding protein 4 concentrations in lean, glucose-tolerant women with PCOS. *Reprod Biomed Online* 2010;20(1):150–5.

57. Roh SG, Song SH, Choi KC, et al. Chemerin–a new adipokine that modulates adipogenesis via its own receptor. *Biochem Biophys Res Commun* 2007;362(4):1013–18.

58. Takahashi M, Takahashi Y, Takahashi K, et al. Chemerin enhances insulin signaling and potentiates insulin-stimulated glucose uptake in 3T3-L1 adipocytes. *FEBS Lett* 2008;582(5):573–8.

59. Shaw LJ, Bairey Merz CN, Azziz R, et al. Postmenopausal women with a history of irregular menses and elevated androgen measurements at high risk for worsening cardiovascular event-free survival: results from the National Institutes of Health–National Heart, Lung, and Blood Institute sponsored Women's Ischemia Syndrome Evaluation. *J Clin Endocrinol Metab* 2008;93(4):1276–84.

60. Dilbaz B, Ozkaya E, Cinar M, et al. Cardiovascular disease risk characteristics of the main polycystic ovary syndrome phenotypes. *Endocrine* 2011;39(3):272–7.

61. Yilmaz M, Isaoglu U, Delibas IB, Kadanali S. Anthropometric, clinical and laboratory comparison of four phenotypes of polycystic ovary syndrome based on Rotterdam criteria. *J Obstet Gynaecol Res* 2011;37(8):1020–6.

62. Morgan CL, Jenkins-Jones S, Currie CJ, Rees DA. Evaluation of adverse outcome in young women with

polycystic ovary syndrome versus matched, reference controls: a retrospective, observational study. *J Clin Endocrinol Metab* 2012;97(9):3251–60.

63. Schmidt J, Landin-Wilhelmsen K, Brännström M, Dahlgren E. Cardiovascular disease and risk factors in PCOS women of postmenopausal age: a 21-year controlled follow-up study. *J Clin Endocrinol Metab* 2011;96(12):3794–803.

64. Guzick DS, Talbott EO, Sutton-Tyrrell K, et al. Carotid atherosclerosis in women with polycystic ovary syndrome: initial results from a case-control study. *Am J Obstet Gynecol* 1996;174(4):1224–9; discussion 1229–32.

65. Talbott EO, Guzick DS, Sutton-Tyrrell K, et al. Evidence for association between polycystic ovary syndrome and premature carotid atherosclerosis in middle-aged women. *Arterioscler Thromb Vasc Biol* 2000;20(11):2414–21.

66. Meyer ML, Malek AM, Wild RA, et al. Carotid artery intima-media thickness in polycystic ovary syndrome: a systematic review and meta-analysis. *Hum Reprod Update* 2012;18(2):112–26.

67. Aydogdu A, Uckaya G, Tasci I, et al. The relationship of epicardial adipose tissue thickness to clinical and biochemical features in women with polycystic ovary syndrome. *Endocr J* 2012;59(6):509–16.

68. Cakir E, Doğan M, Topaloglu O, et al. Subclinical atherosclerosis and hyperandrogenemia are independent risk factors for increased epicardial fat thickness in patients with PCOS and idiopathic hirsutism. *Atherosclerosis* 2013;226(1):291–5.

69. Orio F Jr, Palomba S, Spinelli L, et al. The cardiovascular risk of young women with polycystic ovary syndrome: an observational, analytical, prospective case-control study. *J Clin Endocrinol Metab* 2004;89(8):3696–701.

70. Celik O, Sahin I, Celik N, et al. Diagnostic potential of serum N-terminal pro-B-type brain natriuretic peptide level in detection of cardiac wall stress in women with polycystic ovary syndrome: a cross-sectional comparison study. *Hum Reprod* 2007;22(11):2992–8.

71. Yarali H, Yildirir A, Aybar F, et al. Diastolic dysfunction and increased serum homocysteine concentrations may contribute to increased cardiovascular risk in patients with polycystic ovary syndrome. *Fertil Steril* 2001;76(3):511–16.

72. Christian RC, Dumesic DA, Behrenbeck T, et al. Prevalence and predictors of coronary artery calcification in women with polycystic ovary syndrome. *J Clin Endocrinol Metab* 2003;88(6):2562–8.

73. Talbott EO, Zborowski JV, Rager JR, et al. Evidence for an association between metabolic cardiovascular syndrome and coronary and aortic calcification among women with polycystic ovary syndrome. *J Clin Endocrinol Metab* 2004;89(11):5454–61.

74. Shroff R, Kerchner A, Maifeld M, et al. Young obese women with polycystic ovary syndrome have evidence of early coronary atherosclerosis. *J Clin Endocrinol Metab* 2007;92(12):4609–14.

75. Sirmans SM, Weidman-Evans E, Everton V, Thompson D. Polycystic ovary syndrome and chronic inflammation: pharmacotherapeutic implications. *Ann Pharmacother* 2012;46(3):403–18.

76. Toulis KA, Goulis DG, Mintziori G, et al. Meta-analysis of cardiovascular disease risk markers in women with polycystic ovary syndrome. *Hum Reprod Update* 2011;17(6):741–60.

77. Rassi A, Veras AB, dos Reis M, Pastore DL, et al. Prevalence of psychiatric disorders in patients with polycystic ovary syndrome. *Compr Psychiatry* 2010;51:599–602.

78. Dokras A. Mood and anxiety disorders in women with PCOS. *Steroids* 2012;77:338–41.

79. Dokras A, Clifton S, Futterweit W, Wild R. Increased risk for abnormal depression scores in women with polycystic ovary syndrome: a systematic review and meta-analysis. *Obstet Gynecol* 2011;117(1):145–52.

80. Dokras A, Clifton S, Futterweit W, Wild R. Increased prevalence of anxiety symptoms in women with polycystic ovary syndrome: systematic review and meta-analysis. *Fertil Steril* 2012;97(1):225–30.

81. Cipkala-Gaffin J, Talbott EO, Song MK, et al. Associations between psychologic symptoms and life satisfaction in women with polycystic ovary syndrome. *J Womens Health* 2012;21(2):179–87.

82. Cinar N, Kizilarslanoglu MC, Harmanci A, et al. Depression, anxiety and cardiometabolic risk in polycystic ovary syndrome. *Hum Reprod* 2011;26(12):3339–45.

83. Deeks AA, Gibson-Helm ME, Paul E, Teede HJ. Is having polycystic ovary syndrome a predictor of poor psychological function including anxiety and depression? *Hum Reprod* 2011;26(6):1399–407.

10

Effects of androgens on female genital tract

Abdulmaged M. Traish and André T. Guay*

Summary

Androgens exert specific effects on the development and function of the female reproductive organs and secondary sexual characteristics and the external genitalia. These include effects on the uterus, vagina, oviduct, preputial gland, clitoris, and the mammary glands [1]. The normal physiology of the female genital tract encompasses complex processes, which are highly dependent on the structural and functional integrity of the genital tissues, involving complex neurovascular mechanisms modulated by numerous local neurotransmitters, vasoactive agents, sex steroid hormones, and growth factors. Regulation of vaginal and clitoral physiology by androgens is an area of great interest, not only in terms of basic understanding of the molecular and cellular bases of vaginal and clitoral function, but also because of clinical implications for use of androgen therapy in women. However, this area of investigation remains, at best, rudimentary and poorly studied. Clinically, testosterone is implicated in maintaining female genital sexual arousal response and sexual desire. Androgens play a key role in maintaining vaginal nerve network fibers and sexual arousal responses and facilitate genital hemodynamics and vaginal lubrication, thus preventing dyspareunia.

Androgens in the development and differentiation of female genitalia

Sexual development is a function of two processes: sex determination and differentiation. Sex determination is the initial decision to direct the development of the bipotential gonadal ridges into either testes or ovaries [2]. This decision of gonadal sex is governed by genetic sex, and is dependent on the integration of a host of molecular pathways that direct and dictate the development of germ cells and their migration to the urogenital ridge. Determination of whether testes or ovaries will develop is dependent on the presence or absence of the Y chromosome. Sex determination sets the conditions for the undifferentiated embryo to develop in a sexually dimorphic manner [3]. Once the gonads have formed, genetic sex becomes unimportant, and gonadal sex directs sexual differentiation into the male or female phenotype [4]. Subsequent sexual differentiation is the sex-specific response of tissues to hormones associated with the type of gonad present [3].

The sex-determining region of the Y gene (SRY) has been established as the testis-determining factor, a single-copy male-specific sequence that has been evolutionarily conserved [2]. Although ovarian development has long been thought of as a default pathway that occurs in the absence of SRY, this default concept was originally intended to describe the female reproductive ducts that develop in the absence of gonads, not ovarian development. The choice between the male or female pathway involves active networks of complex and antagonistic signals [5]. For example, Sox9 and FGF9 promote testis differentiation, while WNT4 represses male genes to promote ovary development. The relative levels of expression of these factors in the bipotential gonad tilts the balance in favor of either the male or female pathway and triggers the differentiation of the testis or ovary [2].

Female sexual differentiation

In XX gonads, the male pathway is antagonized and the female pathway proceeds [6]. Independent of the

Androgens in Gynecological Practice, ed. Leo Plouffe and Botros Rizk. Published by Cambridge University Press. © Cambridge University Press 2015.

* This chapter is dedicated to the memory of my dear friend and colleague Andre T. Guay, MD and for his immense contributions to the advancement of clinical endocrinology, especially sexual function in men and women.   Abdulmaged M. Traish

ovaries, absence of anti-müllerian hormone (AMH) and androgen action [7] results in feminization of the internal and external genitalia. Lack of testosterone results in regression of the wolffian ducts while lack of AMH results in stabilization of the müllerian ducts in the immature female [2]. Further maturation depends on the effects of estrogens [8].

The time line for sex differentiation in female begins at approximately 7–10 weeks of gestation, in which the müllerian ducts fuse caudally to form the fornix of the vagina. At approximately 11–12 weeks, the germ cell migrates from the dorsal endoderm of the yolk sac to the ovary. By weeks 12–16, the uterus forms in the fused portion of the müllerian duct and the first ovarian follicles appear. By the twentieth week, the peak oocyte number is reached. Subsequently, the primordial follicles with granulosa cells are established and the vagina becomes fully canalized [9].

As discussed above, around the seventh week of gestation, the wolffian ducts involute and the müllerian ducts differentiate into the fallopian tubes, uterus, cervix, and vagina [2]. The cranial müllerian duct regresses in the presence of AMH; thus, the upper female genital tract is under negative control of AMH. In contrast, the caudal müllerian duct is insensitive to AMH and vaginal development is under negative control of androgens [10,11]. In rodent males with testicular feminization (Tfm) and human males with complete androgen insensitivity syndrome (CAIS), the vagina forms irrespective of AMH function, demonstrating the critical role of androgen in suppressing vaginal development. Similarly, in AMH transgenic female mice, the oviduct and uterus fail to form due to the inhibition by AMH, whereas the vagina develops [reviewed in 10].

The cranial portions of the müllerian ducts extend into the peritoneal cavity and form the fallopian tubes. The right and left müllerian ducts grow caudally, and lateral fusion at the midline by week 10 gives rise to the uterus, cervix, and vagina [7,10,12]. The thin uterine septum remaining after fusion eventually resorbs to yield a single cavity, and by the fifth month of gestation müllerian organogenesis is complete [2].

External genitalia

Absence of androgens maintains the appearance of the external genitalia as developed upon the sixth week of gestation. Feminization is achieved with little or no changes; the genital tubercle develops very slightly to form the clitoris, lateral genital (labioscrotal) swellings remain unfused to form the labia majora, and adjacent urethral folds form the labia minora. The vaginal introitus and urethral meatus form between the labia minora [9].

Both the fetal testis and fetal adrenal gland are capable of testosterone biosynthesis during the first trimester. During this time, the presence of androgen excess will result in male sexual differentiation of the external genitalia. From the second trimester on, the potential for further virilization is limited by two key events: downregulation of androgen receptor (AR) expression and fetal expression of the enzyme aromatase (CYP19A1) [13]. AR expression in the external female genitalia is decreased after the first trimester, and is only continuously expressed in the clitoris [13]. Clitoral growth in human females is androgen-responsive in the fetus, as well as in the postnatal and adult female.

In experimental animal models, exposure of the female fetus to testosterone produced a small increment in clitoral length. However, the maximum clitoral length occurred during fetal and postnatal treatment with minimal increase at adulthood. This suggests that clitoral response in the animal model may not resemble that seen in humans, since clitoral growth is noted in female-to-male transsexuals who are treated with testosterone as adults [14].

Fetal aromatase serves as a protector to AR by converting androstenedione and testosterone into estrone and 17β-estradiol, respectively [13]. Aromatase is found in the gonads, brain, adipose tissue, and placenta. In the placenta, it serves a protective effect by converting fetal adrenal androgens such as dehydroepiandrosterone (DHEA), its sulfated derivative (DHEA-S), and 16-hydroxy DHEA-S into estrogens (estrone, estradiol, estriol). Efficient aromatization prevents accumulation of testosterone and ensures that no fetal androgen crosses into the maternal circulation. In the case of placental aromatase deficiency, fetal androgens and placentally derived testosterone will circulate freely to virilize the mother and fetus. If genetically male, sexual differentiation will continue normally, but if genetically female, the fetus will be born with clitoral enlargement and labial fusion [15].

Additionally, prenatal exposure of testosterone in rats has resulted in decreased size of the vaginal orifice, reduced distance from the phallus, persistent cleaving of the phallus, and even absence of the vaginal orifice. Such malformations are similar to underdeveloped reproductive tracts and the early urogenital sinus. Presence of a vaginal thread, or isthmus of tissue across the diameter of the vaginal orifice was also observed in the study conducted by Wolf et al. [16], in which testosterone propionate treatment resulted in permanently increased anogenital distance.

In an immunohistochemical study of androgen, estrogen, and progesterone receptors in the vulva and vagina, Hodgkins et al. [17] reported an increase in ARs and decrease in estrogen and progesterone receptors upon the transition from vagina to vulva. ARs are especially abundant in epidermal keratinocytes and dermal fibroblasts of the labia majora and are also expressed in sebaceous glands, sweat glands, hair follicles, dermal fibroblasts, and epithelial cells and stromal fibroblasts of the vagina. Estrogen receptors are also found in relative abundance in vulval epidermis, whereas progesterone receptors are not seen in the vulva [17]. Development of pubic hair appears to be under the control of androgens [18].

Androgens maintain female genital structure and function

Androgens modulate the physiology of female reproductive organs and female secondary sexual characteristics. These organs include the uterus, the vagina, the oviduct, the preputial gland, the clitoris, and the mammary glands [1]. In animal studies, androgen treatment of immature animals resulted in premature opening and cornification of the vagina. The loss of vaginal tissue weight subsequent to ovariectomy was partially restored by testosterone treatment. Androgen treatment produced mucification, but not cornification of the vaginal mucosa, and it was suggested that androgens oppose the effects of estrogens on vaginal cornification [1]. In ovariectomized animal studies, treatment with testosterone, Δ^5androstenediol (Δ^5Adiol), DHEA, or androsterone was shown to result in the enlargement and development of the preputial gland. Also, the paraurethral glands (Skene's ducts) in the female were more sensitive to androgens, but not estrogens.

It is evident that androgens continue to have an important role in regulating the structure and function of female genital tissues well beyond the prenatal and early postnatal periods. This section will focus on the effects of androgens in the adult clitoris and vagina. The normal physiology of the female genital tract involves maintenance of tissue structure and complex neurovascular processes modulated by local neurotransmitters,

vasoactive agents, sex steroid hormones, and growth factors. Regulation of these processes by androgens is an area of great interest, because of the clinical implications for use of androgen therapy in women.

Effects of androgens on clitoral and vaginal structure

In animal studies, ovariectomy produces profound changes in the vaginal epithelium and, to a limited extent, the muscularis and lamina propria layers (Figure 10.1a and b) [19]. Two weeks post-ovariectomy, estradiol treatment restored changes in vaginal tissue structure. In contrast, testosterone treatment alone did not restore growth of the vaginal epithelium (Figure 10.1c), but produced partial restoration of the muscularis fiber bundles (Figure 10.1c). Combined treatment with estradiol and testosterone restored the muscularis fiber bundle similar to that noted in vaginal tissues of control intact animals (Figure 10.1d). Interestingly, testosterone treatment of ovariectomized animals significantly enhanced the density of the nerve network fibers, as assessed by protein gene product (PGP) 9.5, a general nerve marker (Figure 10.2a and b). Also, testosterone treatment increased the density of the adrenergic nerve marker tyrosine hydroxylase (TH) and PGP 9.5 in the vagina (Figure 10.3a and b). These changes were not noted in the tissue from animals treated with estradiol [19], observations suggesting differential regulation of vaginal innervation by androgens.

Pelletier et al. [20] reported that ovariectomy produced significant reduction in the thickness of the lamina propria area and treatment with DHEA restored the thickness to approximately 69% of the intact value. In the ovariectomized animals, protein gene product 9.5 (PGP 9.5)-immunopositive fibers and muscularis layer thickness were reduced. These reductions were prevented with DHEA treatment. In addition, DHEA treatment resulted in increased density of TH-containing nerve fibers in the vagina. These findings paralleled those reported by us [19] and suggest that DHEA is acting as an androgen or androgen precursor in the vagina.

As in the prenatal and early postnatal periods, androgens are known to modulate clitoral growth in adolescents and adults. This is supported by clinical observations in women with disorders of sex development. Mutations in steroidogenic factor NR5A1 in patients with 46 XY genotype often produce sufficient testosterone for spontaneous virilization during puberty. Phenotypic females with NR5A1 mutations may present

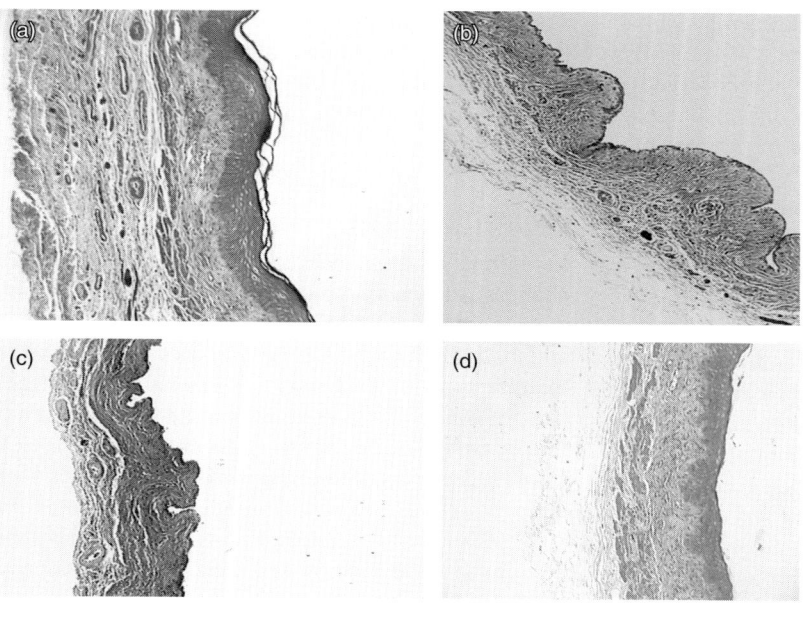

Figure 10.1 Effects of ovariectomy and estrogen or androgen treatment on vaginal tissue structure (a–d). Female rats were left intact (control; a) or ovariectomized and after 2 weeks were infused with vehicle (b) or testosterone (c) or testosterone and estradiol (d). Hormone treatment was maintained for 2 weeks. Vaginal tissue was fixed, sectioned, and subjected to Masson's trichrome staining as described in references 19 and 26. A black and white version of this figure will appear in some formats. For the color version, please refer to the plate section.

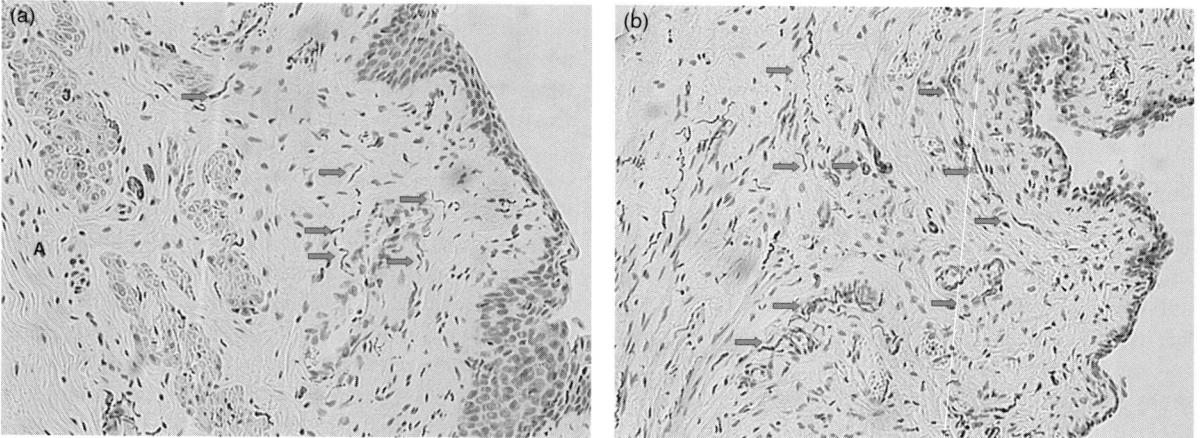

Figure 10.2 Effects of ovariectomy and androgen treatment on vaginal tissue nerve fiber network density. Female rats were left intact (control; a) or ovariectomized and after 2 weeks were infused with testosterone (b). Hormone treatment was started 2 weeks after ovariectomy and was maintained for 2 weeks. Vaginal tissue was fixed and paraffin-embedded and vaginal tissue sections subjected to immunostaining procedures using anti-PGP 9.5 primary antibody and counterstained with Gill's hematoxylin. In vaginal tissue from intact rats (a), nerve fibers extending from the lamina propria into the epithelial layer were occasionally observed – see arrows in (a). PGP-positive fibers were commonly observed surrounding blood vessels (arrows) as depicted in tissue sections from intact control (a) and testosterone-infused (11 μg/day for 14 days), ovariectomized animals (b). Details are given in references 19 and 26. A black and white version of this figure will appear in some formats. For the color version, please refer to the plate section.

with clitoromegaly at puberty [21]. In another study, female-to-male transsexuals treated with testosterone exhibited signs of polycystic ovarian disease, but no noticeable endometrial pathologies. However, increases in clitoral size, libido, body, and facial hair, deepened voices, and decline in breast size were all noted [22,23]. In utero, virilization of the fetus can occur due to loss of activity of 17α-hydroxylase and 21-hydroxylases

that shift the pathway to androgen biosynthesis by the adrenal gland, resulting in clitoral growth.

Androgens regulate key vasomotor and trophic enzymes in genital tissue

Blood flow in the vagina is critical for vaginal engorgement. This hemodynamic process is thought to be

(a)

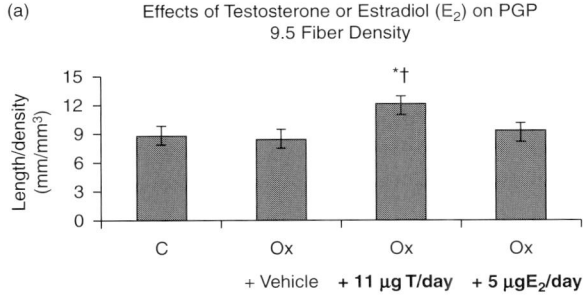

(b)

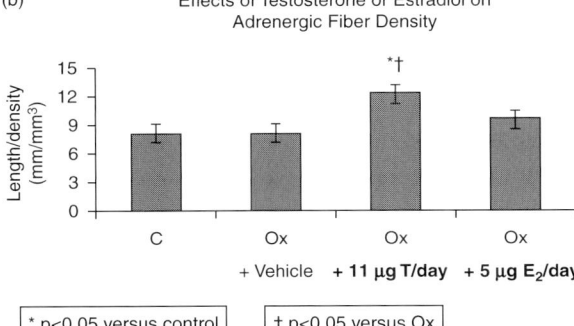

* p<0.05 versus control   † p<0.05 versus Ox

Figure 10.3 Effects of ovariectomy (Ox), and testosterone (T) or estradiol (E$_2$) on PGP 9.5 density (a) and adrenergic fiber length-density (b). Rats were ovariectomized for 2 weeks, followed by 2 weeks of continuous infusion of vehicle or hormone replacement using subcutaneous osmotic pumps. Animals were sacrificed and vaginal tissue was removed, fixed, and embedded as described above. Immunostaining with the appropriate antibody to PGP 9.5 or tyrosine hydroxylase (a marker for the adrenergic nerve fiber) was employed. The density of the nerve fibers was assessed as described in reference 19. *P <0.001 versus control, †P <0.01 versus Ox.

important for vaginal sexual arousal. Nitric oxide (NO), one of the key mediators of vaginal blood flow, is thought to be the non-adrenergic non-cholinergic neurotransmitter and the vasoactive dilator. In female rats, Kim et al. [24,25] demonstrated that administration of inhibitors of nitric oxide synthase (NOS) and inhibitors of guanylyl cyclase diminished vaginal blood flow in response to pelvic nerve stimulation. These observations implicate NO/cGMP as one of the key mediators of vaginal smooth muscle relaxation and pelvic nerve-stimulated increase in blood flow [26]. Treatment of ovariectomized rabbits with testosterone resulted in increased expression and activity of total NOS in the proximal vagina, but not in the distal vagina [27]. Treatment with 5α dihydrotestosterone (5α-DHT) resulted in increased expression and

activity of total NOS in both the proximal and distal vagina.

Arginase hydrolyzes arginine, the main co-substrate for NOS [28]. Treatment of ovariectomized animals with DHEA, Δ^5 A-diol, or 5α-DHT resulted in decreased arginase activity in the proximal and distal vagina. Interestingly, treatment with testosterone increased arginase activity in the proximal vagina, but not in the distal vagina. These observations suggest that, in vaginal tissue, arginase activity is modulated by androgens in a tissue- and region-specific manner [27]. Δ^5 A-diol produced effects similar to those obtained with 5α-DHT. These observations indicate that Δ^5 A-diol acted specifically as an androgen in the rabbit vagina.

Androgens modulate female genital sexual arousal responses

Peripheral sexual arousal consists of increased pelvic blood flow that causes genital vasocongestion and production of plasma transudate and mucin in the vagina. These vascular events are also accompanied by increased sensitivity of afferent nerves and changes in the tissue properties of the vaginal canal and the clitoris, among other genital tissues. Thus, genital sexual responses are in part regulated by the vascular smooth muscle in the vaginal blood vessels of the submucosa and the clitoral corpora cavernosa, as well as the tone of the non-vascular smooth muscle within the vaginal muscularis. Androgens may play an important role in vaginal hemodynamics and in maintaining the vaginal sexual arousal response. Vaginal blood flow in response to pelvic nerve stimulation in the animal model was significantly diminished in ovariectomized rats infused with vehicle, while testosterone treatment of ovariectomized animals enhanced vaginal blood flow responses (Figures 10.4 and 10.5) [29]. Given the histological and biochemical data presented in the previous sections, it is likely that neuronal structure/function, smooth muscle health, and expression of NOS and arginase provide a mechanistic basis for the effects of androgens on vaginal blood flow.

In studies with ovariectomized animals, atrophy and inflammatory changes were noted in the vagina. DHEA, an androgen precursor, stimulated vaginal epithelium growth with mucous cells typical of an androgenic effect, produced epithelium mucification, and increased muscularis thickness [30,31]. Interestingly, treatment of ovariectomized animals with Δ^5 A-diol increased total NOS activity in proximal and distal vagina, similar to

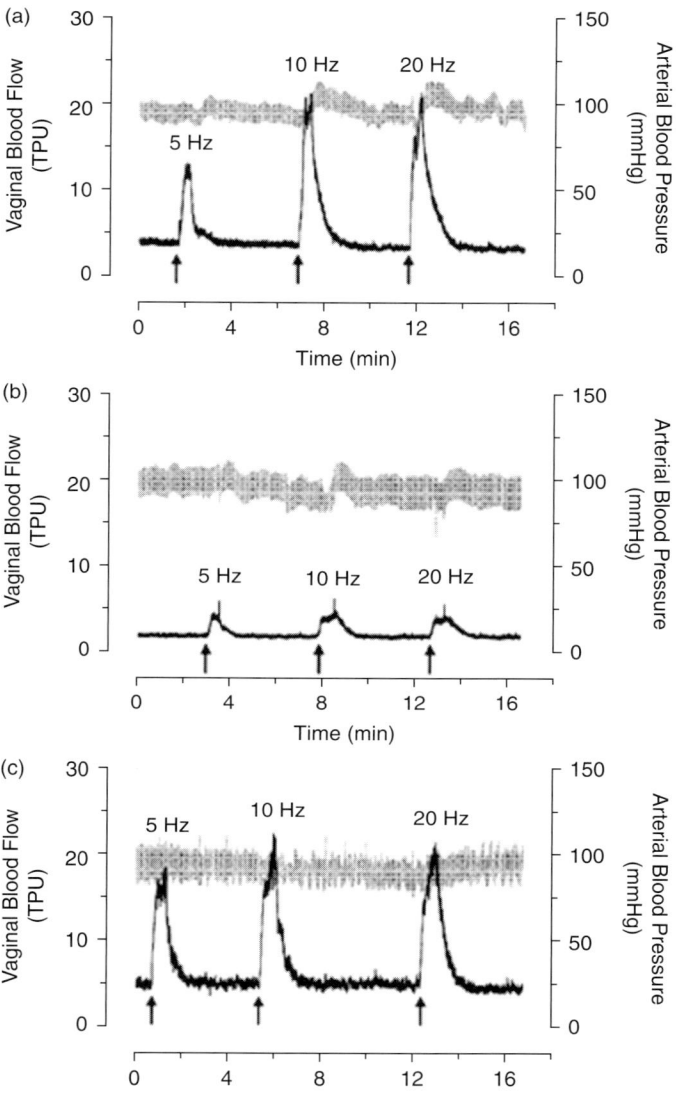

Figure 10.4 Vaginal blood flow recordings. After 2 weeks of vehicle or testosterone infusion, vaginal blood flow in response to pelvic nerve stimulation at varying frequencies was assessed by laser Doppler flowmetry (black). Systemic blood pressure was recorded simultaneously (gray). Shown are representative vaginal blood flow recordings from intact control (a) and ovariectomized rats infused with vehicle (b) or 55 µg/day of testosterone (c). Data derived from such recordings were then subjected to statistical analyses, as shown in Figure 10.5. TPU, tissue perfusion units. (With permission from reference 29.)

that obtained with 5α-DHT [27]. These observations suggest that Δ⁵A-diol may be a unique hormone with specific activity in the vagina [32–34] or simply conversion into 5α-DHT, and modulates NOS activity in the vagina via an androgenic mechanism.

It has been suggested that androgens are important in vaginal mucification [34,35]. While moisture production for lubrication is thought to be estrogen dependent by increasing the moisture content in the vagina, androgens are thought to be important for the production of mucin. Lubrication encompasses secretion of

sialo-glycoproteins, which provide the lubricant substance, and is modulated by androgens. Vaginal mucin production, as assessed by tissue sialic acid content, has been reported to be androgen dependent. Testosterone treatment of ovariectomized rabbits restored sialic acid to that of the vehicle-treated group (Traish et al., unpublished data). These data are consistent with the findings of Kennedy and Armstrong, who have shown that androgens increase vaginal mucification in the animal model [34,35]. In a separate study, treatment of ovariectomized rats with topical DHEA resulted in complete

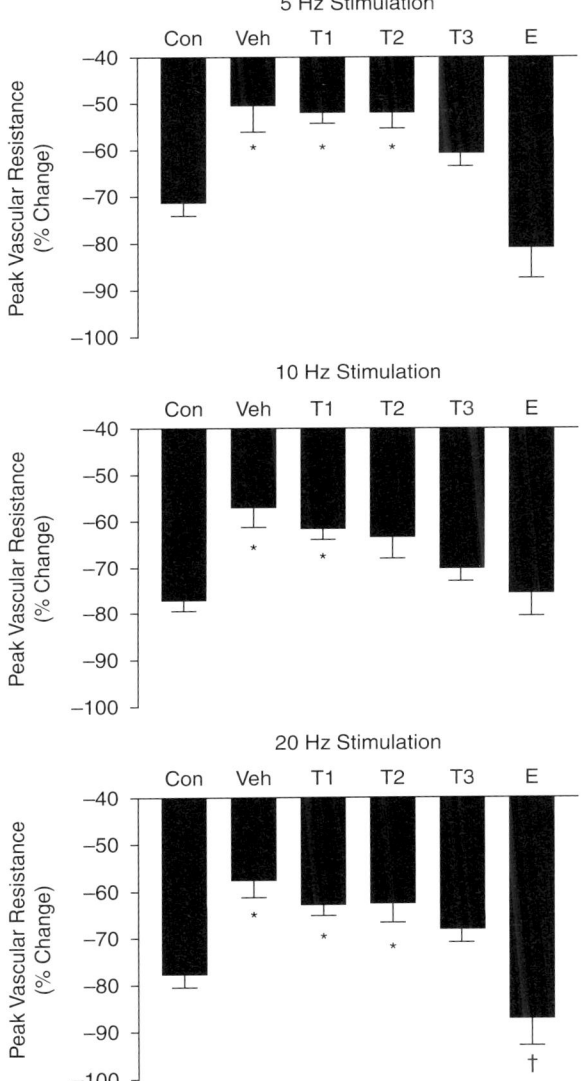

Figure 10.5 Effect of testosterone (T) on vaginal vascular resistance. Instantaneous vascular resistance values were calculated from vaginal blood flow and systemic blood pressure data (see Figure 10.4). The percent increase in mean peak vascular resistance over baseline vascular resistance was determined for each animal. For comparison, data from ovariectomized rats infused with a physiological dose of estradiol (E; 15 μg/day, resulting in a plasma concentration of 31 ± pg/mL) have been included. These data were derived from a previous study regarding the effects of estradiol on vaginal blood flow. Values are the mean ± SEM for each treatment group (see Figure 10.1). For each frequency of pelvic nerve stimulation, comparisons between groups were analyzed by one-way analysis of variance (ANOVA) and Tukey's post hoc test (*P <0.05 vs. control group; †P <0.05 vs. group T3). (With permission from reference 29.)

reversal of vaginal atrophy and stimulated proliferation and mucification of the vaginal epithelium [36]. Furthermore, DHEA treatment resulted in vaginal epithelial mucification [31,32].

Testosterone has also been shown to regulate vaginal muscularis function. Ovariectomy reduces vaginal muscularis relaxation to electric field stimulation and to exogenous vasointestinal polypeptide (VIP) in organ bath studies [37]. Estrogen treatment of animals failed to improve the relaxation response to either electrical stimulation or VIP. In contrast, testosterone treatment enhanced relaxation to electrical stimulation and restored VIP-induced relaxation, suggesting that androgens may modulate neurotransmitter function. Further, 5α-DHT, Δ5A-diol, and DHEA had similar effects to that of testosterone, indicating that several androgen hormones are equally effective in facilitating neurogenic or VIP-induced relaxation. These observations may be explained by increased neurogenic input or upregulation of neurotransmitter receptors mediating relaxation. Although the main neurotransmitter mediating relaxation of the rabbit vaginal muscularis is yet to be identified [37], similar mechanisms have been demonstrated in other tissues with VIP. Taken together, these data suggest that androgens facilitate vaginal smooth muscle relaxation, while estrogens attenuate this response [37]. In addition, tissue from testosterone-infused ovariectomized animals developed significantly greater contractile force to norepinephrine, suggesting that testosterone may be an important regulator of vaginal non-vascular smooth muscle contractility. It must be noted that these studies were carried out with supraphysiological levels of hormones, so this may be a pharmacologic rather than physiologic effect. Yet, Labrie et al. [38,39] recently provided clinical data that intravaginal DHEA improved vaginal sexual arousal and had beneficial effects on orgasm in postmenopausal women.

Clinical implications of androgen deficiency or excess in female genital physiology

Androgen excess

Congenital adrenal hyperplasia (CAH)

Conditions of CAH are generally diagnosed by neonatologists and pediatricians and are not pertinent, with limited exceptions, to the practice of the adult office gynecologist. Therefore, the treatment of these conditions will not be considered in this short treatise.

Polycystic ovary syndrome (PCOS)

PCOS is a pertinent exception to the group of disorders that are congenital in origin, because gynecologists will see these conditions quite regularly in the office. Approximately 10% of women are thought to exhibit androgen excess in adolescence and/or in adulthood. The excess androgens have numerous effects on the female genital tract. The ovaries are large and thickened and may present with pelvic discomfort. Women will present with hirsutism and acne, but also with menstrual irregularity and infertility. Of serious concern is the effect of prolonged amenorrhea on endometrial hyperplasia and the risk of endometrial cancer [40].

Although there is no direct effect on the female genital tract, practicing gynecologists should be aware that PCOS is linked to increased cardiovascular (CV) risk with considerable disagreements among studies as to the actual increase in CV events. Treatment of women with PCOS and infertility still may involve the use of clomiphene stimulation, wedge resection of the ovaries by a variety of techniques, and more sophisticated gonadotropin stimulation of the ovaries. Younger women present to gynecologists for hirsutism, acne, seborrhea, and often with concurrent menstrual disorders. Often, estrogen-progestin therapies are used to lower androgen levels by directly suppressing the mid-cycle surge of androgens, along with raising sex hormone-binding globulin (SHBG) levels to decrease free testosterone [41,42].

Androgen deficiency

One of the main problems with diagnosing androgen deficiency is that women produce about one-tenth of the amount of testosterone produced by men. Although free testosterone assays are the most accurate (measured in picogram amounts), they have been developed for men and the accuracy of the results is questioned in women. Despite this, it was thought that the current assays were adequate for the diagnosis of androgen deficiency. One approach is the measurement of DHEA-S, a main precursor to testosterone, and the main precursor in postmenopausal women. DHEA is the most abundant steroid produced by men and women, and the measurement is more accurate, as it is reported in microgram quantities. Measuring DHEA-S along with free testosterone lessens the potential decreased accuracy of the free testosterone assay. Davison et al. [43] investigated the levels of androgens in 1400 women from ages 18 to 75 years, with self-reported normal sexual function, and presented data listed by decade, along with a differentiation of women with normal and surgical menopause. These data (Table 10.1) provide the most comprehensive baseline level of androgens in women to date [44].

Premenopausal women

There are numerous possible causes for androgen deficiency in younger women [44]. These include surgical menopause, radiation and/or chemotherapy, hyperprolactinemia, destruction of the hypothalamic/pituitary centers from tumors or Sheehan's postpartum necrosis, functional hypothalamic amenorrhea, GnRH agonists or antagonists for treatment of endometriosis, AR blockers used in PCOS, corticoid treatments for asthma, lupus, colitis, etc., and use of oral contraceptives, which can raise SHBG. Other drugs may also raise SHBG, such as anticonvulsants. As SHBG increases, free testosterone decreases.

Oral contraceptives

Androgen deficiency in younger premenopausal women most often seen in the gynecologist's office is associated with the use of oral contraceptives, and most frequently is associated with sexual symptoms of decreased desire, but there may also be decreased arousal/orgasm, and occasionally genital pain. It is interesting that, according to good epidemiological data, the presenting sexual symptom in women of all ages is decreased sexual desire. It has been shown that women on oral contraceptives, and who complain of decreased libido, have lower androgen levels than similar women not on oral contraceptives [reviewed in 44]. These drugs prevent ovulation, but also decrease the mid-cycle surge of androgens seen in normal menstrual cycles. They also increase SHBG, which lowers free testosterone, even that which comes from adrenal production of androgens. The obvious option is to have the woman stop the oral contraceptive and monitor the androgen levels. However, this may not be acceptable to women for a variety of reasons: fear of pregnancy, return of menstrual irregularity and pain, or even suppression of endometriosis. In these cases, androgen replacement should be considered after a careful discussion with the patient, especially indicating that this

Table 10.1 Normal ranges of androgens for women aged 18–75 years, and who went through natural or surgical menopause

	18–24 years	25–34 years	35–44 years	45–54 years	55–64 years	65–75 years
Total T (nmol/L)						
Mean	1.58	1.11	0.92	0.81	0.66 (0.38 SM)	0.71 (0.39 SM)
10th percentile	0.86	0.58	0.50	0.40	0.20	0.30
90th percentile	2.47	1.70	1.40	1.30	1.25	1.10
cFree T (pmol/L)						
Mean	23.61	17.25	13.67	11.62	10.81 (5.54 SM)	9.76 (6.06 SM)
10th percentile	12.91	8.17	5.80	5.25	3.69	3.43
90th percentile	38.64	31.70	23.52	21.28	20.88	17.26
DHEA-S (μmol/L)						
Mean	7.49	4.72	4.31	3.42	2.36 (1.89 SM)	1.76 (1.13 SM)
10th percentile	4.03	2.20	1.86	1.30	0.70	0.50
90th percentile	10.78	7.90	7.31	6.20	5.30	3.10
A-dione (nmol/L)						
Mean	8.46	6.44	5.15	4.17	3.14 (2.15 SM)	3.07 (2.93 SM)
10th percentile	4.86	3.00	2.64	2.00	1.30	1.30
90th percentile	13.72	9.46	8.48	6.29	5.10	5.40

cFree T, calculated free T; SM, surgical menopause; A-dione, androstenedione;
To convert nmol/L to ng/dL, or pmol/L to pg/dL, divide by 0.0347.
To convert μmol/L to μg/dL, divide by 0.027.
Adapted from reference 43.

is an off-label treatment. Androgen treatment is not sanctioned in the United States, although it is in the European Union.

Idiopathic premenopausal androgen deficiency

Some women have lower androgen and precursor levels than one would expect from their age and menopausal status. Sexual symptoms are the usual presenting complaint, especially decreased libido without any history of recent oral contraceptive or other medication usage. Davis et al. [45] studied 261 premenopausal women, aged 35–46 years, who presented with decreased sexual satisfaction and free testosterone levels ≤1 pg/mL, a level that is considered low for younger women. These women were healthy, but being on oral contraceptives was not an exclusion; <30% of the women on treatment were on these drugs, and no good reason was given why the others had such a low level of free testosterone. The androgen-treated groups, as well as the placebo group, increased the number of satisfactory sexual experiences, consistent with other studies. The lack of difference in satisfactory sexual experiences probably reflects the vague nature of this endpoint. In this study, the group of women with the higher dose

of transdermal testosterone treatment did show a significant difference over placebo. Although the data are scanty, a practicing gynecologist will undoubtedly see women complaining of sexual dysfunction who have hormonal levels lower than expected for age or menopausal status.

Postmenopausal women

The menopausal period is hallmarked by decreased androgen and estrogen production. Androgen replacement therapy for women has never been approved in the United States. Previously, it was thought that the body needed a certain amount of estrogen in order for testosterone to work. This is why the prior studies on the testosterone patch in surgically menopausal women included the concomitant use of estrogen. This has not been found to be true, especially for women with sexual dysfunction. Davis et al. [45] replaced testosterone in postmenopausal women without estrogen therapy. The results showed a significant increase in satisfying sexual experiences in the women who received testosterone replacement in the normal range. This is important for women who do not wish to take estrogen for whatever reason, commonly seen when women have a positive family history of breast cancer. Also, it is standard practice to withhold estrogen replacement therapy in women who have been treated for breast cancer, especially with estrogen receptor-positive tumors. These women may be treated for atrophic vaginitis symptoms with intravaginal estradiol. Gynecologists have used intravaginal estrogen and testosterone for decades for relief of menopausal vulvovaginal atrophy symptoms. An encouraging study by Raghunandan et al. [46] found that local intravaginal therapy, with or without testosterone, was highly effective in relieving symptoms of urogenital atrophy. Standard oral testosterone is not given because it is aromatized to estradiol, but methyltestosterone may be considered as it is not aromatized to estrogen. It should be stressed that methyltestosterone would have to be made in a compounding pharmacy, and women need to be told that this is an off-label therapy.

Androgens in postmenopausal women

The postmenopausal ovary ceases making any significant amount of estrogen early in the menopausal transition, but continues to make androgens for a significant amount of time, up to approximately 8 years, because of the high level of circulating pituitary hormones – luteinizing hormone and follicle-stimulating hormone. Eventually, many women will become androgen deficient and this affects libido, lubrication, and arousal. In a clinical study with 272 naturally menopausal women, in which some were on standard menopausal HRT and others were on androgen replacement therapy, irrespective of the type of HRT, both groups showed significant increase over placebo for sexual desire, arousal, and satisfying sexual experiences. This further supports the view that testosterone does not require estrogen to have effectiveness concerning vulvovaginal and systemic sexual symptoms [reviewed in 45,48].

DHEA in postmenopausal women

Lee et al. [47] have shown that DHEA can potentiate relaxation of rabbit clitoral smooth muscle, thereby facilitating blood flow and arousal. Labrie et al. [38,39] have created an intravaginal DHEA for use in postmenopausal women. The levels of DHEA, and its metabolites, have been shown to remain in the normal range. This treatment has been shown to help vaginal atrophy and, in turn, counteract decreased libido, as well as decreased arousal and lubrication [38,39]. Genital

sensation is also helped with this therapy, perhaps because DHEA therapy has been shown to increase intravaginal nerve fiber density [20]. This is consistent with the theory that sex steroid deficiency in men and women may be related to nerve degeneration with loss of sensory input.

Safety of androgen therapy in women

The safety and efficacy of testosterone therapy in women has been evaluated by an expert panel, looking at testosterone replacement with and without estrogens [45,48]. It was concluded that physicians could be reassured regarding CV, breast, and endometrial outcomes, as long as the levels of hormones examined are in the normal physiological range. Data from several clinical trials demonstrated that administration of testosterone patch for up to 4 years in 1094 surgically menopausal women with hypoactive sexual desire disorder did not produce any significant adverse events. The main side effects were increased hair growth and acne [49]. Further, no clinically significant changes were noted in serum chemistry, hematology, lipid profile, carbohydrate metabolism, renal and liver function, or coagulation parameters through 4 years of therapy [49]. There were no reports of adverse CV, breast, and endometrial outcomes. Interim data from a long-term phase III safety trial of a testosterone gel demonstrate a continued low rate of CV events and breast cancer in postmenopausal women at increased CV risk.

In studies with supraphysiological doses of testosterone in female-to-male transsexuals, there were no serious adverse events reported, suggesting that testosterone at doses producing physiological levels in women with sexual dysfunction is expected to produce limited and minimal adverse effects. Another issue has arisen about the use of testosterone replacement in postmenopausal women, especially but not exclusively, in women who use androgen replacement in conjunction with standard HRT. No data are available to support the contention that testosterone increases the risk of breast cancer [reviewed in 50], and a search in the literature did not provide any evidence of such risk, especially if the data in the various studies are adjusted for estrogen levels [50]. In fact, testosterone may decrease the risk of breast cancer by inhibiting cell growth induced by estrogens [reviewed in 50]. Testosterone and DHT inhibited proliferation and increased apoptosis in breast cell cultures, and opposed estradiol-stimulated breast cell proliferation

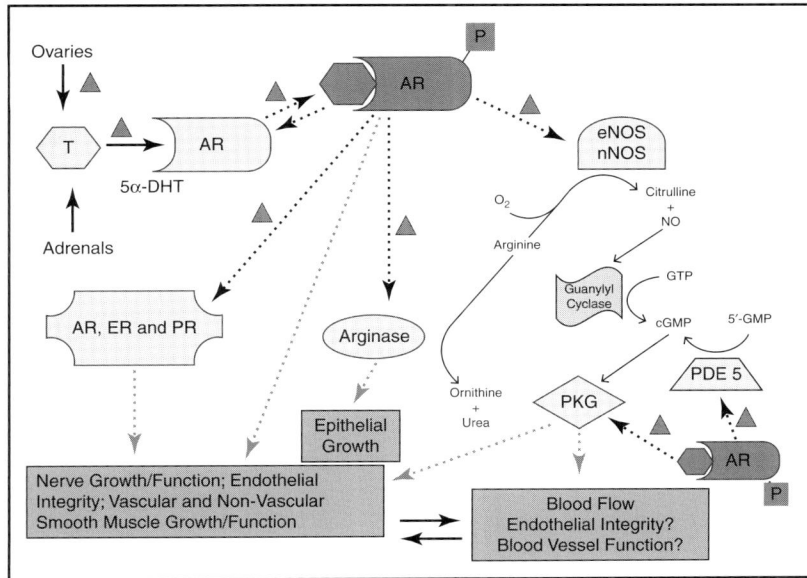

Figure 10.6 Role of androgens in modulating female genital physiology. Androgens produced by the ovaries and the adrenals interact with AR in various genital tissues (vagina, clitoris, labia, etc.). Androgens regulate vascular and non-vascular smooth muscle growth and function, endothelial and nerve function as well as blood vessels, either directly or via regulating estrogen and progesterone receptor function. Androgens regulate nitric oxide synthase, arginase, and phosphodiesterase type 5 enzyme pathways, and are implicated in regulation of genital hemodynamics.

and cell survival in an androgen receptor-dependent manner. Data published by Davis et al. [45,48] support this contention.

Conclusion

Androgens play an important role in sexual differentiation and in genital physiology. There is reasonable pre-clinical and clinical evidence to suggest that androgens are important in the function of various domains of female sexual function including, desire, arousal, lubrication, orgasm, and overall sexual satisfaction [45,48]. As depicted in Figure 10.6, we propose that testosterone secreted from the ovaries and the adrenals is converted, in part, to 5α-DHT via 5α-reductases. Testosterone and/or 5α-DHT bind to the AR in the various genital organs (vagina, clitoris, labia majora, labia minora, etc.) with high affinity. This binding activates a host of biochemical signaling pathways involved in vascular and non-vascular smooth muscle, epithelial, and endothelial cellular function. Furthermore, androgens regulate the activity of a host of enzymes, such as NOS, guanylyl cyclase, arginase, and protein kinase G, which are involved in facilitating genital hemodynamics in these organs. In addition, AR may regulate expression and activity of estrogen and progesterone receptor function in these tissues, thus regulating the synthesis and function of nerve and vascular endothelial growth factors and maintaining nerve

network and endothelia cellular integrity, vascular and non-vascular smooth muscle function, and blood vessel growth and function [26]. The pre-clinical and clinical data available to date suggest that androgens play an important role in maintaining female genital tissue physiology and genital sexual function. Androgen deficiency has been linked to female sexual dysfunction and several clinical trials have provided reasonable evidence that androgen treatments are important in women's sexual health [45,48]. These studies have also provided reasonable evidence relating to the safety of androgens in the treatment of female sexual dysfunction [45,48,49].

The concept of androgen deficiency in women remains controversial, and inadequately studied. There is not even an International Classification of Diseases (ICD-9) diagnostic code for female androgen deficiency. There is an urgent need to develop more sensitive assays for total and free testosterone in women, which would provide clinical evidence to validate this diagnosis. More research is needed to clarify the physiological role of adrenal androgens in women, which account for half of all circulating androgens in women. Better physiological clinical endpoints in androgen treatment are needed in clinical studies. Circulating androgen levels reach a critical low point in postmenopausal women and we foresee the day when standard hormone replacement in menopause will include androgens as well as estrogens and progestins.

References

1. Dorfman RI, Shipley RA. *Androgens: Biochemistry, Physiology and Clinical Significance.* New York: Wiley; 1956.

2. Strauss JF, Barbieri RL. *Yen & Jaffe's Reproductive Endocrinology: Physiology, Pathophysiology, and Clinical Management*, Sixth Edition. Philadelphia, PA: Saunders; 2009.

3. MacLaughlin DT, Donahoe PK. Sex determination and sexual determination. *N Engl J Med* 2004; 350: 367–378.

4. Johnson MH, Everett BJ. *Essential Reproduction*, Sixth Edition. Malden, MA: Blackwell; 2007.

5. Bhandari RK, Haque MM, Skinner MK. Global genome analysis of the downstream binding targets of testis determining factor SRY and SOX9. *PLoS ONE* 2012; 7: 1–17.

6. Cederroth CR, Pitetti J, Papaioannou MD, Nef S. Genetic programs that regulate testicular and ovarian development. *Mol Cell Endocrinol* 2007; 265–266: 3–9.

7. Rey R, Picard J. Embryology & endocrinology of genital development. *Baillières Clin Endocrinol Metab* 1998; 12: 17–33.

8. Healey A. Embryology of the female reproductive tract. In: Mann GS, Blair JC, Garden AS, editors. *Imaging of Gynecological Disorders in Infants and Children*. Berlin Heidelberg: Springer-Verlag; 2012.

9. Diamond DA, Yu RN. Sexual differentiation: normal and abnormal. In: Kavoussi LR, Novick AC, Partin AW, Peters CA, editors. *Pediatric Urology*. Philadelphia, PA: Saunders; 2012.

10. Cai Y. Revisiting old vaginal topics: conversion of the Müllerian vagina and origin of the "sinus" vagina. *Int J Dev Biol* 2009; 53: 925–934.

11. Drews U, Sulak O, Schenck PA. Androgens and the development of the vagina. *Biol Reprod* 2002; 67: 1353–1359.

12. Pajkrt E, Chitty LS. Prenatal gender determination and the diagnosis of genital anomalies. *BJU International* 2004; 93: 12–19.

13. Krone N, Hanley NA, Arlt W. Age-specific changes in sex-steroid biosynthesis & sex development. *Best Pract Res Clin Endocrinol Metab* 2007; 21: 393–401.

14. Welsh M, MacLeod DJ, Walker M, et al. Critical androgen-sensitive periods of rat penis & clitoris development. *Int J Androl* 2010; 33: e144–152.

15. Warne GL, Kanumakala S. Molecular endocrinology of sex differentiation. *Semin Reprod Med* 2002; 20: 169–180.

16. Wolf CJ, Hotchkiss A, Ostby JS, et al. Effects of prenatal testosterone propionate on the sexual development of male and female rats: a dose-response study. *Toxicol Sci* 2002; 65: 71–86.

17. Hodgins MB, Spike RC, Mackie RM, MacLean AB. An immunohistochemical study of androgen, oestrogen and progesterone receptors in the vulva and vagina. *Br J Obstet Gynaecol* 1998; 107: 216–222.

18. Hutchinson KA. Androgens & sexuality. *Am J Med* 1995; 98: 111S–115S.

19. Pessina MA, Hoyt RF Jr, Goldstein I, Traish AM. Differential effects of estradiol, progesterone, and testosterone on vaginal structural integrity. *Endocrinology* 2006; 147: 61–69.

20. Pelletier G, Ouellet J, Martel C, Labrie F. Effects of ovariectomy and dehydroepiandrosterone (DHEA) on vaginal wall thickness and innervation. *J Sex Med* 2012; 9: 2525–2533.

21. Tantawy S, Lin L, Akkurt I, et al. Testosterone production during puberty in two 46,XY patients with disorders of sex development and novel NR5A1 (SF-1) mutations. *Eur J Endocrinol* 2012; 167: 125–130.

22. Mueller A, Kiesewetter F, Binder H, et al. Long-term administration of testosterone undecanoate every 3 months for testosterone supplementation in female-to-male transsexuals. *J Clin Endocrinol Metab* 2007; 92: 3470–3475.

23. Mueller A, Haeberle L, Zollver H, et al. Effects of intramuscular testosterone undecanoate on body composition and bone mineral density in female-to-male transsexuals. *J Sex Med* 2010; 7: 3190–3198.

24. Kim SW, Jeong SJ, Munarriz R, et al. Role of the nitric oxide-cyclic GMP pathway in regulation of vaginal blood flow. *Int J Impot Res* 2003; 15: 355–361.

25. Kim SW, Jeong SJ, Munarriz R, et al. An in vivo rat model to investigate female vaginal arousal response. *J Urol* 2004; 171: 1357–1361.

26. Traish AM, Botchevar E, Kim NN. Biochemical factors modulating female genital sexual arousal physiology. *J Sex Med*. 2010; 7: 2925–2946.

27. Traish AM, Kim NN, Huang YH, et al. Sex steroid hormones differentially regulate nitric oxide synthase and arginase activities in the proximal and distal rabbit vagina. *Int J Impot Res* 2003; 15: 397–404.

28. Cama E, Colleluori DM, Emig FA, et al. Human arginase II: crystal structure and physiological role in male and female sexual arousal. *Biochemistry* 2003; 42: 8445–8451.

29. Traish AM, Kim SW, Stankovic M, et al. Testosterone increases blood flow and expression of androgen and estrogen receptors in the rat vagina. *J Sex Med* 2007; 4: 609–619.

30. Berger L, El-Alfy M, Martel C, Labrie F. Effects of dehydroepiandrosterone, Premarin and Acolbifene on histomorphology and sex steroid receptors in the rat vagina. *J Steroid Biochem Mol Biol* 2005; 96: 201–215.

31. Berger L, El-Alfy M, Labrie F. Effects of intravaginal dehydroepiandrosterone on vaginal histomorphology, sex steroid receptor expression and cell proliferation in the rat. *J Steroid Biochem Mol Biol* 2008; 109: 67–80.

32. Traish AM, Huang YH, Min K, et al. Binding characteristics of [3H]delta(5)-androstene-3b,17b-diol to a nuclear protein in the rabbit vagina. *Steroids* 2004; 69: 71–78.

33. Shao TC, Castaneda E, Rosenfield RL, Liao S. Selective retention and formation of a delta5-androstenediol-receptor complex in cell nuclei of the rat vagina. *J Biol Chem* 1975; 250: 3095–3100.

34. Kennedy TG. Vaginal mucification in the ovariectomized rat in response to 5alpha-pregnane-3,20-dione, testosterone and 5alpha-androstan-17beta-ol-3-one: test for progestogenic activity. *J Endocrinol* 1974; 61: 293–300.

35. Kennedy TG, Armstrong DT. Induction of vaginal mucification in rats with testosterone and 17beta-hydroxy-5alphaandrostan-3-one. *Steroids* 1976; 27: 423–430.

36. Sourla A, Flamand M, Belanger A, Labrie F. Effect of dehydroepiandrosterone on vaginal and uterine histomorphology in the rat. *J Steroid Biochem Mol Biol* 1998; 66: 137–49.

37. Kim NN, Min K, Pessina MA, et al. Effects of ovariectomy and steroid hormones on vaginal smooth muscle contractility. *Int J Impot Res* 2004; 16: 43–50.

38. Labrie F, Archer D, Bouchard C, et al. Effect of intravaginal dehydroepiandrosterone (Prasterone) on libido and sexual dysfunction in postmenopausal women. *Menopause* 2009; 16: 923–991.

39. Labrie F, Archer D, Bouchard C, et al. Intravaginal dehydroepiandrosterone (Prasterone), a physiological and highly efficient treatment of vaginal atrophy. *Menopause* 2009; 16: 907–922.

40. Coulam CB, Annegers JF, Kranz JS. Chronic anovulation syndrome and associated neoplasia. *Obstet Gynecol* 1983; 61: 403–407.

41. Ehrmann DA. Polycystic ovary syndrome. *N Engl J Med* 2005; 352: 1223–1236.

42. Norman RJ, Dewailly D, Legro RS, Hickey TE. Polycystic ovary syndrome. *Lancet* 2007; 370: 685–697.

43. Davison SL, Bell R, Donath S, et al. Androgen levels in adult females: changes with age, menopause, and oophorectomy. *J Clin Endocrinol Metab* 2005; 90: 3847–3853.

44. Guay A. Traish A. Testosterone therapy in women with androgen deficiency: its time has come. *Curr Opinion in Invest Drugs* 2010; 11: 1116–1126.

45. Davis SR, Moreau M, Kroll R, et al; the APHRODITE Study Team. Testosterone for low libido in postmenopausal women not taking estrogen. *N Engl J Med* 2008; 359: 2005–2017.

46. Raghunandan C, Agrawal S, Dubey P, et al. A comparative study of the effects of local estrogen with or without testosterone on vulvovaginal and sexual dysfunction in postmenopausal women. *J Sex Med* 2010; 7: 1284–1290.

47. Lee SY, Myung SC, Lee MY, et al. The effects of dehydroepiandrosterone (DHEA) / DHEA-Sulfate (DHEAS) on the contraction responses of the clitoral cavernous smooth muscle from female rabbits. *J Sex Med.* 2009; 6: 2653–2660.

48. Davis SR, Braunstein GD. Efficacy and safety of testosterone in the management of hypoactive sexual desire disorder in postmenopausal women. *J Sex Med.* 2012; 9: 1134–1148.

49. Nachtigall L, Casson P, Lucas J, et al. Safety and tolerability of testosterone patch therapy for up to 4 years in surgically menopausal women receiving oral or transdermal oestrogen. *Gynecol Endocrinol* 2011; 27: 39–48.

50. Traish AM, Fetten K, Miner M, et al. Testosterone and risk of breast cancer: appraisal of existing evidence. *Horm Mol Biol Clin Invest* 2010; 2: 177–190.

Congenital adrenal hyperplasia in females

Candice P. Holliday and Botros R. M. B. Rizk

Introduction

The first documented case of congenital adrenal hyperplasia (CAH) occurred in 1865 with Italian anatomist Luigi De Crecchio's description of a cadaver with male external genitalia, female internal organs, and very large adrenal glands [1]. The cadaver was Giuseppe Marzo, who was born in June 1820 and named Josephine [2]. At the age of 4, Josephine was declared male and then given the name "Joseph." By age 18, Joseph had developed a deep voice and beard, and began to have "adventures with women" [2]. At 25, Joseph fell in love and prepared to marry; but when he discovered his birth certificate said "female," he began to procrastinate. Unfortunately, he procrastinated so long that his fiancée found a new lover and Joseph then began to drown his sorrows in alcohol and tobacco. He had repeated episodes of vomiting and diarrhea, and died at 43 during such an episode [2,3]. Upon autopsy, it was discovered that Joseph had a 6-cm long penis with hypospadias, no testes, a normal vagina, a normal uterus, normal fallopian tubes, normal ovaries, and enlarged adrenal glands [2].

CAH is a common autosomal recessive disorder that results from an enzymatic defect that affects cortisol biosynthesis [1,4]. The resulting glucocorticoid deficiency can occur with concomitant excess or deficiency in mineralocorticoid and/or sex steroids. The type and severity of the enzyme defect is associated with a specific presentation of CAH. Severe enzyme defects tend to present in neonates and infants with symptoms of virilized and ambiguous genitalia, i.e., females have masculinized external genitalia. Milder enzyme defects tend to present in childhood, adolescence, or adulthood with symptoms of androgen excess.

Of the various types of enzyme defects (Figure 11.1), approximately 90% of CAH is due to a 21-hydroxylase (P450c21) deficiency (Figure 11.2)

from a CYP21A2 mutation on chromosome 6p21.3 [1,4,5]. Approximately 5% of CAH is due to a defect in 11β-hydroxylase [6]. The remaining 5% of defects are in 3β-hydroxysteroid dehydrogenase, 17α-hydroxylase /17,20-lyase, steroidogenic acute regulatory (StAR) protein, and P450 oxidoreductase [4]. There are two types of 21-hydroxylase deficiency presentations: classic 21-hydroxylase deficiency CAH and nonclassic 21-hydroxylase deficiency CAH.

The incidence of classic 21-hydroxylase deficiency CAH ranges from 1 in 5000 to 1 in 16 000 live births [1,7]. Approximately 1 in 60 people is a carrier for classic CAH [1]. Interestingly, CAH is more common in white and Hispanic people than in African Americans in the USA [1,7]. The incidence of nonclassic 21-hydroxylase deficiency CAH, however, is higher than that of classic 21-hydroxylase deficiency, and the incidence is even as high as 3.7% in Ashkenazi Jews and 1.9% in Hispanics [8].

Patients with CAH will need lifelong treatment and care. For newborns and infants, the treatment goals are to prevent an adrenal crisis, to assign a gender with surgery in the more severe cases of CAH, and to protect linear growth [5]. For children and adolescents, the treatment goals are to protect linear growth, to promote a normal pubertal development, and to maintain an appropriate body weight. Lastly, the treatment goals of adults include fertility, prevention of metabolic syndrome, and prevention of osteoporosis.

Pathophysiology

Although CAH is an autosomal recessive disorder, its pathophysiology is more complex than most. There are three main steroids produced by the adrenal: glucocorticoids, mineralocorticoids, and sex steroids. Production is controlled by adrenocorticotropic hormone

Androgens in Gynecological Practice, ed. Leo Plouffe and Botros Rizk. Published by Cambridge University Press. © Cambridge University Press 2015.

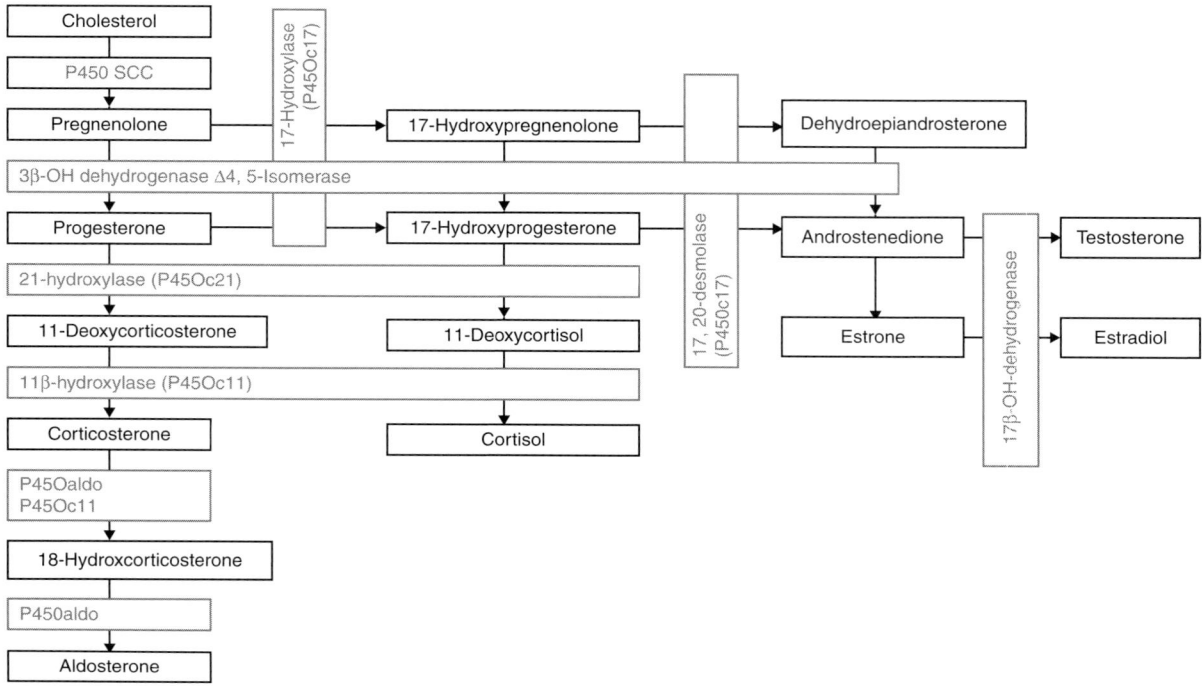

Figure 11.1 Types of enzyme blocks and their consequences in CAH. The inherited defect in steroid biosynthesis of cortisol decreases cortisol, which leads to an increase in ACTH. This can result in compensated cortisol block and precursors (androgens, corticoids).

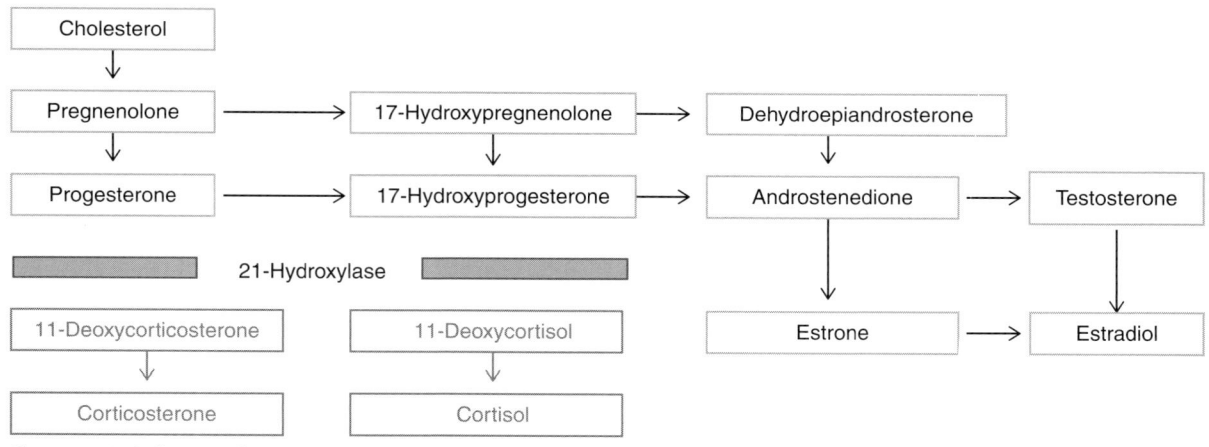

Figure 11.2 Deficiency of 21-hydroxylase enzyme defect. The potential spectrum of salt loss and shock, virilization, amenorrhea, and infertility.

(ACTH). ACTH is regulated by the central nervous system, and its production is increased during times of physiological stress with corticotropin-releasing factor (CRF) from the hypothalamus. ACTH and CRF are then controlled by negative feedback from circulating plasma levels of cortisol. ACTH augments StAR protein function, which moves free cholesterol to the inner mitochondrial membrane so that pregnenolone is produced and steroidogenesis can occur.

Thus, when cortisol is not present because of CAH, ACTH continues to be produced and steroidogenesis continues unchecked in the metabolic pathways not affected by CAH and with an accumulation of precursor molecules in the metabolic pathways affected by CAH. The precursor molecules (such as 17-hydroxyprogesterone, progesterone, and androstenedione) that have accumulated due to an enzyme deficiency from CAH are then used in the metabolic pathways not affected by CAH and more androgens are created. Additionally, the excess ACTH causes hypertrophy in the adrenal glands of the zona fasciculata and zona reticularis, and thus, the name CAH.

Clinical presentation

The internal genitalia of a woman with CAH remain normal because she does not have müllerian-inhibiting hormone from testicular Sertoli cells [4]. Her external genitalia, however, can appear normal at birth or ambiguous/virilized, depending on the type and severity of her CAH (Figure 11.3a–c) [9]. Additionally, excess ACTH can cause increased pigmentation of nipples and genitalia [5]. If the patient's CAH is not appropriately treated, she will continue to have excess exposure to androgens. This continued excess exposure can lead to progressing penile/clitoral enlargement, premature development of pubic and axillary hair, acne, and apocrine odor (Figure 11.3a–c) [1,4]. While an individual with CAH may appear tall as a child, she will often be short as an adult due to premature epiphyseal fusion from accelerated skeletal maturation [1].

If a CAH patient is successfully treated from birth, she will likely have a relatively normal pubertal age with normal appearance of secondary sex characteristics [4].

If she is not successfully treated, hyperandrogenism can result. Hyperandrogenism in adult females can appear as alopecia, acne, infertility, hirsutism (Figure 11.3a–c), and absent/irregular menses [1,4]. Infertility in these female patients can be due to various etiologies, including anovulation, irregular menses, abnormal introitus, and uncontrolled progesterone levels. There is also some research demonstrating an association of CAH with polycystic ovary syndrome (PCOS) [10].

21-Hydroxylase deficiency

Female patients with the classic form of 21-hydroxylase deficiency will suffer from the more severe hyperandrogenism symptoms in utero, virilization of external female genitalia, as a result of androgen exposure from about the sixth week of gestation (Figure 11.4) [1,4]. This virilization can result in genital ambiguity in females upon birth, ranging from clitoromegaly to a penile urethra. There are two subtypes of classic 21-hydroxylase deficiency CAH: a simple virilizing form and a salt-wasting form, the latter related to the adequacy of aldosterone production (Figure 11.2). Over 75% of classic 21-hydroxylase deficiency CAH presents with the salt-wasting subtype, which is the more dangerous as it subjects the newborn to risk of adrenal crises [4,7]. For patients with salt-wasting classic 21-hydroxylase deficiency, less than 1% of enzyme activity remains [5].

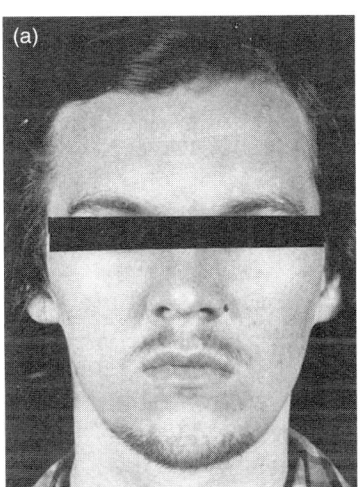
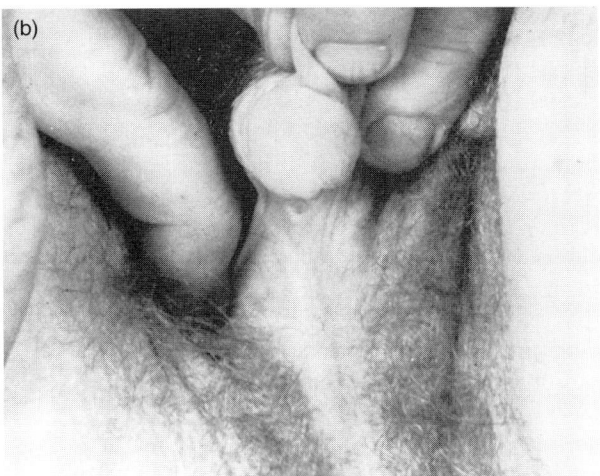
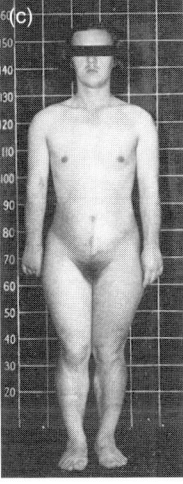

Figure 11.3 (a) Patient N.W. at age 16 after 3 days without shaving. (b) Patient N.W.'s external genitalia before surgery. (c) Patient N.W. after 8 months of cortisone therapy; notice the breast development. Reproduced with permission from Jones HW Jr., Scott WW. Female Intersexuality with Adrenal Hyperplasia. In: Jones HW Jr., Scott WW (Eds.) Hermaphroditism, Genital Anomalies and Related Endocrine Disorders. The Williams and Wilkins Company; 1971: Ch. 9.

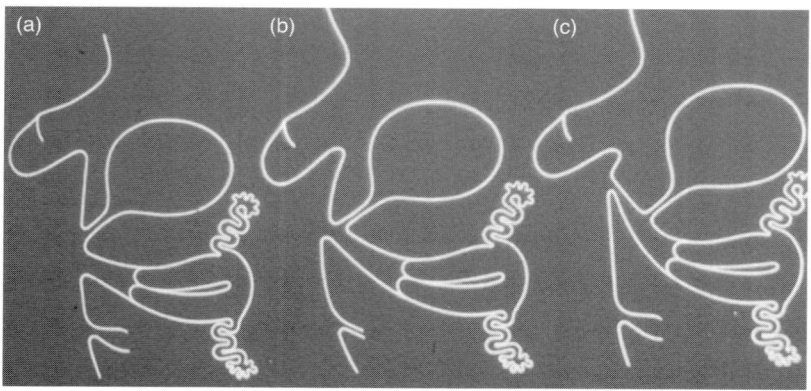

Figure 11.4 External genitalia as dependent on exposure to androgens. (a) Female external genitalia exposed to testosterone after 12th week of pregnancy, leading to clitoral hypertrophy. (b) Earlier exposure between 7 and 12 weeks, leading to progressive masculinization of urogenital sinus. (c) Exposure at 7 weeks leading to penile urethra.

Patients with the nonclassic form of 21-hydroxylase deficiency CAH are only mildly affected by their enzymatic deficiency and tend to present as adults with signs of hyperandrogenism, if they present with signs at all [4]. They can present with premature development of pubic hair, advanced bone age, severe cystic acne, infertility, alopecia, hirsutism, voice deepening, and male habitus. Their menarche tends to be normal or delayed, although secondary amenorrhea frequently occurs. Both classic and nonclassic 21-hydroxylase deficiency patients can exhibit poor growth if attempts at treating them with glucocorticoids exceed the appropriate physiological levels [4].

Diagnosis

Early diagnosis of CAH is important to improve morbidity and mortality. Female infants with classic 21-hydroxylase deficiency are often diagnosed by ambiguous external genitalia at birth and/or salt-wasting (i.e., failure to thrive, hypovolemia, hypotension, hyponatremia, and hyperkalemia) [5]. After identifying the ambiguous genitalia, pelvic ultrasound and genitograms are appropriate to evaluate the newborn [11]. Children with virilizing classic 21-hydroxylase can present with precocious adrenarche and accelerated growth. Women with nonclassic 21-hydroxylase deficiency tend to present the same as women with PCOS. For that reason, in patients who present with hyperandrogenism and are diagnosed with PCOS, it is necessary to consider CAH as well [4].

Hormonal diagnosis

CAH can be diagnosed by evaluating a female's hormone levels with radioimmunoassays, enzyme-linked immunosorbent assays, and time-resolved fluoroimmunoassays.

For classic 21-hydroxylase deficiency, a high concentration of 17-hydroxyprogesterone (17-OHP), which is the precursor of the enzyme that is defective, is diagnostic [1,4]. Blood should be drawn in the early morning for this assay [12].

For nonclassic 21-hydroxylase deficiency, a corticotropin stimulation test (CST), also known as the acute ACTH stimulation test, is diagnostic. For a CST, 250 μg of cosyntropin (Cortrosyn) is given intravenously, then 17-OHP and Δ^4-androstenedione are measured at baseline and then again at 30–60 minutes. Because it may be impractical to do a CST on all women with hyperandrogenism signs, it has been suggested that unstimulated 17-OHP levels (170–300 ng/dL), measured in the morning and in the follicular phase of the menstrual cycle, be used to screen women [1]. Additionally, plasma renin activity should be measured in order to determine whether the patient also has a mineralocorticoid deficiency.

Molecular genetics

Although genetic testing may be useful, it is not the gold standard for diagnosis of 21-hydroxylase deficiency and should only be used when 17-OHP and CST results are equivocal [12]. There are over 120 mutations on *CYP21A2* associated with CAH, which result in partial to total loss of enzyme activity [1]. The various forms of CAH tend to have specific genotypes; however, phenotypes do not always correlate with the genotype [1,4]. For CAH, a recessive disease, usually it is the least deficient mutation of the two alleles that governs. Thus, for classic 21-hydroxylase deficiency, usually two alleles with severe mutations occur together. For nonclassic 21-hydroxylase deficiency, either the patient has two mild mutations or one severe and one mild mutation on her alleles. The screening panels that are commercially available tend to only measure the 10–12 most common mutations [1]. Genetic analysis can be limited because of the complexity of CAH gene duplications, gene deletions, and gene rearrangements within chromosome 6p21.3 [12].

Newborn screening

All 50 states in the USA screen for CAH [12]. Screening for CAH involves measuring 17-OHP, and often detects newborns with classic 21-hydroxylase deficiency. Of note, female newborns with classic 21-hydroxylase deficiency will often have genital ambiguity due to virilization as another indicator of CAH. The Endocrine Society recommends that all newborns are screened for 21-hydroxylase deficiency with an initial immunoassay and then further evaluation of positive immunoassays with liquid chromatography/tandem mass spectrometry [12]. Nonclassic 21-hydroxylase deficiency is often not detected by newborn screening because their 17-OHP levels tend to be lower [1]. There has been some research advancing the use of ultrasonography to screen newborns for enlarged adrenal glands in order to achieve more rapid diagnostic results than with biochemical assays [13].

Treatment

Medical treatment

Medical treatment is focused on addressing the cortisol deficiency and the androgen overproduction that creates symptoms in CAH patients (Figure 11.3). Glucocorticoids will decrease the stimulation of androgen production and are the current gold standard of treatment. For classic 21-hydroxylase deficiency, about 10–15 mg/m^2 per day of hydrocortisone is given in 2–3 doses per day [4,14]. For adults with CAH, longer-acting glucocorticoids, alone or with hydrocortisone, may be used. In children that dose is usually 6–15 mg/m^2 per day and is given in 3 daily doses [1,14]. As usual, the goal is to use the lowest dose necessary for the patient's age, such as titrated doses yielding 17-OHP levels under 1000 mg/dL in children [4,14]. Patients with nonclassic 21-hydroxylase deficiency can sometimes be treated with lesser amounts, such as 0.25–0.5 mg of dexamethasone at night [4,14]. Additionally, therapy to replace mineralocorticoids usually involves 9α-fludrocortisone acetate to get the patient to a low normal plasma renin activity for their

age [1]. Antiandrogen therapy (flutamide, cyproterone acetate, or finasteride) can also be used in adult women who need further suppression of hyperandrogenism [4]. For CAH patients who also have PCOS, oral contraceptives can be used to help control symptoms [4].

During times of physiological stress or illness, the dosage of corticosteroid should be increased to 2–3 times the usual dose, or 45 mg/m^2 per day, although dexamethasone should not be used for stress doses due to its delayed onset of action [1,4,14]. For surgeries, the dose should be increased to 5–10 times the usual dose or 100 mg parenteral hydrocortisone (Solu-Cortef), IV or IM, can be used [1,4,14]. Along the same lines, patients should be issued emergency injection kits of hydrocortisone (25 mg for infants, 50 mg for children, and 100 mg for adults) and taught how to use them [4,14].

As patients with CAH age, the amount of glucocorticoids that they need to decrease their excess androgen production diminishes. Excess use of glucocorticoids to treat CAH can result in Cushing's syndrome or growth suppression in children [4]. Thus, monitoring of treatment is necessary, including every 3–4 months in children and less often as the children age [4]. These guidelines for treatment of CAH, however, are not easily met. In 2010, the United Kingdom Congenital Adrenal Hyperplasia Adult Study Executive (CaHASE) reviewed the management of 203 CAH patients [15]. After finding that only a low percentage of CAH patients were under an endocrinologist's care and had adequately controlled androgen levels, among other issues, CaHASE deemed the current health status of adult CAH patients in the United Kingdom to be poor. Similar concerns regarding the inadequacy of CAH patient management and treatment in Norway have been raised [16].

Surgical treatment

Previously, early corrective surgery on neonates with ambiguous genitalia was the norm. Recently, however, such routine surgery has come under fire because of a dearth of research on long-term functional outcomes of the surgery [4]. Thus, when a child is born with ambiguous genitalia, the parents should be fully informed of the controversy over early corrective surgery and advised of all available options for the child.

Surgical treatment for CAH is often considered when there is concern regarding lower tract and upper tract genitourinary malformations because of virilization, which could lead to frequent urinary tract infections [11]. The guidelines of the Endocrine Society currently provide that surgery is usually appropriate for severely virilized infant girls and when there is an experienced surgeon in a facility with similarly experienced pediatric endocrinologists, mental health professionals, etc. [12].

A multidisciplinary and individualized approach should be used when corrective surgery is contemplated. Typically, when corrective surgery is chosen, redundant erectile tissue is removed (while preserving the glans clitoris), fused labia are separated, and a normal vaginal orifice is created [4,17]. The goal of such surgery is to construct female-appearing external genitalia to aid in normal psychosexual development and to construct a functional vagina for menstruation and sexual activity [17]. Not only is the decision to use surgery controversial, but the best time to do vaginoplasty (early one-stage procedure vs. two-stage with delayed vaginoplasty until after puberty) is also a matter of much debate [17]. One study of women diagnosed with classic 21-hydroxylase deficiency found that the women who had undergone corrective surgery suffered impaired clitoral sensitivity and sexual functioning, including problems with penetration and anorgasmia [18].

Adrenalectomy should be considered only when medical therapy has failed [12]. The cautionary use of adrenalectomy is based on concerns of surgical risk and the loss of protective functions from the adrenal glands [19]. It is often in adult females who have salt-wasting classic 21-hydroxylase deficiency and infertility that adrenalectomy is considered due to failure of medical therapy [20]. Bilateral adrenalectomy for CAH, however, remains controversial.

Other considerations

All CAH patients (and all patients on corticosteroids) should wear medical bracelets that identify their appropriate treatment. Furthermore, adults with CAH should be monitored for bone mineral density, obesity, and cardiovascular issues [4]. Because bone mineral density is adversely affected by both androgen excess and glucocorticoid excess, women with 21-hydroxylase deficiency demonstrate decreased bone mineral density [21]. Although there is limited research supporting the proposition, appropriate levels of vitamin D should be maintained in CAH patients [1]. Obesity is associated with long-term glucocorticoid treatment, which is in addition to excess androgens. CAH patients' cardiovascular issues tend to be significant once the patient is over 30 years of age [3].

For CAH women who are sexually active, but not on a very effective contraceptive, or for women who plan to become pregnant, they should take hydrocortisone, prednisone, or prednisolone [1]. In women who want to become pregnant, appropriate suppression of progesterone has been shown to be essential for conception [22]. For women with CAH who are pregnant, many who have classic 21-hydroxylase deficiency will need or electively choose a cesarean section [23]. Moreover, CAH patients who are on glucocorticoids should be given a stress dose of medication for the physiological stresses of labor and delivery [1,4].

Prenatal treatment

As an initial matter, the Endocrine Society strongly recommends genetic counseling for parents of a CAH child and for CAH patients considering fertility [12]. Prenatal treatment of CAH is currently considered experimental by the Endocrine Society due to the lack of data and uncertainty over whether dexamethasone treatment is safe when started early during pregnancy [12,24]. When there is a significant risk of classic 21-hydroxylase deficiency and should prenatal treatment be deemed advisable, dexamethasone (0.02 mg/kg per day, based on patient's pre-pregnancy weight, divided into 2–3 doses) is appropriate [4]. Dexamethasone should be begun before 8–10 weeks of gestation and used until amniocentesis or chorionic villus sampling can be done to confirm the diagnosis [1,4]. Prenatal treatment with dexamethasone is experimental, however, and a risk exists of Cushingoid features and glucose intolerance in the mother [1]. When dexamethasone is used at or before 9 weeks of gestation, studies have shown a reduction in female virilization [25,26].

Conclusion

CAH is a common autosomal recessive disorder with a wide spectrum of effects. It results from inherited defects in enzymes for adrenal steroids. The most common type of CAH is 21-hydroxylase deficiency. The presentation of females with symptomatic CAH ranges from signs of mild hyperandrogenism in adulthood to ambiguous genitalia at birth. CAH can be treated with appropriate hormone replacement; however, patients need to be correctly dosed and monitored.

Acknowledgments

The authors would like to thank and acknowledge Dr. Theodore Baramki for his advice.

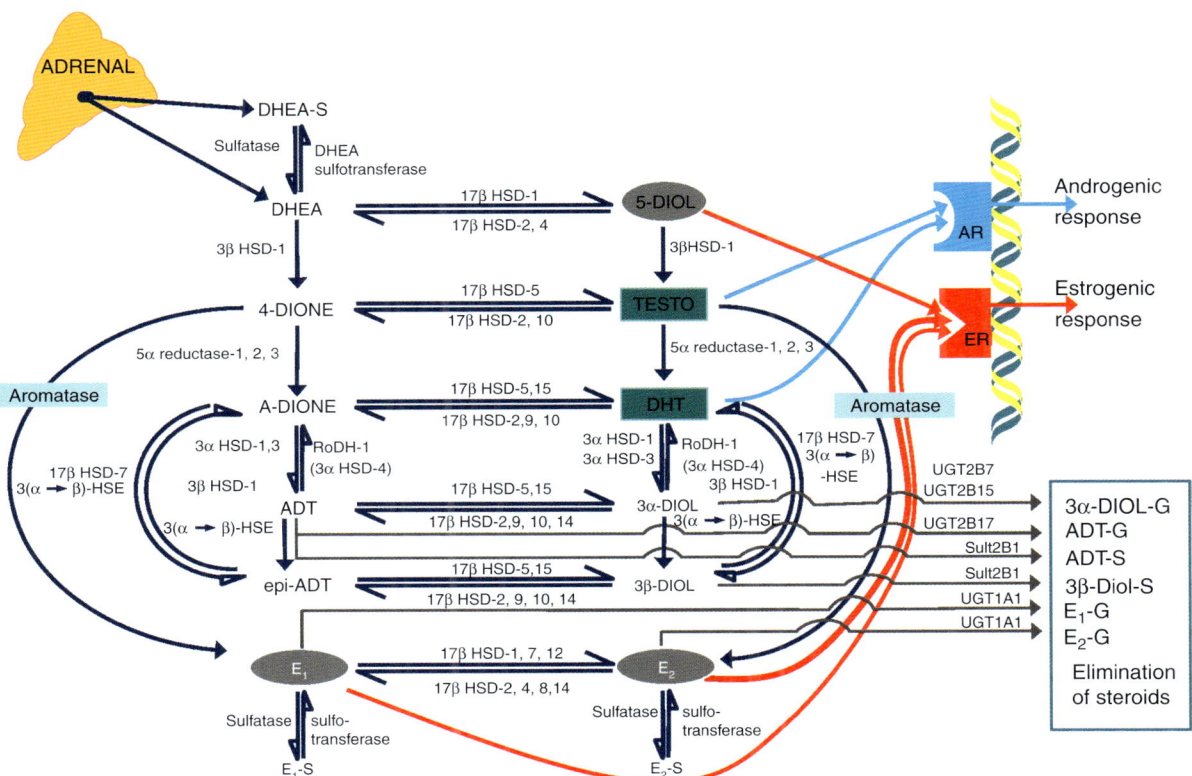

Figure 6.1 Human steroidogenic and steroid-inactivating enzymes in peripheral intracrine tissues. 4-dione, androstenedione; A-dione, 5α-androstane-3,17-dione; ADT, androsterone; epi-ADT, epiandrosterone; E_1, estrone; E_1-S, estrone sulfate; E_2, 17β-estradiol; E_2-S, estradiol sulfate; 5-diol, androst-5-ene-3α, 17β-diol; HSD, hydroxysteroid dehydrogenase; testo, testosterone; RoDH-1, Ro dehydrogenase 1; ER, estrogen receptor; AR, androgen receptor; UGT2B28 family (including UGT2B7, UGT2B15, and UGT2B17), uridine glucuronosyl transferase 2B28; Sult2B1, sulfotransferase 2B1; UGT1A1, uridine glucuronosyl transferase 1A1.

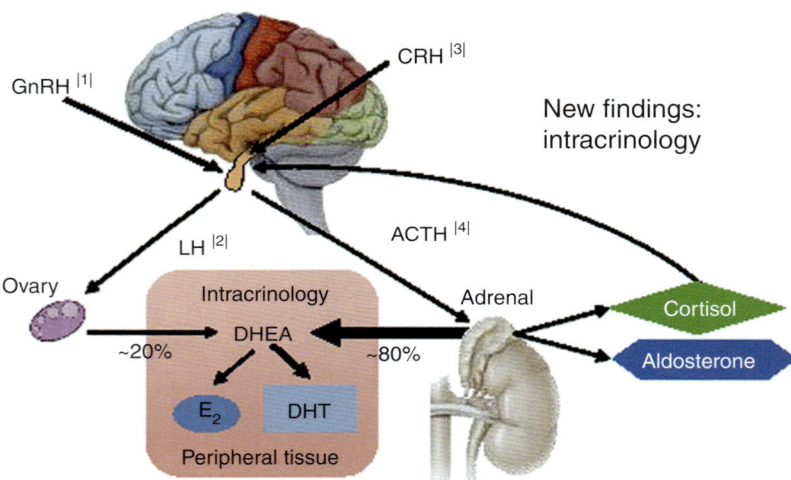

After menopause

GnRH [1]   CRH [3]

New findings:
intracrinology

LH [2]   ACTH [4]

Ovary   Adrenal   Cortisol

Intracrinology

DHEA

~20%   ~80%   Aldosterone

E₂   DHT

Peripheral tissue

(1) Gonadotropin-releasing hormone.
(2) Luteinizing hormone.
(3) Corticotropin-releasing hormone.
(4) Adrenocorticotropic hormone.

Figure 6.2 Schematic representation of dehydroepiandrosterone (DHEA) acting as the unique source of sex steroids in women after menopause. Approximately 80% of circulating DHEA is of adrenal origin while about 20% is released from the ovary [4]. Accordingly, after menopause, all androgens and all estrogens are made locally from DHEA in peripheral target tissues. The amount of sex steroids made depends upon the level of the steroid-forming enzymes specifically expressed in each cell in each tissue. GnRH, gonadotropin-releasing hormone; LH, luteinizing hormone; CRH, corticotropin releasing hormone; ACTH, adrenocorticotropic hormone; E₂, 17β-estradiol; DHT, dihydrotestosterone. From Labrie et al. [4].

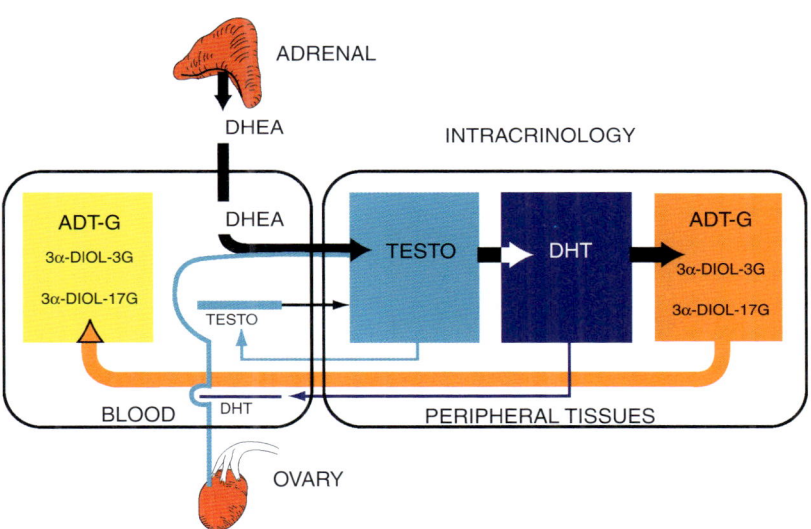

Figure 6.3 Schematic representation of the role of the precursor DHEA of both adrenal (~80%) and ovarian (~20%) origins in total androgenic activity in postmenopausal women. A very small proportion of the active androgens testosterone and DHT made intracellularly by the steroidogenic enzymes of the intracrine pathway diffuse into the circulation, thus avoiding systemic effect. The height of the bars is proportional to the concentration of each steroid. For androgens, the mechanisms of intracrinology [2] make it possible to keep intracellularly the formation and inactivation of androgens with only some leakage at very low levels in the circulation. In the circulation, the biologically significant and representative parameters are the precursor DHEA and the glucuronidated metabolites of androgens (ADT-G and 3α-diol-3G and 3α-diol-17G).

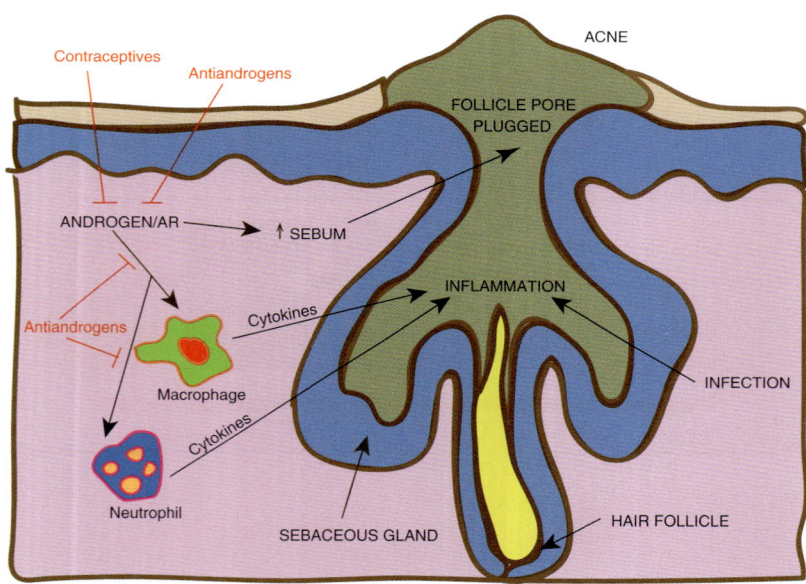

Figure 8.3 The role of androgens/AR in acne formation and progression. Acne formation results from excessive sebum formation accompanied by excessive inflammation and infections in the hair follicle. Macrophages and neutrophils are recruited to the inflamed follicles and secrete cytokines and other factors that promote inflammation and infection clearance. However, the inflammatory response also damages the normal tissues in the follicles. AR can promote the inflammatory response mediated by macrophages and neutrophils as well as directly promote sebum production that plugs the follicle pore. Current treatments for acne include antiandrogens and contraceptives, which reduce androgen levels, thereby reducing sebum production and suppressing the function of macrophages and neutrophils to dampen the inflammatory response. Adapted from "The role of androgen and androgen receptor in skin-related disorders," by J. J. Lai, P. Chang, K. P. Lai, L. Chen, and C. Chang, 2012, *Arch Dermatol Res, 304*, p. 504.

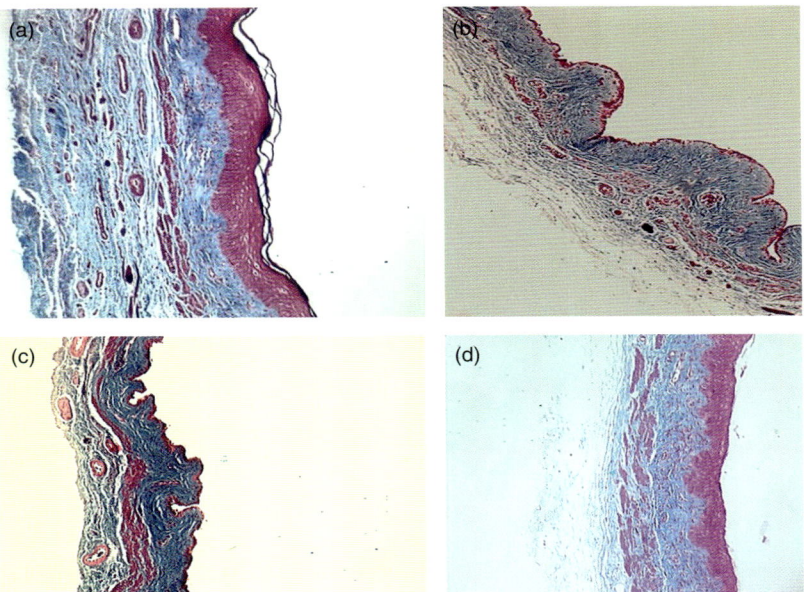

Figure 10.1 Effects of ovariectomy and estrogen or androgen treatment on vaginal tissue structure (a–d). Female rats were left intact (control; a) or ovariectomized and after 2 weeks were infused with vehicle (b) or testosterone (c) or testosterone and estradiol (d). Hormone treatment was maintained for 2 weeks. Vaginal tissue was fixed, sectioned, and subjected to Masson's trichrome staining as described in references 19 and 26.

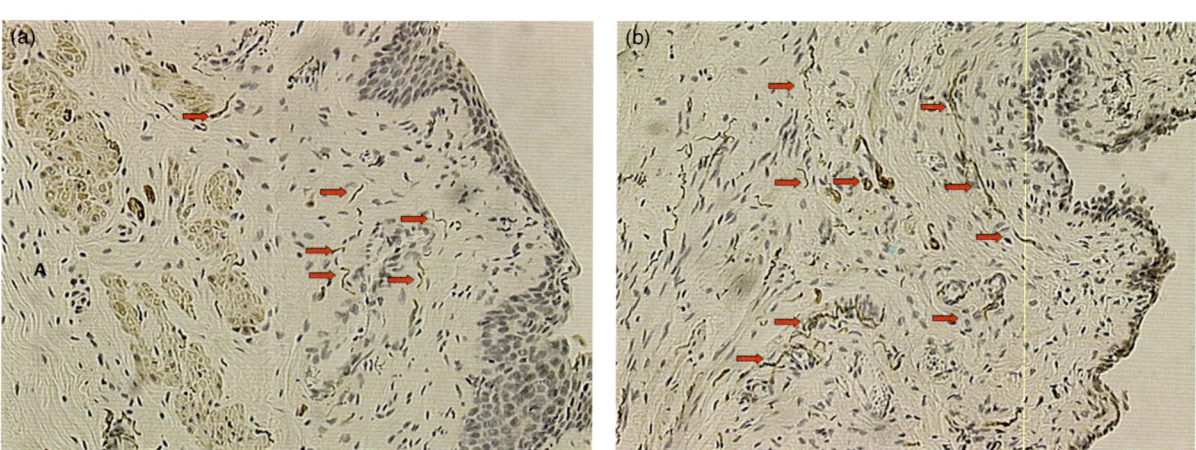

Figure 10.2 Effects of ovariectomy and androgen treatment on vaginal tissue nerve fiber network density. Female rats were left intact (control; a) or ovariectomized and after 2 weeks were infused with testosterone (b). Hormone treatment was started 2 weeks after ovariectomy and was maintained for 2 weeks. Vaginal tissue was fixed and paraffin-embedded and vaginal tissue sections subjected to immunostaining procedures using anti-PGP 9.5 primary antibody and counterstained with Gill's hematoxylin. In vaginal tissue from intact rats (a), nerve fibers extending from the lamina propria into the epithelial layer were occasionally observed – see arrow in (a). PGP-positive fibers were commonly observed surrounding blood vessels (arrows) as depicted in tissue sections from intact control (a) and testosterone infused (11 µg/day for 14 days), ovariectomized animals (b). Details are given in references 19 and 26.

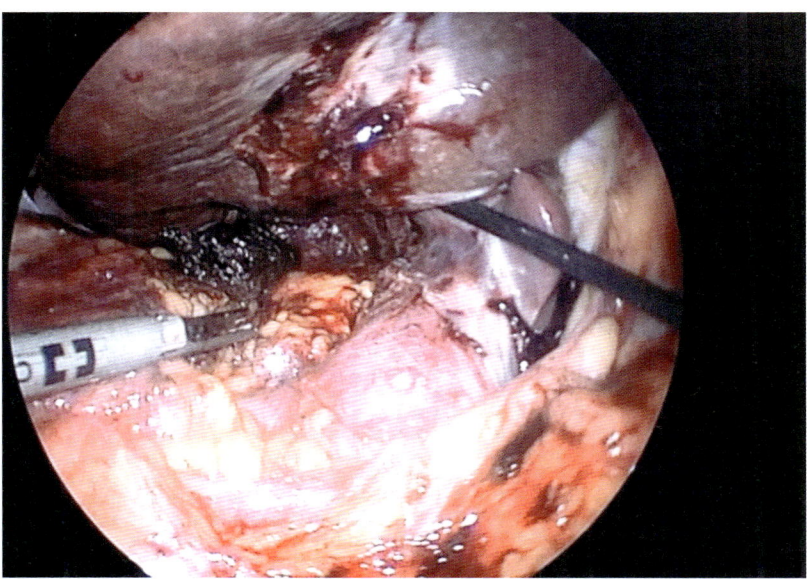

Figure 12.4 An adrenal nodule in a woman with Cushing's syndrome visualized by laparoscopy. In this case, the adrenal nodule was metastatic from a sarcoma.

References

1. Witchel SF, Azziz R. Congenital adrenal hyperplasia. *J Pediatr Adolesc Gynecol* 2011;24(3):116–26.

2. Jones HW Jr, Scott WW. Female intersexuality with adrenal hyperplasia. In: Jones HW Jr., Scott WW (Eds.) *Hermaphroditism, Genital Anomalies and Related Endocrine Disorders.* The Williams and Wilkins Company; 1971: Ch. 9: 197–269.

3. Falhammar H, Thorén M. Clinical outcomes in the management of congenital adrenal hyperplasia. *Endocrine* 2012;41(3):355–73.

4. Nimkarn S, Lin-Su K, New MI. Steroid 21 hydroxylase deficiency congenital adrenal hyperplasia. *Pediatr Clin North Am* 2011;58(5):1281–300, xii.

5. Marumudi E, Khadgawat R, Surana V, et al. Diagnosis and management of classical congenital adrenal hyperplasia. *Steroids* 2013;78(8):741–6.

6. Forest MG. Recent advances in the diagnosis and management of congenital adrenal hyperplasia due to 21-hydroxylase deficiency. *Human Reprod Update* 2004;10(6):469–85.

7. Therrell BJ, Berenbaum S, Manter-Kapanke V, et al. Results of screening 1.9 million Texas newborns for 21-hydroxylase-deficient congenital adrenal hyperplasia. *Pediatrics* 1998;101:583–90.

8. Speiser PW, Dupont B, Rubinstein P, et al. High frequency of nonclassical steroid 21-hydroxylase deficiency. *Am J Hum Genet* 1985;37:650–75.

9. Carlson AD, Obeid JS, Kanellopoulou N, et al. Congenital adrenal hyperplasia: update on prenatal diagnosis and treatment. *J Steroid Biochem Mol Biol* 1999;69(1–6):19–29.

10. Barnes RB, Rosenfield RL, Ehrmann DA, et al. Ovarian hyperandrogynism as a result of congenital adrenal virilizing disorders: evidence for perinatal masculinization of neuroendocrine function in women. *J Clin Endocrinol Metab* 1994;79:1328–33.

11. Nabhan ZM, Eugster EA. Upper-tract genitourinary malformations in girls with congenital adrenal hyperplasia. *Pediatrics* 2007;120(2):e304–7.

12. Speiser PW, Azziz R, Baskin LS, et al.; Endocrine Society. Congenital adrenal hyperplasia due to steroid 21-hydroxylase deficiency: an Endocrine Society clinical practice guideline. *J Clin Endocrinol Metab* 2010;95(9):4133–60.

13. Hernanz-Schulman M, Brock JW 3rd, Russell W. Sonographic findings in infants with congenital adrenal hyperplasia. *Pediatr Radiol* 2002;32(2):130–7.

14. Fritz MA, Speroff L (Eds.) *Clinical Gynecologic Endocrinology and Infertility.* Philadelphia: Lippincott, Williams and Wilkins, 8th ed. 2010.

15. Arlt W, Willis DS, Wild SH, et al.; United Kingdom Congenital Adrenal Hyperplasia Adult Study Executive (CaHASE). Health status of adults with congenital adrenal hyperplasia: a cohort study of 203 patients. *J Clin Endocrinol Metab* 2010;95(11):5110–21.

16. Nermoen I, Husebye ES, Svartberg J, Løvås K. Subjective health status in men and women with congenital adrenal hyperplasia: a population-based survey in Norway. *Eur J Endocrinol* 2010;163(3):453–9.

17. Gozar H, Pascanu I, Ardelean M, et al. Surgical reconstruction of the genitalia in a 3-year-old infant with a 46XX karyotype: case report. *Aesthetic Plast Surg* 2014;38(3):549–53.

18. Crouch NS, Liao LM, Woodhouse CR, et al. Sexual function and genital sensitivity following feminizing genitoplasty for congenital adrenal hyperplasia. *J Urol* 2008;179:634–8.

19. Van Wyk JJ, Ritzen EM. The role of bilateral adrenalectomy in the treatment of congenital adrenal hyperplasia. *J Clin Endocrinol Metab* 2003;88(7):2993–8.

20. Ogilvie CM, Rumsby G, Kurzawinski T, Conway GS. Outcome of bilateral adrenalectomy in congenital adrenal hyperplasia: one unit's experience. *Eur J Endocrinol* 2006;154(3):405–8.

21. Falhammar H, Filipsson H, Holmdahl G, et al. Fractures and bone mineral density in adult women with 21-hydroxylase deficiency. *J Clin Endocrinol Metab* 2007;92(12):4643–9.

22. Witchel SF. Management of CAH during pregnancy: optimizing outcomes. *Curr Opin Endocrinol Diabetes Obes* 2012;19(6):489–96.

23. Hagenfeldt K, Janson PO, Holmdahl G, et al. Fertility and pregnancy outcome in women with congenital adrenal hyperplasia due to 21-hydroxylase deficiency. *Hum Reprod* 2008;23(7):1607–13.

24. Mercè Fernández-Balsells M, Muthusamy K, Smushkin G, et al. Prenatal dexamethasone use for the prevention of virilization in pregnancies at risk for classical congenital adrenal hyperplasia because of 21-hydroxylase (CYP21A2) deficiency: a systematic review and meta-analyses. *Clin Endocrinol (Oxf)* 2010;73(4):436–44.

25. New M, Carlson A, Obeid J, et al. Update: Prenatal diagnosis for congenital adrenal hyperplasia in 595 pregnancies. *Endocrinologist* 2003;13:233–9.

26. Nimkarn S, New MI. Prenatal diagnosis and treatment of congenital adrenal hyperplasia owing to 21-hydroxylase deficiency. *Nat Clin Pract Endocrinol Metab* 2007;3:405–13.

Chapter

12

Cushing's syndrome in females

Andrew S. Fischer, Christopher B. Rizk, and Sylvia Hsu

Introduction

Cushing's syndrome is a condition that results from excessive, long-term exposure to glucocorticoids. Whereas exogenous glucocorticoids are most commonly responsible, the endogenous etiologies are what continue to both challenge and fascinate even the most adept clinicians. The stark female predominance within these endogenous etiologies, particularly amongst those with Cushing's disease, makes women of particular concern. Patients often present with a constellation of signs and symptoms that encompasses all aspects of the body and overlaps with many common conditions. Therefore, there is continuous research on the diagnosis and management of this condition. Even so, many aspects of this condition remain an enigma.

Pathophysiology

Aberrations in the hypothalamic–pituitary–adrenal (HPA) axis are common to all Cushing's syndromes. Corticotropin-releasing hormone (CRH) is synthesized in the hypothalamus and regulates corticotropin (adrenocorticotropic hormone [ACTH]) release from the anterior pituitary. ACTH stimulates the adrenal cortex to synthesize and release cortisol, which then exerts negative feedback on the hypothalamus and anterior pituitary to down-regulate CRH and ACTH secretion, respectively.

In Cushing's syndrome, there is excessive serum cortisol. This may be the result of tumorigenesis, or more commonly, the exogenous administration of glucocorticoids. If tumorigenesis is responsible, the tumor may be located within the HPA axis or ectopically. Tumors of the adrenal glands autonomously secrete cortisol, whereas tumors outside the adrenals secrete ACTH which signals for the release of cortisol.

The secretion of cortisol or ACTH by tumors may be constant or cyclical in nature. Collectively, the common result is a supraphysiologic level of serum cortisol.

The excessive serum cortisol is not subject to normal endogenous negative feedback on the HPA axis (owing to continued exogenous administration or tumorigenesis). This resistance to the suppressive effects of glucocorticoids can be utilized for the diagnosis of Cushing's syndrome. While the HPA axis becomes resistant, the peripheral tissues remain sensitive to the effects of the glucocorticoids, leading to the progressive clinical features of the disease.

Excessive serum ACTH is present in some types of Cushing's syndrome, termed ACTH-dependent. ACTH is derived from a larger precursor molecule, pro-opiomelanocortin (POMC). Other derivatives of POMC include beta-lipoprotein, beta-endorphin, and melanocyte-stimulating hormone (MSH). States of high ACTH may be additionally marked by skin pigmentation due to excessive MSH. In addition to cortisol, ACTH is also responsible for the synthesis and secretion of the adrenal androgens, dehydroepiandrosterone sulfate (DHEA-S) and androstenedione, as well as 11-deoxycorticosterone. DHEA-S and androstenedione are peripherally converted to testosterone, leading to a state of androgen excess. Moreover, ACTH induces the local production of growth factors, such as insulin-like growth factors 1 and 2 (IGF-1, IGF-2) and basic fibroblast growth factor (bFGF). These stimulate the growth of the adrenal gland and its vasculature resulting in bilateral adrenal hyperplasia.

Etiology of Cushing's syndrome

Cushing's syndrome is first categorized by the source of the glucocorticoids, exogenous or endogenous.

Androgens in Gynecological Practice, ed. Leo Plouffe and Botros Rizk. Published by Cambridge University Press. © Cambridge University Press 2015.

Table 12.1 Etiology of endogenous Cushing's syndrome

	Frequency (%)
ACTH-dependent	
ACTH-secreting pituitary adenoma (Cushing's disease)	70
Ectopic ACTH syndrome	10
Ectopic CRH syndrome	<1
ACTH-independent	
Adrenal adenoma	10
Adrenal carcinoma	5
Macronodular adrenal hyperplasia	<2
Micronodular adrenal hyperplasia (including the Carney complex)	<2
McCune–Albright	<2
Adapted from [1,2].	

Exogenous, or iatrogenic, Cushing's syndrome is caused by the chronic administration of high-dose glucocorticoids. Endogenous Cushing's syndrome is further categorized into conditions that are dependent on ACTH and conditions that are not dependent on, or are independent of, ACTH. The etiology and frequency of endogenous Cushing's syndrome is provided in Table 12.1. Over 80% of endogenous Cushing's syndrome is ACTH-dependent, and over 80% of ACTH-dependent Cushing's syndrome is due to an ACTH-secreting pituitary adenoma (Cushing's disease). The remaining cases are due to ectopic ACTH or CRH syndromes. Cases of ACTH-independent Cushing's syndrome are more commonly due to adrenal tumors (adenomas and carcinomas) which autonomously secrete cortisol. Very rarely ACTH-independent Cushing's syndrome is due to macro- or micronodular adrenal hyperplasia or McCune–Albright syndrome.

Iatrogenic Cushing's syndrome

Glucocorticoids are commonly used for neoplastic, inflammatory, and autoimmune disorders. Long-term, high-dose glucocorticoid therapy of any preparation can lead to iatrogenic Cushing's syndrome that may be virtually indistinguishable from an endogenous etiology on exam. Therefore, the clinician must have a high clinical suspicion. The development of symptoms is related more to the dose and duration than the potency, so even low-potency glucocorticoids can cause clinical features if given frequently and at high dose [3]. Although very rare, clinicians should also be aware of the surreptitious use of glucocorticoids leading to a factitious Cushing's syndrome [4].

Cushing's disease

ACTH-secreting pituitary adenomas are the most common cause of endogenous Cushing's syndrome. They were first described by Harvey Cushing in 1932 and are now referred to as Cushing's disease [5]. Most are microadenomas less than 10 mm in diameter. Half of these are less than 5 mm, making them virtually invisible on imaging. They may be located anywhere in the anterior pituitary. Macroadenomas are larger than 1 cm in diameter and may extend out of the sella turcica and compress the optic chiasm, or extend laterally into the cavernous sinus. Most pituitary tumors arise spontaneously. However, they may also arise more rarely in the setting of multiple endocrine neoplasia type 1 (MEN1).

Ectopic ACTH syndrome

Nearly any tumor can be an ectopic source of ACTH. Most commonly, the tumors are of neuroendocrine origin: bronchial carcinoids, small cell carcinomas of the lung, pancreatic carcinoids, thymic carcinoids, medullary carcinomas of the thyroid, and pheochromocytomas [6]. Particularly aggressive tumors, like small cell carcinomas, may release unprocessed POMC in addition to ACTH.

Ectopic CRH syndrome

Ectopic CRH production is an extremely rare manifestation of some tumors, such as medullary thyroid carcinoma, pheochromocytoma, and prostate cancer. There is hyperplasia of the pituitary gland with resultant ACTH and cortisol overproduction.

Primary adrenal disorders

Primary adrenal disorders include adenomas, carcinomas, and nodular hyperplasias. Hyperfunctioning benign adrenal adenomas are most common and are often less than 4 cm in diameter. Malignant carcinomas are often larger, greater than 6 cm, and may present as a palpable abdominal flank mass. They may invade into the surrounding kidney, liver, and retroperitoneum. Macronodular adrenal hyperplasia is often the result of ectopic hormones or growth factors. The adrenals are massively enlarged and have characteristic nodules over 1 cm in diameter that may, over time, begin to autonomously secrete cortisol.

With the advent of high-resolution imaging, adrenal adenomas are being discovered incidentally with increasing frequency. The majority are less than 4 cm and non-functional, but some may display varying amounts of autonomous cortisol secretion.

Primary pigmented nodular adrenal disease

Primary pigmented nodular adrenal disease (PPNAD) is the most frequent variant of micronodular adrenal hyperplasia. It may be sporadic, or part of the autosomal dominant Carney complex, and is a rare cause of Cushing's syndrome presenting in late childhood [1]. The adrenals are often normal or slightly enlarged with small, often less than 6 mm, ACTH-secreting nodules. Mutations in the regulatory subunit of R1A of protein kinase A (PRKAR1A) are associated with approximately 45% of cases [7,8]. There is increased expression of the glucocorticoid receptor, leading to a paradoxical rise in cortisol secretion in response to dexamethasone.

McCune–Albright

McCune–Albright syndrome is a sporadic disease associated with hyperfunctioning endocrine glands. Constitutive cortisol production from adrenal glands can lead to Cushing's syndrome, presenting in the first few years of life. An activating mutation of GNAS1 leads to increased activation of a Gs protein in the adrenal cortex, resulting in increased cAMP formation and cortisol synthesis [9]. It is commonly associated with the classic triad of precocious puberty, polyostotic dysplasia, and café-au-lait skin pigmentation.

Epidemiology

Cushing's syndrome is rare. The true incidence and prevalence in the population is difficult to assess for many reasons – the syndrome may remain incipient for many years, the clinical features are shared by common diseases (type 2 diabetes, metabolic syndrome, polycystic ovary syndrome), and the diagnostic journey is rather arduous. Population-based studies estimate an annual incidence of 0.7–2.4 per million and a prevalence of 39–79 cases per million [10– 12].

There exists a growing population of patients diagnosed with subclinical or occult Cushing's syndrome. These patients have a low level of autonomous cortisol production, often from an adrenal adenoma. They lack the overt features of Cushing's syndrome, but often have one or more features of metabolic syndrome

[13]. Screening studies in type 2 diabetics, osteoporotics, and hypertensives have revealed an unsuspected Cushing's syndrome in up to 10% of these patients [14–18]. These studies highlight that Cushing's syndrome truly exists on a spectrum, with a wide range of clinical phenotypes, making categorization and quantification difficult.

Female predominance

Cushing's syndrome has a marked female predominance, similar to many endocrine disorders. The female-to-male ratio of Cushing's disease and adrenal adenomas (accounting for roughly 80% of Cushing's syndrome) is 3–4:1 [19]. The female-to-male ratio of Cushing's disease alone is even higher, 3–8:1 [20]. Interestingly, there is no female predominance observed in adolescents with Cushing's disease. In fact, there is a prepubertal male predominance, which equalizes during puberty. The female predominance does not emerge until adulthood [21]. Few studies have looked at gender-related differences in the presentation, diagnosis, and clinical course of the disease. Two studies, looking at Cushing's disease, found that women often present with less severe clinical features, lower serum ACTH, and lower urinary cortisol levels than their male counterparts [20,22].

It is unknown why there is a female preponderance. Likely, gender-related differences in pituitary and adrenal secretions are at play [22]. Some propose that a change in hormones, namely an increase in estrogen during and after puberty, is responsible [21]. Further, estrogen receptor β (ERβ) is expressed in corticotroph pituitary tumors, further supporting a role for estrogen in the pathogenesis of this disease [23].

Clinical presentation

The clinical presentation of Cushing's syndrome is broad (Figures 12.1 and 12.2) and represents the widespread action of cortisol on human tissue. Whereas a florid presentation of Cushing's syndrome is unmistakable, mild cases may be much less obvious. A number of signs and symptoms of excess cortisol are also common amongst women in the general population – such as obesity, depression, diabetes, hypertension, and menstrual irregularities – further confusing the clinical picture. Since the disease tends to be progressive, the presence of multiple features is more suggestive than any one single feature. Table 12.2 provides the clinical features of Cushing's syndrome with their

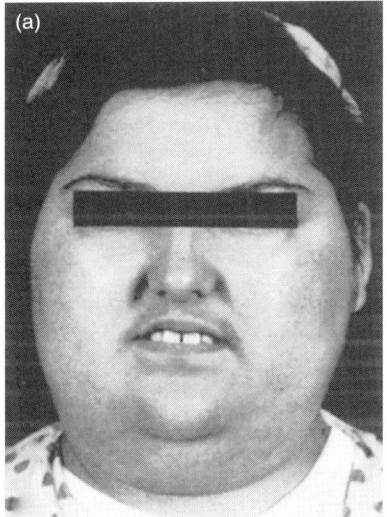
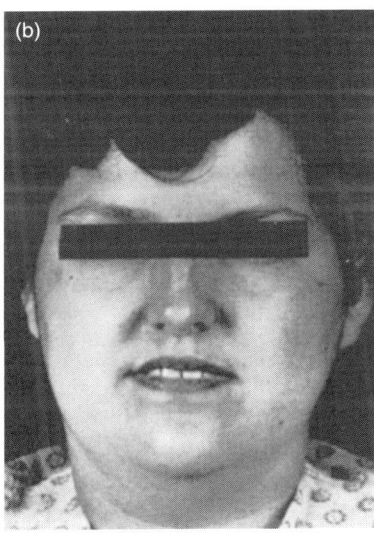

Figure 12.1 A patient with Cushing's syndrome before (a) and after (b) adrenalectomy. Note the typical round "moon" face in (a). Picture courtesy of [24].

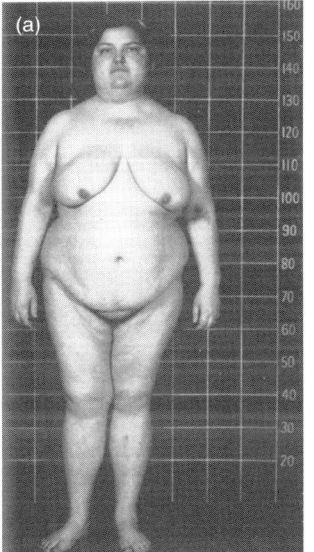
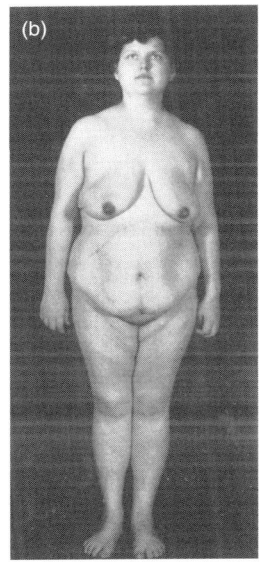

Figure 12.2 Patient with Cushing's syndrome before (a) and after (b) adrenalectomy. This patient depicts the Cushingoid habitus in (a) with central obesity, peripheral wasting, and a round "moon" face. Picture courtesy of [24].

Table 12.2 Clinical features of women with Cushing's syndrome (the discriminatory features are often less common, whereas the less discriminatory features are often more common)

More discriminatory		Less discriminatory	
Feature	*Frequency (%)*	*Feature*	*Frequency (%)*
Facial plethora	90	Obesity/weight gain	95
Thin skin/ striae[a]	70–90	Rounded face	90
Easy bruising	65	Reduced libido	90
Proximal myopathy[b]	60	Menstrual irregularity	80
Osteoporosis	50	Hypertension	75
		Hirsutism	75
		Psychiatric symptoms	70
		Impaired glucose tolerance	60

[a] Striae are typically reddish purple and wider than 1 cm.
[b] Sixty percent of patients complain of generalized weakness.
Adapted from [1,25].

respective frequencies, separated into those that are more and less discriminatory.

Cortisol exerts an anabolic effect on the liver, stimulating gluconeogenesis and glycogenesis, and a catabolic effect on muscle and adipose tissue, stimulating proteolysis and lipolysis. The result is the classic Cushingoid habitus: central fat deposition with peripheral wasting.

There is progressive fat deposition in the abdomen, face (including retro-orbital), and neck. Fat accumulation in the face gives rise to a round face (the "moon face"), while fat accumulation in the neck, specifically the supraclavicular and dorsocervical fat pads, is termed a "buffalo hump." Central obesity is the most sensitive manifestation, and often the first. Obesity is only one component of a metabolic syndrome seen in many patients characterized additionally by hypertension, diabetes mellitus type 2, and hyperlipidemia. Acanthosis nigricans is a possible manifestation of insulin resistance.

The signs and symptoms of protein catabolism are of high discriminatory value to the clinician, as they are most specific [26–28]. Patients may complain of muscle weakness, specifically proximally, which may manifest as the inability to climb stairs or rise from a seated position. Decreased bone formation, increased bone resorption, and decreased intestinal and renal calcium reabsorption may lead to osteoporosis manifesting as fractures. A cortisol-mediated inhibition of collagen synthesis results in skin atrophy that appears translucent and is susceptible to tearing. Poor dermal support and reduced elasticity of the vessel walls make patients especially susceptible to bruising in response to minor trauma, which manifests as ecchymosis. Dermal tears lead to atrophic scarring and violaceous striae, which are pathognomonically more than 1 cm wide and located on the abdomen and flank. Facial plethora is a consequence of a thinned epidermis.

Hypogonadism results from cortisol's inhibitory effect on the release of gonadotropin-releasing hormone from the hypothalamus. It may manifest clinically as menstrual irregularities (menorrhagia, amenorrhea), infertility, and reduced libido. A hyperandrogenic state, common to ACTH-dependent Cushing's syndromes, may present as hirsutism, acne, and temporal hair loss.

A wide range of emotional disturbances is associated with cortisol excess, including, but not limited to, depression, hypomania, frequent mood swings, psychosis, insomnia, and short-term memory loss. Depression is most common, and residual psychiatric symptoms may persist even with normalization of cortisol [29].

Glucocorticoids inhibit the synthesis of arachidonic acid, a key molecule in the inflammatory cascade, and suppress cell-mediated immunity. Cushing patients are at increased risk for infections, particularly invasive and cutaneous mycoses.

Patients with aggressive ACTH-secreting tumors may present without the classic Cushingoid appearance. High levels of cortisol overwhelm 11β-hydroxysteroid dehydrogenase type 2 in the renal tubules and exert their effect on the mineralocorticoid receptors. These patients present with the acute onset of severe hypertension, hypokalemia, hyperglycemia, and myopathy. There may be no weight gain.

Iatrogenic Cushing's syndrome will present similarly to endogenous Cushing's syndrome, but without

hyperandrogenism and with an increased incidence of glaucoma, cataracts, and avascular necrosis [3].

Morbidity and mortality

Women with Cushing's syndrome have a more than twofold increased mortality risk [30,31]. While this risk has been well established for patients with Cushing's disease, it is currently unclear whether the same applies for those with an adrenal Cushing's syndrome as well [32,33]. There are currently no good data on patients with malignant Cushing's syndrome, such as adrenal carcinomas and ectopic ACTH syndrome, but these patients generally have poor prognoses [10].

The cardiovascular complications of Cushing's syndrome – coronary artery disease, congestive heart failure, and myocardial infarction – contribute greatly to morbidity and mortality. These are likely the result of a hypercortisol-related metabolic syndrome. Further, there is often some persistence of the metabolic syndrome and cardiovascular risk factors in patients with long-term normalization of cortisol [34,35]. It is inconclusive at this time whether patients in remission from Cushing's syndrome still have increased mortality. However, persistence or recurrence of disease is a main determinant of mortality in this population [31,33]. Collectively, these conclusions form the rationale for early diagnosis, prompt treatment, and thorough follow-up.

Differential diagnosis

Table 12.3 presents conditions that are associated with mild elevations in cortisol, but are not considered Cushing's syndrome per se. Women with physiologic elevations in cortisol may present with nonspecific features of a true Cushing's syndrome – such as obesity, diabetes, and depression. Along with biochemical evidence of hypercortisolemia, it becomes very difficult to distinguish these patients from those with mild Cushing's syndrome. For example, a depressed patient may present with weight gain, menstrual irregularities, reduced libido, mild hirsutism, and a urinary cortisol above the upper limit of normal. The clinician is then left to determine whether the depression led to the excess cortisol, or the excess cortisol led to the depression. Physiologic hypercortisolism is thought to result from over-activation of the HPA axis in response to a condition or stressor. It should resolve once the condition is reversed (i.e., when the depression is treated). It is also paramount to remember

Table 12.3 Conditions associated with hypercortisolemia in women

May present with clinical features of Cushing's syndrome	Unlikely to present with clinical features of Cushing's syndrome
Pregnancy	Physical stress
Depression/psychiatric conditions	Malnutrition/anorexia nervosa
Alcoholism	Intense chronic exercise
Glucocorticoid resistance	Hypothalamic amenorrhea
Morbid obesity	Corticosteroid-binding globulin (CBG) excess
Diabetes mellitus (poorly controlled)	
Adapted from [25].	

that while these conditions are relatively common in women, true Cushing's syndrome is rare.

Diagnosis

The diagnosis of Cushing's syndrome begins with a thorough history and physical searching for the clinical features outlined in Table 12.2 [29]. A comprehensive drug history helps investigate the possibility of iatrogenic Cushing's syndrome. Glucocorticoids are found in many preparations (oral, rectal, inhaled, topical, injected) and may be found in many over-the-counter skin creams, bleaching agents, and herbal medications [25]. It is prudent to recognize that iatrogenic Cushing's syndrome, as it is most common, can be easily reversed with withdrawal of the drug, and saves the patient from unnecessary testing.

Endogenous Cushing's syndrome, on the other hand, is rare and the diagnostic workup is complex, expensive, and taxing. Therefore, the clinician must have a high index of suspicion before recommending such testing. The non-discriminate screening of patients with relatively common conditions such as diabetes mellitus, obesity, and depression is not recommended, owing to a high false-positive rate [25,29]. The current recommendation is to screen the following patient populations: [25]

1. Patients with features that are unusual for age, such as osteoporosis and hypertension in a young woman
2. Patients with multiple and progressive features, specifically those features which are more discriminatory (Table 12.2)
3. Children with decreasing height percentile and increasing weight

4. Patients with incidentally discovered adrenal adenomas.

Patients with a family history of Carney complex or MEN1 should undergo surveillance screening for the development of Cushing's syndrome.

An algorithm for the diagnosis in a woman who is suspected of having Cushing's syndrome is presented in Figure 12.3.

Biochemical testing

The diagnosis of Cushing's syndrome is established with the following sequential steps:

1. Establishing the presence of hypercortisolism
2. Determining ACTH dependence vs. independence
3. Determining the source of ACTH in ACTH-dependent Cushing's syndrome.

Step 1. Establishing the presence of hypercortisolism

Hypercortisolism is established with one of the following first-line tests [25]:

- 24-hour urine free cortisol (≥ 2)
- Late-night salivary cortisol (≥ 2)
- 1 mg overnight dexamethasone suppression test
- 2 mg/day 48-hour low-dose dexamethasone suppression test.

There are two additional second-line tests used for specific situations:

- 2 mg/day 48-hour low-dose dexamethasone suppression test with CRH
- midnight serum cortisol.

Table 12.4 provides a summary of the screening tests. They are highly sensitive and, with the exception of midnight serum cortisol, can be performed in an outpatient setting. The difficulty in the first step is in differentiating between true Cushing's syndrome and other causes of hypercortisolemia, presented in Table 12.3. Therefore, an abnormal result from a first-line test, or a high clinical suspicion with a normal result, should prompt retesting with a different first- or second-line test. If the results are concordantly positive, Cushing's syndrome is confirmed and further testing to elucidate the cause is indicated (step 2). If the results are concordantly negative, Cushing's syndrome is unlikely and no further evaluation is necessary. If the results are discordant, and there is still a high clinical suspicion of a cyclical disease, then instructing the patient to collect a urinary free cortisol or late-night salivary cortisol, when symptoms arise, may be beneficial.

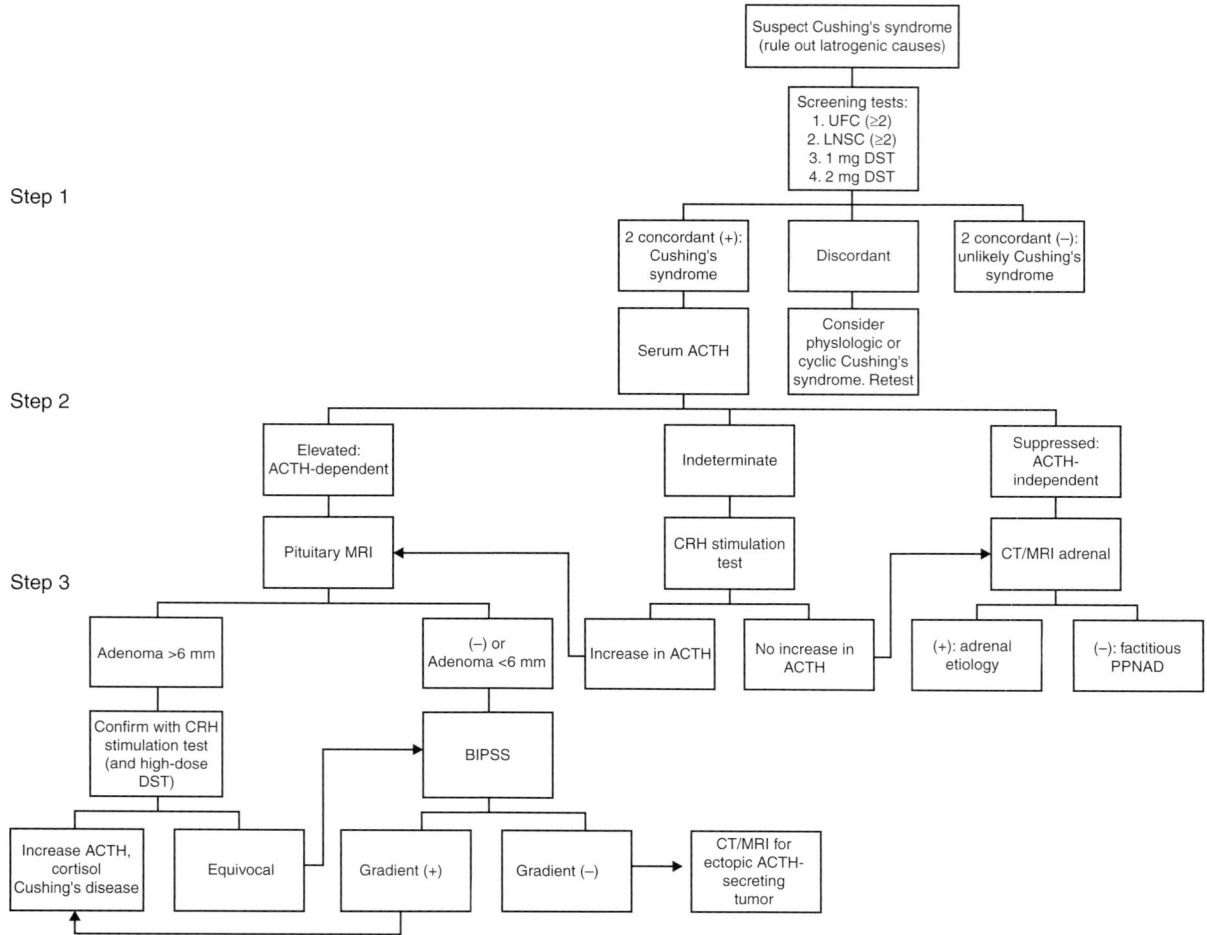

Figure 12.3 Diagnostic algorithm for Cushing's syndrome in females [25,29]. UFC, 24-hour urinary free cortisol; LNSC, late-night salivary cortisol; DST, dexamethasone suppression test; PPNAD, primary pigmented nodular adrenal dysplasia; BIPSS, bilateral inferior petrosal sinus sampling.

None of the tests for diagnosing Cushing's syndrome are perfect, and a wide range of diagnostic thresholds have been proposed – each with different discriminatory powers, on different patient populations, and with different assays. Likely, this reflects that Cushing's syndrome exists on a wide spectrum. Regardless, meta-analysis has shown a similar diagnostic accuracy for urine free cortisol, late-night salivary cortisol, and 1 mg overnight dexamethasone suppression test, as well as various combined strategies [39].

24-hour urinary free cortisol (UFC)

Cortisol circulates both freely and bound to corticosteroid-binding globulin (CBG).

Free cortisol is filtered by the kidneys and excreted into the urine. When measured over a 24-hour period, it provides a good indication of cortisol levels. Free cortisol is not affected by conditions that alter the level of CBG, such as estrogen. Plasma cortisol, on the other hand, measures both bound and free cortisol, and can be falsely elevated in patients on oral contraceptives, for example.

Patients are instructed to discard the first morning void on the first day and to collect subsequent voids throughout the day and night, including the first morning void on the second day. The urine should be kept refrigerated [25].

A value above the assay's upper limit of normal (~80 μg/day) should prompt further testing. At least

Table 12.4 Summary of diagnostic screening tests for suspected Cushing's syndrome

Test	Procedure	Measure	Diagnostic cut-off	Sensitivity (%)	Specificity (%)	False positives	False negatives
24-hour urine free cortisol	Collect voids over 24 h	Cortisol, creatinine	Assay upper limit of normal; ~80 μg/24 h	91–96	91	High fluid intake (>5 L/day) Exogenous steroids, carbamazepine, digoxin, fenofibrate	Renal failure Cyclic Cushing's Mild Cushing's
Late-night salivary cortisol	Collect saliva sample between 11 p.m. and 12 a.m.	Salivary cortisol	145 ng/dL	92	96	Ill or depressed patients Shift-workers Chewing tobacco, cigarette smoking Contamination/user error	Cyclic Cushing's
1 mg DST	1 mg dexamethasone between 11 p.m. and 12 a.m.	Serum cortisol between 8 and 9 a.m.	1.8 μg/dL	98–100	80	Medications inducing CYP 3A4 system Estrogens/pregnancy	Medications inhibiting CYP 3A4 system Liver and/or renal failure
2 mg DST	0.5 mg dexamethasone administered q6h starting at 9 a.m. for 48 h	Serum cortisol at 9 a.m. 6 h after last dose	1.8 μg/dL	96	79	Decreased dexamethasone absorption Alcohol	
2 mg DST with CRH	2 mg DST followed by IV 1 μg/kg CRH 2 h after last dose	Serum cortisol 15 min after CRH	1.4 μg/dL	98	70		
Midnight serum cortisol	Venous sampling from indwelling catheter at 12 a.m.	Serum cortisol	1.8 μg/dL 7.5 μg/dL	99–100 91–96	20 88	Stress from hospitalization Ill or depressed patients	Cyclic Cushing's

DST, dexamethasone suppression test. CRH, corticotropin-releasing hormone.
Adapted from [2] with additional information from [19,25,29,36–38].

two tests should be performed, as values can be variable. However, if several are normal, there is a low probability of Cushing's syndrome. Mild elevations in UFC can be seen with physiological hypercortisolemia. Pregnancy, owing to estrogen, may present with UFC values up to three times normal. UFC values fourfold greater than the upper limit of normal, however, are virtually diagnostic of Cushing's syndrome [29].

In clinical practice, the accuracy of the test is often limited by the adequacy of collection. Expressing free cortisol over urinary creatinine may help verify an adequate urine sample. Further, patients with renal failure and a creatinine clearance less than 60 mL/min may receive a falsely low result. Excessive amounts of fluids (>5 L/day) may falsely increase urinary cortisol. Finally, certain drugs – such as carbamazepine, digoxin, and fenofibrate – may co-elute with cortisol and falsely elevate urinary cortisol [25].

Late-night salivary cortisol

Late-night salivary cortisol is a simple and convenient non-invasive screening test.

Normally, cortisol rises in the early morning, peaks during mid-morning, and then falls to its nadir around midnight. Patients with Cushing's syndrome lose their rhythmic cortisol fluctuation and fail to demonstrate depressed cortisol levels in the late evening. Salivary cortisol concentrations are an accurate representation of the free serum cortisol, independent of the salivary flow rate, and show high concordance with UFC in patients with Cushing's syndrome [40,41].

The main advantage of testing salivary cortisol is in its ease of collection in the outpatient setting. Patients are instructed to collect saliva samples (by drooling into a plastic bag or chewing on a salivette for 1–2 minutes) between 11 p.m. and 12 a.m. on two occasions. The samples are stable at room temperature for up to a week, allowing patients to mail them to the lab. It can also be particularly useful if there is a need for repeated testing, an advantage when the presumed diagnosis is a cyclical Cushing's syndrome.

Most normal patients have a late-night salivary cortisol concentration of less than 145 ng/dL by ELISA and mass spectrometry (LC-MS/MS) [25,42]. Patients should collect saliva while relaxed, as spuriously elevated values can result if patients are stimulated at the time of testing. Further, patients with abnormal circadian rhythms, such as shift workers or those in critical condition, may not be good candidates for testing. Patients should also be particularly careful to avoid contamination with steroid-containing lotions and gels.

Dexamethasone suppression tests

Negative feedback from an exogenous glucocorticoid should suppress serum ACTH and cortisol in patients with a normally functioning HPA axis. Women with Cushing's syndrome, however, lose their HPA axis feedback regulation and do not demonstrate the expected suppression after the administration of a low dose of dexamethasone. There are three versions of the dexamethasone suppression test: the 1 mg overnight dexamethasone suppression test, the 2 mg/day 48-hour low-dose dexamethasone suppression test, and the 2 mg/day 48-hour low-dose dexamethasone suppression test with CRH.

Abnormalities in dexamethasone absorption and metabolism can produce false results. A false negative may result when dexamethasone clearance is slowed, either by hepatic or renal failure, or by drugs that inhibit the hepatic CYP 3A4 enzymes: aprepitant/fosaprepitant, itraconazole, ritonavir, fluoxetine, diltiazem, and cimetidine. Conversely, a false positive may result when dexamethasone clearance is increased by drugs that induce the hepatic CYP 3A4 enzymes: phenobarbital, phenytoin, carbamazepine, primidone, rifampin, rifapentine, ethosuximide, pioglitazone, and alcohol [25]. If there is concern about dexamethasone clearance, the simultaneous measurement of a serum dexamethasone (>0.22 µg/dL) with serum cortisol may be constructive [25]. Additionally, women should withhold oral estrogen-containing drugs for 6 weeks before testing [43]. Oral estrogens, such as oral contraceptives, and pregnancy may increase circulating CBG, yielding a false positive result.

The 1 mg overnight dexamethasone suppression test

The patient is given 1 mg of dexamethasone between 11 p.m. and 12 midnight and serum cortisol is measured between 8 and 9 a.m. the following morning. A cut-off of 1.8 µg/dL by radioimmunoassay achieves a high sensitivity, appropriate for screening [25]. However, the specificity is increased with an increasing serum cortisol suppression cut-off.

The 2 mg/day 48-hour low-dose dexamethasone suppression test

The 2 mg/day 48-hour low-dose dexamethasone suppression test is similar to the 1 mg overnight suppression test but spread over two days. The patient is given 0.5 mg every 6 hours beginning at 9 a.m. on the first day (9 a.m., 3 p.m., 9 p.m., 3 a.m.). This is repeated on the second day. Serum cortisol is then measured at 9 a.m. on the third day, 6 hours after the last dose of dexamethasone.

This test was once thought to provide more accuracy in discriminating between Cushing's syndrome and other conditions of hypercortisolism, when 24-hour urinary free cortisol may be falsely elevated. However, recent studies have shown similar performance to the 1 mg overnight dexamethasone suppression test [25,38]. Additionally, because it offers equivocal diagnostic acumen and is more cumbersome to perform, it has largely fallen out of favor.

The 2 mg/day 48-hour low-dose dexamethasone suppression test with CRH

The 2 mg/day 48-hour low-dose dexamethasone suppression test with CRH was developed to improve the sensitivity of the 2 mg/day 48-hour low-dose dexamethasone suppression test in discriminating between true Cushing's syndrome and states of physiological hypercortisolemia. A small population of patients with Cushing's syndrome suppress cortisol in response to dexamethasone. Administration of CRH should increase ACTH and cortisol in those patients with Cushing's syndrome only, whereas patients with physiological hypercortisolemia often have a blunted response to the administration of CRH.

The test is carried out similarly to the 2 mg low-dose dexamethasone suppression test, but 1 µg/kg CRH is administered intravenously (IV) 2 hours after the last dose of dexamethasone. Dexamethasone is measured at the time of CRH administration and cortisol is measured 15 minutes later. Although initially shown to yield higher sensitivity and specificity than the 2 mg/day 48-hour low-dose dexamethasone suppression test, more recent studies fail to show improved specificity [38]. It may be useful in patients with questionable urine cortisol levels [25].

Midnight serum cortisol

Similar to the late-night salivary cortisol test, the midnight serum cortisol test is based on the premise that women with Cushing's syndrome lose their rhythmic cortisol fluctuation and therefore their nocturnal nadir. The patient is admitted to the hospital and serum cortisol is measured at midnight through an indwelling catheter. The patient is often kept for 48 hours to avoid false elevations from the initial stress of hospitalization. A serum cortisol greater than 1.8 µg/dL in

a sleeping patient is highly sensitive [36]. Its greatest utility may be to exclude Cushing's syndrome when a patient produces an elevated UFC and fails to suppress cortisol with dexamethasone testing, but there is a low clinical suspicion [25]. Specificity is increased if the cut-off is raised to 7.5 µg/dL. If the patient is awake, a serum cortisol greater than 7.5 µg/dL is highly sensitive, but less so than in a sleeping patient. Because of the cost and inconvenience associated with inpatient admission, serum cortisol is generally not used as an initial test.

Step 2. Determining ACTH dependence vs. independence

A morning serum ACTH at 9 a.m. will help determine the cause of an established Cushing's syndrome. Blood is collected at 9 a.m. in a prechilled EDTA tube, placed in an ice water bath, and delivered to the lab quickly, as ACTH is degraded by plasma proteases. A serum ACTH depressed below 10 pg/mL is indicative of an ACTH-independent Cushing's syndrome [29]. The next step is thin-section computed tomography (CT) or magnetic resonance imaging (MRI) of the adrenal glands to search for an adrenal etiology (adrenocortical adenoma, carcinoma, or ACTH-independent macronodular adrenal hyperplasia). If imaging fails to reveal the etiology, then consider PPNAD (either sporadic or part of the Carney Complex) or that the patient is surreptitiously using glucocorticoids.

A serum ACTH elevated above 20 pg/dL suggests an ACTH-dependent cause of Cushing's syndrome [29]. The next step is a gadolinium-enhanced T1-weighted pituitary MRI and biochemical tests to determine the source of ACTH (step 3). For indeterminate tests (serum ACTH 10–20 pg/dL), a CRH stimulation test may be fruitful.

Step 3. Determining the source of ACTH in ACTH-dependent Cushing's syndrome

It is important for treatment purposes to make the distinction between pituitary and ectopic ACTH-secreting tumors, although this is not always possible in clinical practice. Determining the source of ACTH in an ACTH-dependent etiology is often the most challenging aspect in the diagnosis of Cushing's syndrome.

Clinically, patients with either etiology may be virtually indistinguishable. Imaging, although invaluable, fails to visualize a pituitary adenoma in roughly 40% of patients with Cushing's disease [44]. Further, there is a possibility that a pituitary "incidentaloma" may be mistaken for an ACTH-secreting

Table 12.5 Summary of dynamic tests used to determine the source of ACTH in ACTH-dependent Cushing's syndrome

Test	Procedure	Measure	Diagnostic cut-off	Sensitivity (%)	Specificity (%)
High-dose DST	2 mg dexamethasone administered q6h starting at 9 a.m. for 48 h	UFC, serum cortisol at 9 a.m. (both beginning and end)	>50% suppression	81–94	29–67
	8 mg dexamethasone at 11 p.m.	Serum cortisol at 9 am	>50% suppression	77–92	57–100
CRH stimulation test	Administer 1 μg/kg CRH IV	Serum cortisol, ACTH	>35% rise in serum ACTH at 15, 30 min	93	100
			>20% rise in serum cortisol at 30, 45 min	91	88–95
BIPSS	Place venous sampling catheters in the inferior petrosal sinuses, administer 100 μg CRH IV	Petrosal, serum ACTH before and after CRH	Petrosal:peripheral ACTH >2:1	94	94
			Petrosal:peripheral ACTH >3:1 after CRH		

DST, dexamethasone suppression test; CRH, corticotropin-releasing hormone; BIPSS, bilateral inferior petrosal sinus sampling.
Adapted from [2] with additional information from [19,29,44–46].

pituitary adenoma, when in fact, the patient has an ectopic ACTH-secreting tumor. Nearly 10% of the population have incidental pituitary tumors visualized on MRI [44]. This limits the reliance on imaging alone.

Measuring serum ACTH and potassium may clue the clinician into one etiology. Ectopic ACTH-secreting tumors tend to have higher serum ACTH values than pituitary ACTH-secreting tumors, but serum ACTH values are rarely able to definitively distinguish between the two etiologies [19]. Hypokalemia is common in patients with ectopic ACTH production. Excessive cortisol saturates 11β-hydroxysteroid dehydrogenase and interacts with the mineralocorticoid receptor in the renal tubules. This, however, is more often a reflection of high cortisol levels than specifically ectopic ACTH production. Notably, 10% of patients with Cushing's disease may also have hypokalemia [19].

Dynamic testing is frequently used to determine a pituitary versus non-pituitary cause of Cushing's syndrome. A summary of these tests is provided in Table 12.5. Testing can be non-invasive (high-dose dexamethasone suppression test, CRH stimulation test) or invasive (inferior petrosal sinus sampling). The clinical utility of each test is judged against the pretest probability that 85–90% of women with ACTH-dependent Cushing's syndrome will have Cushing's disease [19].

High-dose dexamethasone suppression test

Eighty to ninety percent of ACTH-secreting pituitary adenomas will partially suppress ACTH secretion in response to high-dose dexamethasone [29]. Ectopic ACTH-secreting tumors, though, are generally resistant to the negative feedback of high-dose glucocorticoids.

There are multiple versions of the high-dose dexamethasone suppression test [19].

In the 48-hour high-dose dexamethasone suppression test, 2 mg dexamethasone is administered every 6 hours starting at 9 a.m. for 48 hours. A 24-hour urinary free cortisol or serum cortisol (between 8 and 9 a.m.) is measured on the first and third days. Because the multi-day test is quite cumbersome for the patient, an 8 mg overnight dexamethasone suppression test was developed. In this test, 8 mg dexamethasone is administered at 11 p.m. Serum cortisol is measured at 8 a.m., before and after dexamethasone administration. To avoid problems with patient compliance and dexamethasone absorption, an intravenous version was developed. Dexamethasone is infused at 1 mg/hour for 4–7 hours and serum cortisol is measured.

Traditionally, 50% suppression of serum or urine cortisol in response to high-dose dexamethasone was evidence of a pituitary source. However, effectiveness analysis of the 48-hour high-dose and 8 mg overnight dexamethasone suppression test has shown a sensitivity and specificity of 81% and 67%, respectively [45].

Increasing the cut-off of serum and urine cortisol suppression to greater than 80–90% can help optimize specificity, but without complete differentiation between pituitary and non-pituitary sources.

It is not completely understood why some ectopic ACTH-secreting tumors suppress ACTH in response to dexamethasone, but it may be due to the presence of functional glucocorticoid receptors [47]. Regardless, given that the data suggest that the diagnostic accuracy of the high-dose suppression tests is no better than the pre-test probability of a patient having Cushing's disease, most clinicians to do not solely rely on them. However, they may be useful as an adjunct to the CRH stimulation test.

CRH stimulation test

CRH stimulates neoplastic cells in the pituitary gland to synthesize and secrete ACTH and raise serum cortisol. An intravenous bolus of 1 μg/kg CRH is administered. Serum ACTH and cortisol are measured 15 minutes before the CRH bolus, at the time of the CRH bolus, and then at 15, 30, 45, 60, 90, and 120 minutes after CRH. The rise in serum ACTH and cortisol is measured as a percentage increase from baseline after CRH administration to control for variable basal levels. There is no consensus on how to interpret the results, but a rise in serum ACTH greater than 35% or cortisol greater than 20% has been shown to be both sensitive and specific [46]. CRH can, albeit rarely, produce a rise in ACTH from ectopic ACTH-secreting tumor cells. Therefore, complete reliance on this test alone is not recommended either [29].

Bilateral inferior petrosal sinus sampling (BIPSS)

Bilateral inferior petrosal sinus sampling is used to determine the source of ACTH in patients with proven ACTH-dependent Cushing's syndrome when there is discordant or equivocal non-invasive testing or imaging [29]. It is widely considered to be the gold standard for differentiating between pituitary and non-pituitary sources of ACTH. It has also been used to help lateralize pituitary tumors in order to guide neurosurgery, although this is not routinely recommended [19,29].

Venous outflow from the pituitary drains through the cavernous sinuses to the petrosal sinuses and jugular bulb. If a pituitary adenoma is responsible for ACTH production, there will be a detectable ACTH gradient from the petrosal sinus to the periphery. Both inferior petrosal sinuses are catheterized and serum ACTH is obtained initially from each sinus and a peripheral

vein. One hundred micrograms of CRH is administered IV and ACTH is measured at 3, 5, and 10 minutes. A petrosal sinus ACTH to peripheral ACTH ratio of greater than 2:1 at baseline, or greater than 3:1 after CRH administration, confirms a pituitary source [29].

The procedure is invasive and technically challenging. Although rare, it carries the risks of deep vein thrombosis, pulmonary emboli, and vascular brainstem damage [29]. False readings may result from improper set-up and anomalous drainage. Venography can help confirm venous anatomy and catheter placement.

The search begins for an ectopic ACTH-secreting tumor if inferior petrosal sinus sampling does not reveal a pituitary source. Axial imaging with thin-cut multislice CT of the chest, abdomen, and pelvis plus MRI of the thorax and pelvis is the next best initial step [6,29]. Imaging alone may fail to reveal the source in up to 50% of patients, so adjunct modalities, such as ^{111}In-octreotide scintigraphy and [18F]fluorodeoxyglucose positron emission tomography (FDG-PET), are often used in conjunction, but with only variable benefit [6].

Treatment

Treatment of Cushing's syndrome is largely dependent on the source of cortisol. The aim is to normalize cortisol levels, reverse clinical features, and prevent recurrence [48]. Slow glucocorticoid withdrawal is appropriate for iatrogenic Cushing's syndrome. Surgery is first line for all endogenous Cushing's syndromes and may be solely sufficient for ACTH-independent adrenal tumors and ACTH-dependent ectopic tumors. Cushing's disease is more challenging to manage, often requiring a combination of surgical, radio-, and medical therapies.

Iatrogenic Cushing's syndrome

Withdrawal from chronic glucocorticoid therapy comes with a potentially life-threatening risk of adrenal insufficiency. There is a great amount of individual variation in the recovery of the HPA axis after the withdrawal of glucocorticoids, with prolonged adrenal suppression lasting up to several months in select patients [49]. Neither the glucocorticoid dose nor the duration of treatment can reliably predict the extent of suppression [49]. Therefore, there is currently a lack of evidence to recommend any single withdrawal regimen over another. Generally, the goal is to reduce supraphysiologic doses of glucocorticoids to the physiologic dose (5–7.5 mg/day prednisone, 15–20 mg/day hydrocortisone). Often, the glucocorticoid is then exchanged for hydrocortisone because of its short half-life. At this point, the drug is gradually tapered. The patient must pay special attention to the signs and symptoms of adrenal insufficiency (weakness, dizziness, nausea, joint pain), with glucocorticoid replacement instituted if there is a strong clinical suspicion [3,49].

ACTH-independent Cushing's syndrome

ACTH-independent Cushing's syndrome requires adrenalectomy, usually unilateral for single adrenal adenomas or carcinomas, partial bilateral for bilateral adenomas, and bilateral for bilateral micronodular or macronodular adrenal hyperplasia. Laparoscopic adrenalectomy is preferred. This approach is appropriate for adrenal tumors less than 8 cm and without invasion into contiguous structures [50]. Figure 12.4 shows an adrenal nodule visualized during laparoscopy. In these cases, the cure rate is nearly 100% and the overall complication rate is low (9.5%) [51]. The most frequent complication is intra- and postoperative bleeding. If there is invasion, as is often the case for adrenal carcinomas, en bloc resection with an open anterior abdominal approach is preferred. High-dose mitotane can help control metastatic adrenal carcinoma.

Women undergoing unilateral adrenalectomy require postoperative glucocorticoid replacement until the HPA axis recovers. After bilateral adrenalectomy, patients require lifelong glucocorticoid and mineralocorticoid replacement.

Figures 12.1(b) and 12.2(b), show patients with Cushing's syndrome several months after adrenalectomy. There is a marked improvement in their appearance.

ACTH-dependent Cushing's syndrome

Medical management

Medical therapy has a limited role in the management of Cushing's syndrome.

It is used preoperatively to correct severe complications, as an adjuvant therapy after surgery or radiotherapy, and as palliative treatment in patients who are not candidates for surgery [48]. Table 12.6 provides a list of commonly used drugs, their mechanisms, doses, and side effects. These drugs inhibit cortisol synthesis in the adrenal gland.

Ketoconazole, an antifungal agent, and metyrapone have a rapid onset of action and effectively reduce

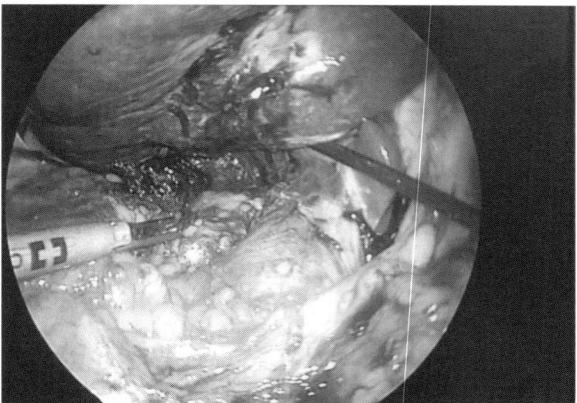

Figure 12.4 An adrenal nodule in a woman with Cushing's syndrome visualized by laparoscopy. In this case, the adrenal nodule was metastatic from a sarcoma. A black and white version of this figure will appear in some formats. For the color version, please refer to the plate section.

cortisol. However, their use is often plagued by the "escape" phenomenon whereby there is a compensatory rise in ACTH in response to hypocortisolemia. This, in addition to their side effects, limits their use as long-term therapeutic options. Aminoglutethimide is similar to ketoconazole and metyrapone, albeit less effective. Mitotane (o,p′-DDD) is an adrenolytic agent predominantly reserved for severe cases of Cushing's disease or malignancy. Although it may take weeks or months to reach full activity, it is effective for long-term cortisol suppression and there is no risk of the "escape" phenomenon. Etomidate is an anesthetic agent and can be administered parenterally for the acute reduction of cortisol levels.

All patients on cortisol-reducing drugs should be monitored to assure that cortisol production is maintained within a normal range. A UFC within the normal range or a cortisol day curve with a mean between 5 and 10 μg/dL indicates normal cortisol production [48].

Surgical management

This section focuses on the surgical management of Cushing's disease. Although ectopic ACTH-secreting tumors are primarily managed surgically, they may be located anywhere and a different approach is required for each location.

Trans-sphenoidal microadenectomy and hypophysectomy

Assuming the tumor can be located, selective adenomectomy is the first-line treatment for Cushing's disease. When performed by an experienced surgeon,

Table 12.6 Medical therapies to lower cortisol in ACTH-dependent Cushing's syndrome

Drug	Mechanism	Dose range	Side effects
Ketoconazole	Inhibits 17–20 lyase, 11β-hydroxylase, cholesterol side-chain cleavage enzyme	200 mg BID to 400 mg TID	Hepatotoxicity, GI symptoms
Metyrapone	Inhibits 11β-hydroxylase	250–1500 mg QID	Hyperandrogenism, hypertension, hypokalemia
Aminoglutethimide	Inhibits cholesterol side-chain cleavage enzyme	250 mg BID to 250 mg TID	Gastrointestinal symptoms, lethargy, ataxia, hypothyroidism, headache, bone marrow suppression, rash
Mitotane (o,p′-DDD)	Adrenolytic; inhibits 11β-hydroxylase, cholesterol side-chain cleavage enzyme	500–3000 mg TID	Gastrointestinal symptoms, weakness, lethargy, leukopenia, hypercholesterolemia
Etomidate	Inhibits 11β-hydroxylase	Bolus of 0.03 mg/kg IV, then infusion starting at 0.1 mg/kg per hour, up to 0.3 mg/kg per hour	Sedation, myoclonus, pain at injection site

BID, twice daily; TID, three times daily; QID, four times daily.
Adapted from [2] with additional information from [48].

patients are often given a long-lasting cure and relief from symptoms. Patients tend to experience significant reductions in blood pressure within 6 weeks, and in body mass index up to 5 years postoperatively [32].

For microadenomas, selective adenomectomy has a remission rate between 65% and 90%, with 5–10% experiencing disease recurrence at 5 years and 10–20% at 10 years [48]. Macroadenomas, unfortunately, have lower initial remission rates (<65%) and higher recurrence rates (12–45%) with a faster onset [48]. If the adenoma cannot be located by sellar exploration, a complete or partial hypophysectomy may offer relief, but at the expense of additional pituitary tissue and higher complication rates.

Following surgery, 30–70% of patients may experience some degree of hypopituitarism [30,52,53]. This may be an especially important complication in patients who are still of reproductive age. For these women, the aggressiveness of removal should be weighed against the preservation of pituitary tissue. Pituitary function should be thoroughly evaluated before and after surgery. Other complications include a transient diabetes insipidus, cerebrospinal fluid leaks, meningitis, sinusitis, and nasal septum perforations.

Following the surgical resection of a pituitary adenoma, a deficiency of cortisol emerges as the HPA axis recovers. This is used as a measure of surgical success in the early postoperative period. Remission is indicated by a morning serum cortisol below 5 µg/dL or UFC <20 µg/24 hours within the first postoperative week [48]. Patients should thereafter be maintained on physiologic doses of glucocorticoids while the HPA axis recovers. Preoperative medical therapy to lower cortisol may preclude an accurate postoperative cortisol assessment, and these patients will not have a true marker of surgical success.

A postoperative morning serum cortisol >5 µg/dL or UFC >100 µg/24 hours warrants further investigation for disease persistence [48]. Small adenomas that cannot be visualized during surgery or locally invasive tumors in the dura of the sella turcica or cavernous sinus may be responsible for persistent disease. Pathologic confirmation of a corticotroph adenoma from the surgical specimen is the first step. Repeat imaging and inferior petrosal sinus sampling, if not done before the initial surgery, may help locate the tumor. Confirmed disease persistence postoperatively requires reoperation, with re-exploration of the sella turcica and walls of the cavernous sinus. Repeat surgery should be done as soon as possible, ideally within the first postoperative week. Reoperation is successful in 50–70% of cases. Patients are more likely to experience hypopituitarism postoperatively, likely owing to more aggressive tissue removal.

Adrenalectomy

Bilateral adrenalectomy will produce an immediate and definitive reduction in cortisol in nearly all patients (Figures 12.1 and 12.2). It is typically reserved for patients who have persistent disease despite repeat trans-sphenoidal surgery and radiotherapy. It may also be appropriate for patients who have intolerance to medical therapy or who wish to maintain fertility [48]. It is becoming a more popular option, as laparoscopic surgery (Figure 12.4) continues to have good outcomes with low associated morbidity. After surgery, patients require lifelong glucocorticoid and mineralocorticoid replacement.

Nelson's syndrome is a locally aggressive pituitary adenoma that will arise in 8–29% of patients with Cushing's disease who undergo bilateral adrenalectomy [54]. Onset is typically within 10 years of surgery, although it has been reported more than 20 years later. High levels of ACTH and hyperpigmentation characterize its onset and patients may present with vision loss from compression of the optic chiasm by the tumor.

Unfortunately, no measures have been validated for predicting the onset of this feared sequela. Starting 3–6 months postoperatively, patients should be monitored for its occurrence with regular serum ACTH and pituitary MRIs [48]. Trans-sphenoidal surgery is the most effective treatment option if the adenoma is caught early and before invasion. Radiotherapy is a helpful adjunct if there is invasion. There is no medical therapy to control Nelson's syndrome. Prophylactic radiotherapy after bilateral adrenalectomy is not currently recommended although at least one study has shown its benefit in reducing the risk and delaying the onset of Nelson's syndrome [48,55].

Pituitary irradiation

Although once a first-line treatment for Cushing's disease, radiotherapy is often now used only for persistent disease after surgery or Nelson's syndrome. It may be an apposite option for locally invasive tumors [48]. As the full effects of radiotherapy are not realized for many months, patients are often placed on medical therapy in the interim period. Additionally, there is a significant risk for hypopituitarism and thus may not be a suitable option for patients wishing to maintain fertility. Other potential complications include optic neuropathy, necrosis, vasculopathies, and a small risk of neoplasia.

The two main methods of radiation are fractionated radiotherapy and stereotactic radiosurgery. Remission rates hover just over 50% with an average time to remission of 3 to 5 years [48]. Fractionated

external beam radiotherapy has historically been the method of choice for Cushing's disease. Radiation is delivered in multiple, fractionated doses daily for 5 to 6 weeks. More than half of patients will experience hypopituitarism afterward. Stereotactic radiosurgery uses a higher, often single, dose of radiation to more precisely target the adenoma. Theoretically, the surrounding tissue has limited exposure to the radiation. Remission rates are similar to conventional radiotherapy with an average time to remission of 12 months. Hypopituitarism will occur in about one-quarter of patients [56]. Radiosurgery with the gamma knife or proton beam has become the more popular method of radiotherapy today because it offers the convenience of a single session, a lower risk of hypopituitarism, and a faster time to remission. Relapse rates up to 20% have been reported [57]. Retreatment is an option for these women.

Conclusion and future directions

Cushing's syndrome is a debilitating condition caused by excessive, long-term exposure to glucocorticoids. Whether the etiology is ACTH-independent or ACTH–dependent, there is increased morbidity and mortality as long as cortisol remains elevated. The diagnosis can be challenging and is often delayed due to the gradual development of symptoms and the clinical overlap with features of a metabolic syndrome. Current research is aimed at refining screening tests to more accurately discriminate true Cushing's syndrome from benign etiologies. Treatment of Cushing's disease is particularly difficult, plagued with both recurrence and undesirable side effects. Advances in pituitary surgery such as endoscopic approaches and peudocapsular dissection promise minimal invasion with lower recurrence rates. Current pharmacological research is focused on developing treatments that will target the pituitary directly. These include somatostatin analogs, dopamine agonists, and peroxisome proliferator-activated receptor gamma (PPARγ) ligands. There is also active research on the blockade of glucocorticoid receptors for the reversal of clinical symptoms. Future studies will help elucidate the long-term efficacy and safety of these drugs, both individually and in combination, for the treatment of Cushing's syndrome. With the advent of personalized medicine, a more individualized approach to the treatment of Cushing's syndrome will likely emerge, with the tailoring of medical and surgical therapies to a woman's specific symptoms.

References

1. Newell-Price J, Bertagna X, Grossman AB, Nieman LK. Cushing's syndrome. *Lancet* 2006;367(9522):1605–1617.

2. Goodwin S, Silverstein J. Cushing's syndrome. In: Henderson KE, ed. *The Washington Manual Endocrinology Subspecialty Consult*. 3rd ed. Philadelphia: Wolters Kluwer Health/Lippincott Williams & Wilkins; 2009.

3. Hopkins RL, Leinung MC. Exogenous Cushing's syndrome and glucocorticoid withdrawal. *Endocrinol Metab Clin North Am* 2005;34(2):371–384, ix.

4. Villanueva RB, Brett E, Gabrilove JL. A cluster of cases of factitious Cushing's syndrome. *Endocr Pract* 2000;6(2):143–147.

5. Cushing H. The basophil adenomas of the pituitary body and their clinical manifestations (pituitary basophilism). 1932. *Obes Res* 1994;2(5):486–508.

6. Alexandraki KI, Grossman AB. The ectopic ACTH syndrome. *Rev Endocr Metab Disord* 2010;11(2):117–126.

7. Kirschner LS, Carney JA, Pack SD, et al. Mutations of the gene encoding the protein kinase A type I-alpha regulatory subunit in patients with the carney complex. *Nat Genet* 2000;26(1):89–92.

8. Groussin L, Jullian E, Perlemoine K, et al. Mutations of the PRKAR1A gene in Cushing's syndrome due to sporadic primary pigmented nodular adrenocortical disease. *J Clin Endocrinol Metab* 2002;87(9):4324–4329.

9. Weinstein LS, Shenker A, Gejman PV, Merino MJ, Friedman E, Spiegel AM. Activating mutations of the stimulatory G protein in the McCune-Albright syndrome. *N Engl J Med* 1991;325(24):1688–1695.

10. Lindholm J, Juul S, Jorgensen JO, et al. Incidence and late prognosis of Cushing's syndrome: a population-based study. *J Clin Endocrinol Metab* 2001;86(1):117–123.

11. Etxabe J, Vazquez JA. Morbidity and mortality in Cushing's disease: an epidemiological approach. *Clin Endocrinol (Oxf)* 1994;40(4):479–484.

12. Bolland MJ, Holdaway IM, Berkeley JE, et al. Mortality and morbidity in Cushing's syndrome in New Zealand. *Clin Endocrinol (Oxf)* 2011;75(4):436–442.

13. Terzolo M, Pia A, Reimondo G. Subclinical Cushing's syndrome: definition and management. *Clin Endocrinol (Oxf)* 2012;76(1):12–18.

14. Catargi B, Rigalleau V, Poussin A, et al. Occult Cushing's syndrome in type-2 diabetes. *J Clin Endocrinol Metab* 2003;88(12):5808–5813.

15. Chiodini I, Torlontano M, Scillitani A, et al. Association of subclinical hypercortisolism with type 2 diabetes mellitus: a case-control study in hospitalized patients. *Eur J Endocrinol* 2005;153(6):837–844.

16. Leibowitz G, Tsur A, Chayen SD, et al. Pre-clinical Cushing's syndrome: an unexpected frequent cause of poor glycaemic control in obese diabetic patients. *Clin Endocrinol (Oxf)* 1996;44(6):717–722.

17. Chiodini I, Mascia ML, Muscarella S, et al. Subclinical hypercortisolism among outpatients referred for osteoporosis. *Ann Intern Med* 2007;147(8):541–548.

18. Chiodini I, Morelli V, Masserini B, et al. Bone mineral density, prevalence of vertebral fractures, and bone quality in patients with adrenal incidentalomas with and without subclinical hypercortisolism: an Italian multicenter study. *J Clin Endocrinol Metab* 2009;94(9):3207–3214.

19. Newell-Price J, Trainer P, Besser M, Grossman A. The diagnosis and differential diagnosis of Cushing's syndrome and pseudo-Cushing's states. *Endocr Rev* 1998;19(5):647–672.

20. Zilio M, Barbot M, Ceccato F, et al. Diagnosis and complications of Cushing's disease: gender-related differences. *Clin Endocrinol (Oxf)* 2014;80(3):403–410.

21. Storr HL, Isidori AM, Monson JP, Besser GM, Grossman AB, Savage MO. Prepubertal Cushing's disease is more common in males, but there is no increase in severity at diagnosis. *J Clin Endocrinol Metab* 2004;89(8):3818–3820.

22. Pecori Giraldi F, Moro M, Cavagnini F, Study Group on the Hypothalamo-Pituitary-Adrenal Axis of the Italian Society of Endocrinology. Gender-related differences in the presentation and course of Cushing's disease. *J Clin Endocrinol Metab* 2003;88(4):1554–1558.

23. Chaidarun SS, Swearingen B, Alexander JM. Differential expression of estrogen receptor-beta (ER beta) in human pituitary tumors: functional interactions with ER alpha and a tumor-specific splice variant. *J Clin Endocrinol Metab* 1998;83(9):3308–3315.

24. Cushing's disease and syndrome. In: Jones H, Scott W, eds. *Hermaphroditism, Genital Anomalies and Related Endocrine Disorders*. 2nd edition. Chapter 24. Baltimore: The Williams & Wilkins Company, 1971; 491–506

25. Nieman LK, Biller BM, Findling JW, et al. The diagnosis of Cushing's syndrome: an Endocrine Society clinical practice guideline. *J Clin Endocrinol Metab* 2008;93(5):1526–1540.

26. Nugent CA, Warner HR, Dunn JT, TYLER FH. Probability theory in the diagnosis of Cushing's syndrome. *J Clin Endocrinol Metab* 1964;24:621–627.

27. Pecori Giraldi F, Pivonello R, Ambrogio AG, et al. The dexamethasone-suppressed corticotropin-releasing hormone stimulation test and the desmopressin test to

distinguish Cushing's syndrome from pseudo-Cushing's states. *Clin Endocrinol (Oxf)* 2007;66(2):251–257.

28. Ross EJ, Linch DC. Cushing's syndrome–killing disease: discriminatory value of signs and symptoms aiding early diagnosis. *Lancet* 1982;2(8299):646–649.

29. Arnaldi G, Angeli A, Atkinson AB, et al. Diagnosis and complications of Cushing's syndrome: a consensus statement. *J Clin Endocrinol Metab* 2003;88(12):5593–5602.

30. Dekkers OM, Horvath-Puho E, Jorgensen JO, et al. Multisystem morbidity and mortality in Cushing's syndrome: a cohort study. *J Clin Endocrinol Metab* 2013;98(6):2277–2284.

31. Graversen D, Vestergaard P, Stochholm K, Gravholt CH, Jorgensen JO. Mortality in Cushing's syndrome: a systematic review and meta-analysis. *Eur J Intern Med* 2012;23(3):278–282.

32. Hassan-Smith ZK, Sherlock M, Reulen RC, et al. Outcome of Cushing's disease following transsphenoidal surgery in a single center over 20 years. *J Clin Endocrinol Metab* 2012;97(4):1194–1201.

33. Clayton RN, Raskauskiene D, Reulen RC, Jones PW. Mortality and morbidity in Cushing's disease over 50 years in Stoke-on-Trent, UK: audit and meta-analysis of literature. *J Clin Endocrinol Metab* 2011;96(3):632–642.

34. Barahona MJ, Sucunza N, Resmini E, et al. Persistent body fat mass and inflammatory marker increases after long-term cure of Cushing's syndrome. *J Clin Endocrinol Metab* 2009;94(9):3365–3371.

35. Geer EB, Shen W, Strohmayer E, Post KD, Freda PU. Body composition and cardiovascular risk markers after remission of Cushing's disease: a prospective study using whole-body MRI. *J Clin Endocrinol Metab* 2012;97(5):1702–1711.

36. Pecori Giraldi F, Ambrogio AG, De Martin M, Fatti LM, Scacchi M, Cavagnini F. Specificity of first-line tests for the diagnosis of Cushing's syndrome: assessment in a large series. *J Clin Endocrinol Metab* 2007;92(11):4123–4129.

37. Carroll T, Raff H, Findling JW. Late-night salivary cortisol for the diagnosis of Cushing syndrome: a meta-analysis. *Endocr Pract* 2009;15(4):335–342.

38. Nieman L. Editorial: The dexamethasone-suppressed corticotropin-releasing hormone test for the diagnosis of Cushing's syndrome: what have we learned in 14 years? *J Clin Endocrinol Metab* 2007;92(8):2876–2878.

39. Elamin MB, Murad MH, Mullan R, et al. Accuracy of diagnostic tests for Cushing's syndrome: a systematic review and metaanalyses. *J Clin Endocrinol Metab* 2008;93(5):1553–1562.

40. Read GF, Walker RF, Wilson DW, Griffiths K. Steroid analysis in saliva for the assessment of endocrine function. *Ann NY Acad Sci* 1990;595:260–274.

41. Doi SA, Clark J, Russell AW. Concordance of the late night salivary cortisol in patients with Cushing's syndrome and elevated urine-free cortisol. *Endocrine* 2013;43(2):327–333.

42. Baid SK, Sinaii N, Wade M, Rubino D, Nieman LK. Radioimmunoassay and tandem mass spectrometry measurement of bedtime salivary cortisol levels: a comparison of assays to establish hypercortisolism. *J Clin Endocrinol Metab* 2007;92(8):3102–3107.

43. Qureshi AC, Bahri A, Breen LA, et al. The influence of the route of oestrogen administration on serum levels of cortisol-binding globulin and total cortisol. *Clin Endocrinol (Oxf)* 2007;66(5):632–635.

44. Invitti C, Pecori Giraldi F, de Martin M, Cavagnini F. Diagnosis and management of Cushing's syndrome: results of an Italian multicentre study. study group of the Italian Society of Endocrinology on the pathophysiology of the hypothalamic-pituitary-adrenal axis. *J Clin Endocrinol Metab* 1999;84(2):440–448.

45. Aron DC, Raff H, Findling JW. Effectiveness versus efficacy: the limited value in clinical practice of high dose dexamethasone suppression testing in the differential diagnosis of adrenocorticotropin-dependent Cushing's syndrome. *J Clin Endocrinol Metab* 1997;82(6):1780–1785.

46. Nieman LK, Oldfield EH, Wesley R, Chrousos GP, Loriaux DL, Cutler GB, Jr. A simplified morning ovine corticotropin-releasing hormone stimulation test for the differential diagnosis of adrenocorticotropin-dependent Cushing's syndrome. *J Clin Endocrinol Metab* 1993;77(5):1308–1312.

47. de Keyzer Y, Lenne F, Auzan C, et al. The pituitary V3 vasopressin receptor and the corticotroph phenotype in ectopic ACTH syndrome. *J Clin Invest* 1996;97(5):1311–1318.

48. Biller BM, Grossman AB, Stewart PM, et al. Treatment of adrenocorticotropin-dependent Cushing's syndrome: a consensus statement. *J Clin Endocrinol Metab* 2008;93(7):2454–2462.

49. Dinsen S, Baslund B, Klose M, et al. Why glucocorticoid withdrawal may sometimes be as dangerous as the treatment itself. *Eur J Intern Med* 2013;24(8):714–720.

50. Young WF, Jr, Thompson GB. Laparoscopic adrenalectomy for patients who have Cushing's syndrome. *Endocrinol Metab Clin North Am* 2005;34(2):489–499, xi.

51. Assalia A, Gagner M. Laparoscopic adrenalectomy. *Br J Surg.* 2004;91(10):1259–1274.

52. Dekkers OM, Biermasz NR, Pereira AM, et al. Mortality in patients treated for Cushing's disease is increased,

compared with patients treated for nonfunctioning pituitary macroadenoma. *J Clin Endocrinol Metab* 2007;92(3):976–981.

53. Rees DA, Hanna FW, Davies JS, Mills RG, Vafidis J, Scanlon MF. Long-term follow-up results of transsphenoidal surgery for Cushing's disease in a single centre using strict criteria for remission. *Clin Endocrinol (Oxf)* 2002;56(4):541–551.

54. Assie G, Bahurel H, Coste J, et al. Corticotroph tumor progression after adrenalectomy in Cushing's disease: a reappraisal of nelson's syndrome. *J Clin Endocrinol Metab* 2007;92(1):172–179.

55. Jenkins PJ, Trainer PJ, Plowman PN, et al. The long-term outcome after adrenalectomy and prophylactic pituitary radiotherapy in adrenocorticotropin-dependent Cushing's syndrome. *J Clin Endocrinol Metab* 1995;80(1):165–171.

56. Ding D, Starke RM, Sheehan JP. Treatment paradigms for pituitary adenomas: defining the roles of radiosurgery and radiation therapy. *J Neurooncol* 2013; 117(3):445–457.

57. Jagannathan J, Sheehan JP, Pouratian N, Laws ER, Steiner L, Vance ML. Gamma knife surgery for Cushing's disease. *J Neurosurg* 2007;106(6):980–987.

Androgen-producing ovarian tumors

Patricia Carney

Introduction

As the attending physician covering the resident gynecology clinic, it was not unusual for a student or a resident to present the case of a woman complaining of hirsutism. When I would ask the presenter for a differential diagnosis, almost always, one of the first responses would be that she could have an androgen-secreting tumor. While true, statistically, this was probably the least likely etiology.

Hyperandrogenism affects 7% of women [1] and can be both physically and psychologically disturbing. Androgen-secreting tumors, however, are an extremely rare cause of this disorder and tend to receive a disproportionate share of attention in the differential diagnosis. Ovarian tumors account for only 0.2% of cases of hyperandrogenism [2].

Androgen-secreting tumors can also arise in the adrenal gland. Briefly, in terms of the adrenal, approximately 60% of adrenocortical carcinomas present with symptoms of hormone excess. The most common presentation is hypercortisolism, though virilization can occur due to secretion of excess androgens. These tumors are extremely rare, with an incidence of only 0.5 to 2 per one million persons per year [3]. The vast majority of adrenal tumors are benign, hormonally inactive adenomas, often found by accident. Generally small (average <4 cm) and unilateral, they are very common and found in approximately 10% of postmortem exams [4]. The majority (85%) are non-functional [5]. There are reported cases of adenomas inducing virilization, but this is extremely unusual [6]. Generally, the biggest risk of these neoplasms is that they can prompt needless surgical exploration.

Moving back to the ovary, a wide variety of tumors may be hormonally active, and even some non-hormonally active tumors can induce the surrounding ovarian stroma to produce androgens [7]. These will be discussed in more detail below. Despite their rarity, these tumors still must be considered, and in most cases eliminated, in the differential diagnosis. How does an androgen-secreting tumor present? What clues do we have? What evaluation should we perform? Before answering these questions, it must be kept in mind that there is no presentation that is *always* indicative of tumor. A detailed history, including information about family history and social behaviors, along with a thorough physical exam are the most valuable tools when determining the etiology of hirsutism/androgen excess [8]. This is no different when evaluating cases caused by tumors.

In this chapter, we will outline the symptoms and signs useful in making the diagnosis of an androgen-secreting tumor. The value of diagnostic testing, including simple lab work, imaging, and invasive procedures, will be reviewed. Next, we will review the wide variety of tumors capable of producing androgens. And finally, we will discuss treatment, which in the majority of cases, is straightforward.

History and physical examination

Algorithms often point to the rapidity and severity of symptoms as clues to an underlying tumor. According to Speroff et al. [9], "a woman who develops hirsutism after the age of 25 and demonstrates very rapid progression or masculinization over several months" has an increased likelihood of an androgen-secreting tumor. Given this presentation, a full evaluation for the presence of a tumor should be performed regardless of serum testosterone levels. Even with this presentation, however, the majority of cases will not be caused by a

Androgens in Gynecological Practice, ed. Leo Plouffe and Botros Rizk. Published by Cambridge University Press. © Cambridge University Press 2015.

tumor. It should also be considered that the concept of an "acute, rapid course" is a relative one. A patient may have a distorted recollection of the process, not noticing the slowly progressive nature of her disorder until suddenly, a certain threshold is reached. In addition, some tumors produce only a moderate amount of androgens; their presentation may be much more indolent and symptoms may develop over several years [10].

Tumors may present any time during the reproductive or post-reproductive years. Very rarely, an ovarian tumor can present with virilization, with or without isosexual precocity, in a child. In contrast, most functional causes of hyperandrogenism show a peripubertal onset [9]. The severity of virilization may support the presence of a tumor. Androgen-secreting tumors can cause severe virilization and "defeminization," with signs such as clitoromegaly, balding, voice change, and decreased breast size [11]. Even marked signs of virilization, however, while suggestive of a tumor, are not pathognomonic. In addition, if a tumor produces only moderate amounts of androgens, the virilization may not be as severe.

A thorough and careful pelvic examination is essential in the evaluation of the patient. The majority of functional ovarian tumors are palpable on examination [6,9]. There can be very small (<2 cm) tumors, particularly those that develop in the hilus of the ovary, that are non-palpable; these may require additional diagnostic testing for localization.

Diagnostic testing

Laboratory

Some guidelines use a serum testosterone level greater than 200 ng/dL as an indication of a potential tumor [7]. This number is by no means absolute, as many androgen-producing tumors are associated with testosterone levels below this level [9,12,13] and the majority of women with levels this high do not have a tumor. This value needs to be placed into context with the additional information obtained from the woman (e.g., the rapidity of symptom onset) and not used as a stand-alone criterion. Postmenopausal women secrete lower levels of androgens; utilizing a cut-off value of 100 ng/dL may be more appropriate in these women [14]. Levens et al. [15] reviewed laboratory data on 136 women under investigation for androgen-secreting tumors. They found the optimal cut-off point for

peripheral serum testosterone in their sample was ≥130 ng/dL. At this level, they found a sensitivity of 93.8% and a specificity of 77.8%.

Measurement of free testosterone is a more sensitive indicator of hyperandrogenism than is total testosterone [8]. For the purposes of screening for an androgen-secreting tumor, however, a routine total testosterone assay is sufficient [9]. The majority of tumors present with clearly abnormal levels of testosterone; the fine discrimination offered by the free testosterone determination is unnecessary in this circumstance.

Levens et al. [15] found no discriminating value to measuring androstenedione. A dehydroepiandrosterone sulfate (DHEA-S) level of ≥700 μg/dL is generally accepted as a marker for abnormal adrenal function and potential tumor [7,9]. The clinical usefulness of measuring DHEA-S has been questioned, especially in the absence of any symptoms suggestive of Cushing's syndrome [9]. When testosterone is significantly elevated, imaging of the adrenal glands may be a preferable, and more definitive, evaluation for the presence of an adrenal tumor than measuring DHEA-S.

Radiology

When an androgen-secreting tumor is suspected based on history and physical exam, imaging of the ovaries and the adrenal glands should be performed. For the adrenals, imaging with CT scan is a sensitive method for detecting even small neoplasms. Adrenocortical malignancies tend to be large (>6 cm) with an irregular outline [7,9]. Adenomas tend to be smaller (<4 cm) and well circumscribed. It must be remembered that the majority of small adenomas are not responsible for hormone production and additional evaluation is needed before concluding surgical removal is necessary [6]. Additional discussion is beyond the scope of this chapter.

Transvaginal ultrasound (TVUS) is the method of choice for evaluating the ovaries [5,9] and the majority of tumors are large enough to be easily visualized. The challenge arises when a tumor is so small, it has not yet altered the size of the ovary. Despite this, gray-scale TVUS with a 7.5 MHz probe and a color Doppler flow study can visualize even very small (~0.5 cm) tumors [14,16].

Virilizing tumors are most commonly solid or mostly solid, depending on the tumor type, and calcification is extremely rare [17]. Even very small tumors can have a different echogenicity than the surrounding ovary. In addition, these tumors tend to have low impedance-to-flow values [16]. The majority of some tumor types (e.g., Sertoli–Leydig and Leydig) may be moderately or abundantly vascularized [18,19]. It should be noted that the high diastolic flow on Doppler indicative of a tumor can also be caused by a corpus luteum; this potential confusion is less of an issue in postmenopausal women and in those who are anovulatory due to their high androgen levels [17].

The value of other imaging techniques is unclear. In general, CT scan is not considered as sensitive as TVUS for the detection of small ovarian tumors. MRI is potentially useful [20]. One study found a small ovarian tumor appeared as an area of high intensity on diffusion and T2-weighted imaging [16]. The effectiveness of MRI in the evaluation of small ovarian tumors, however, has not yet been fully demonstrated.

Another imaging technique that has been explored is fluorine-18-deoxyglucose positron emission tomography/computed tomography (FDG-PET/CT). After a short period of fasting, ^{18}F-FDG is administered intravenously and increased uptake may be demonstrated in the affected ovary [21]. One of the problems with this technique is that benign luteal cysts may also show increased uptake. The literature contains studies both supporting, and not supporting, the use of this technique.

Interventional testing

Selective venous catheterization and hormonal sampling (SVCHS) is an invasive procedure that has been advocated for the identification and lateralization of very small ovarian tumors not visualized on imaging studies. The resolution of imaging studies has improved since the 1980s and the cases in which this would be applicable today are extremely limited.

The procedure involves selectively sampling testosterone levels from both adrenal glands and from both ovaries. This presents a number of challenges. The success rate of sampling all four sites is 27–45% [16]. In particular, the right ovarian vein is difficult to access due to anatomic issues [17]. In some women, the left ovarian vein may have reflux and drain into the right ovarian vein, thus obscuring any difference in unilateral ovarian secretion [22]. Rupture of the ovarian vein is a potentially serious complication [23].

If prior imaging of the adrenal glands does not demonstrate a tumor, it has been advocated that the adrenal veins not be sampled [24]. If, however, a small adenoma is found on CT scan, sampling of the adrenal veins may definitively rule out the adrenal gland as the source of increased androgens and prevent unnecessary surgery [18].

Often, it is known that the source of excess androgen is ovarian and the key goal then is to lateralize the hypersecretion [14]. Levens et al. [15] looked at venous catheterization data from 136 cases. They found that unilateral lesions of the ovary were associated with a higher ipsilateral (mean 17 653 ± 4670 ng/dL) than contralateral (mean 761 ± 150 ng/dL) ovarian effluent concentration. This result, however, was not statistically significant as the high mean ipsilateral value was driven by a small number of extremely high values. They looked at a subset of women with peripheral serum testosterone levels ≥130 ng/dL and calculated the right:left and left:right ovarian testosterone ratios. (In this group, 18 women had right-sided lesions, 14 women had left-sided lesions, and 7 women had bilateral lesions.) They correctly identified 90% of women with a right-sided lesion versus 86% of women with a left-sided or bilateral lesion. When combined with the criterion to identify left-sided lesions, overall, 66% of women were correctly categorized.

If SVCHS is technically difficult to perform or the results are inconclusive, some have advocated the performance of intraoperative ovarian vein sampling [24–26]. If a tumor is very small, the ovary may appear grossly normal on examination and this technique can assist in lateralization of the tumor.

Cases where SVCHS is needed or appropriate are few. It is an invasive procedure with risk attached and should be performed only if it alters patient management (e.g., preserves fertility) [27]. It should be performed in units with special expertise in this area [28]. In a postmenopausal woman with a negative adrenal scan, it may be more appropriate to surgically explore the patient and perform a bilateral oophorectomy, than subject her to a catheterization study, which will be followed by surgery anyway [14].

Summary

The combination of history of rapid onset of symptoms, marked symptomology, and significantly elevated serum testosterone increases the likelihood of a tumor. Even under these circumstances, most women will not have a tumor. The clinician must also be aware of atypical presentations and clinical judgment based on the totality of information is essential. For purposes of ruling out a tumor, in the majority of cases, a routine total testosterone assay is sufficient. The majority of androgen-secreting tumors are palpable on examination. If imaging is required, TVUS is the best method for ruling out ovarian pathology, and CT scan is preferred for adrenal masses. If a tumor is suspected despite negative imaging, and location/lateralization of the tumor will change management, SVCHS may be performed either preoperatively or intraoperatively.

We now turn to the variety of ovarian tumors potentially responsible for the development of virilization in women.

Embryological origin of ovarian tumors

Embryologically, the ovaries initially appear as a pair of longitudinal ridges formed by thickenings of the coelomic epithelium. The proliferation of the cells of the genital ridge into the underlying mesenchyme forms the primitive sex cords, or indifferent gonads. At the same time, primordial germ cells from the yolk sac migrate to and invade the genital ridges. Cells from the adjacent mesonephros also invade the mesenchyme and become closely associated with the germ cells [29,30]. The variety of tumors that can occur in the ovary are all derived from the different cell types that exist in the undifferentiated gonad: mesenchymal, mesonephric, coelomic epithelial, and germ cells.

Any one of these cells, alone or in combination, can become part of an ovarian tumor. In addition, because ovaries and testes develop in a similar fashion during early gestation, cells that are typically associated with testicular tissue, such as Sertoli cells and Leydig cells, can also occur in ovarian tumors [29,31]. Knowing the origin and function of the various cell types of the ovary makes it easier to understand which tumors are likely to have androgen-producing capabilities.

According to the International Agency for Research on Cancer (IARC) classification system [see 29], ovarian tumors can be divided into five general categories. The largest group, epithelial cell tumors, account for 60–70% of all tumors and 90–95% of all ovarian malignancies [29,32]. The stromal component of these tumors can, on very rare occasions, secrete androgens, but this is the exception and will be covered under "idiopathic" tumors [31]. Ovarian sex cord-stromal tumors account for 5–10% of all ovarian tumors and approximately 7% of malignant ovarian tumors [29,32,33]. Approximately 15 000–20 000 new ovarian

sex cord-stromal tumors are diagnosed every year in the United States [32]. Most hormonally active tumors, including those producing androgens, are found in this category. Even in this category, however, the majority of tumors are hormonally inactive. Approximately 70% of sex cord-stromal tumors are benign, inactive fibromas. Potentially active tumors in this category, as well as some of their characteristics, are summarized in Table 13.1 and described below.

The remaining categories in the IARC classification system include germ cell tumors (15–20% overall), metastatic tumors (5%), and tumors classified as "other." With a few notable exceptions, which will be reviewed, these tumors generally are not associated with androgen production.

Sex cord-stromal tumors

Adult granulosa cell tumor (AGCT)

Apart from fibromas, AGCTs are the most common tumor in this category, with an incidence of 0.5–1.5/100 000 and accounting for 2–3% of all ovarian tumors [34]. While generally considered estrogenic tumors in 75–85% of cases [33], 10% are clinically androgenic. Interestingly, AGCTs do not contain all the enzymes necessary for complete steroid synthesis and de novo estrogen production. The predominant secretory products may be androgens synthesized in the thecal cells, which are then aromatized to estrogens [33].

AGCTs most commonly occur in peri- and postmenopausal women, with a median age of 50 [34]. Because of their estrogenic potential, they commonly present with postmenopausal or abnormal vaginal bleeding and may lead to endometrial adenocarcinoma in up to 10% of cases. In preadolescent girls, they are a rare cause of isosexual pseudo-precocity. The majority (>95%) are unilateral and they may range from microscopic size to 30 cm. While the average tumor is 12 cm, it is important to note that in 10–15% of cases, the tumor may be too small to be felt on pelvic exam [35].

Approximately 90% of tumors are diagnosed at stage I and have an 86–96% 10-year survival [35]. Tumors presenting at an advanced stage have only a 26–49% 10-year survival. Due to its generally positive prognosis, unilateral salpingo-oophorectomy with complete staging, particularly in younger women, can be the initial approach [34]. In postmenopausal

Table 13.1 Characteristics of ovarian sex cord-stromal tumors [29,33–36]

	Incidence	Androgen secreting	Other hormones	Age	Laterality	Size (diameter)	Malignant	Comments
Granulosa cell tumors (GCT)	2–3% of all ovarian tumors							
Adult	85–95% of granulosa cell tumors	10%	75–85% estrogenic	Median age 50 years; peri- and postmenopause	>95% unilateral	Average 12 cm (range 0–30 cm)	90% stage I (86–96% survival); >stage I (26–49% survival)	
Juvenile	5–15% of granulosa cell tumors	Rare	80% estrogenic	Average 13 years; 97% below age 30	98% unilateral	Average 12 cm (range 2.5–32 cm)	5% malignant; often fatal	80% in children result in isosexual pseudo-precocity
Thecoma	One-third as frequent as GCT	10% if luteinized; rare other	Commonly estrogenic	Mean 59 years; 84% postmenopausal	95% unilateral	5–10 cm (range 0–15 cm)	3% malignant	
Sclerosing stromal	2–6% of sex cord-stromal	Rare	Rarely estrogenic	Mean 27 years; 80% <30 years	Unilateral	1.5–20 cm	Benign	
Sertoli-stromal								
Sertoli–Leydig	<1% of all ovarian tumors	One-third	Occasionally estrogenic	Mean 25 years (15 years, retiform variant; 35 years well differentiated)	98% unilateral	Average 13.5 cm (majority 5–15 cm)	12% malignant	Can present with sudden and severe virilization
Sertoli	Rare; 4% of Sertoli-stromal tumors	Rare	30% estrogenic	Average 30 years, but can occur any age	Unilateral	Avg. 9 cm (range 4–12 cm)	Most benign	
Steroid cell tumors (lipoid, lipid cell)	0.1% of all ovarian tumors							
Stromal luteoma	23% of steroid cell tumors	12%	Most estrogenic	80% postmenopausal (range 26–74 years)	95% unilateral	0.5–2 cm	Most benign	
Leydig cell	20% of steroid cell tumors	80%	Rarely estrogenic or mixed; DHEA-S; androstenedione	Average 58 years (range 32–78 years)	Unilateral	Average 2.4 cm	Rarely malignant	Androgenic signs may develop over many years
Not otherwise specified	58% of steroid cell tumors	50%	Rarely DHEA-S, androstenedione, estrogen, cortisol, aldosterone	Average 43 years (range 2–80 years)		Average 8.4 cm (range 1.2–45 cm)	25–40%; extra-ovarian spread at time of diagnosis 20%	Androgenic signs may develop over many years

women, total abdominal hysterectomy and bilateral salpingo-oophorectomy (TAH BSO) should be considered [35]. Late recurrences are relatively common, with an average time to recurrence of 5 years. Intervals of up to 30 years are known to have occurred [34]. Recurrences tend to be fatal [35].

Juvenile granulosa cell tumor

Juvenile granulosa cell tumors (JGCTs) are a pathologically distinct entity and account for 5–15% of all granulosa cell tumors. The average age of development is 13 years, with 97% of tumors occurring in women below the age of 30, and 40% below the age of 10. As with AGCTs, the majority of tumors produce estrogen and 80% of tumors occurring in children result in isosexual pseudo-precocity [35]. Rare tumors can, however, be androgenic and present with signs of virilization.

The majority of tumors (98%) are unilateral and range from 2.5 to 32 cm in diameter, with an average of 12 cm [34]. The majority of tumors present as stage IA with a survival rate over 90% [34]. Approximately 5% of tumors behave in a malignant fashion [33]. High-stage tumors are often fatal [34]. Recurrences tend to occur earlier with juvenile tumors than with adult tumors, generally within 2 or 3 years [33,35].

Given the fact that the majority of patients are young and the disease is generally unilateral, a unilateral salpingo-oophorectomy and staging is appropriate for patients presenting with stage IA tumors. Debulking and/or chemotherapy are necessary for advanced disease [34].

Thecoma

Thecomas are one-third as common as GCTs and generally occur in older women, with a mean age of 59 years [34]. While they actually can occur at any age, in one series, 84% of tumors were in postmenopausal women and only 10% occurred in women under the age of 30 [35]. Luteinized thecomas (tumors that contain cells resembling luteinized theca and luteinized stromal cells) tend to occur at a somewhat earlier age, with 30% found in women aged 30 or younger [34]. The majority, however, still occur in postmenopausal women.

Thecomas are most commonly estrogenic and can cause endometrial adenocarcinoma in up to one-fourth of cases [35]. Rarely, a thecoma can produce virilization [36]. Luteinized thecomas display a somewhat different hormonal predilection, with approximately half the tumors manifesting estrogenic properties and

approximately 10% androgenic. In one series of four patients, half of the patients had symptoms of virilization [35]. The majority of tumors (95%) are unilateral [34] and they are usually benign [35]. Tumors can range from microscopic size to 15 cm, with the majority 5–10 cm in diameter [34]. In most cases, a unilateral salpingo-oophorectomy can be curative; however, TAH BSO is recommended for post-reproductive aged women [35].

Sclerosing stromal tumor

These tumors are very rare and account for only 2–6% of sex cord-stromal tumors [34]. The average age of occurrence is 27 [35], and 80% of tumors occur in women under the age of 30 [34]. They can very rarely produce either estrogen or androgens. Tumors are generally unilateral and benign, and can range in size from 1.5 to 20 cm. Due to these characteristics, and the generally young age at presentation, a unilateral oophorectomy is usually appropriate treatment [34].

Sertoli-stromal cell tumors

Tumors within this category contain varying amounts of Sertoli and Leydig cells. Due to their cellular composition, these tumors are also referred to as "androblastomas" [34]. Despite this name, many of these tumors are hormonally inactive [34]. They are divided into two major categories: Sertoli–Leydig cell tumors and Sertoli cell tumors.

Sertoli–Leydig cell tumors represent less than 1% of all ovarian tumors [34,35], though they are the tumors often considered to be responsible for virilization. Indeed, virilization can be sudden and severe, leading to amenorrhea, hirsutism, clitoromegaly, deepening of the voice, loss of female secondary sexual characteristics, and even erythrocytosis [35]. Approximately one-third of tumors can result in virilization [34]. Occasionally, tumors are estrogenic, but 50% have no endocrine manifestations [34,35].

There are five major histologic categories of Sertoli–Leydig cell tumor, which affect prognosis and age of occurrence. Across all categories, the average age of occurrence is 25 years. Overall, 98% of tumors are unilateral [34] and can range from microscopic size to very large; the majority of tumors are 5–15 cm in diameter with an average of 13.5 cm [35]. Approximately 12% of all tumors are malignant [34], though the majority present as stage I [29]. Tumors presenting as stage IA, particularly in younger women, can be treated with a unilateral salpingo-oophorectomy [35]. Recurrences tend to occur early, with 66% presenting within 1 year and only 6.6% after 5 years [35]. Recurrences tend to be confined to the pelvis and abdomen [35]. In a series of 64 tumors of intermediate and poor differentiation, the corrected 10-year survival rate was 92% [33]. According to Chen et al. [29], however, the death rate for tumors presenting at greater than stage I is close to 100%.

After removal of the tumor, normal menses usually resume in approximately 4 weeks. Hirsutism will diminish to some extent. Unfortunately, clitoromegaly and deepening of the voice are less likely to resolve [35].

Sertoli cell tumors account for approximately 4% of Sertoli-stromal tumors. Tumors contain cell proliferations that resemble the rete testes and rete ovarii [29]. These tumors are usually non-functional; however, 30% may be estrogenic and can result in isosexual precocity in young girls. Tumors may rarely be androgenic [34], though this may be debatable [33]. Tumors can occur at any age; however, the average is 30 years. Most tumors are clinically benign [34] and present as stage I [29]. The average size is 9 cm in diameter, with a range of 4–12 cm [34]. Unilateral salpingo-oophorectomy is usually appropriate.

Steroid cell tumors

Tumors in this group are also referred to as lipid cell tumors or lipoid cell tumors [35]. They are extremely rare and account for only 0.1% of all ovarian tumors [34]. Tumors are composed entirely of hormone-producing cells. These can include cells that resemble lutein cells (stromal luteoma), testicular Leydig cells (Leydig cell tumor), or adrenal cortical cells [29,34].

Stromal luteomas represent 23% of all steroid cell tumors. They are typically centered in the ovarian parenchyma [34] and 90% of cases are associated with stromal hyperthecosis in the same or in the contralateral ovary [35]. Most of the tumors are estrogenic and 60% of cases present with abnormal bleeding [35]. It is not known whether the tumor directly secretes estrogen or is in fact secreting androgens that are peripherally converted [35]. Approximately 12% of tumors are considered androgenic and, in rare cases, can be associated with acanthosis nigricans [34].

Most stromal luteomas are unilateral (95%) and benign. Approximately 80% occur in postmenopausal women, though they have an age range of 26–74 years. While tumor excision or unilateral salpingo-oophorectomy is considered curative, these

small tumors (0.5–2 cm in diameter) may be difficult to locate [34].

Leydig cell tumors represent approximately 20% of all steroid cell tumors. There are two types of tumor: hilus cell tumor, which accounts for the majority of Leydig cell tumors, and nonhilar tumors [34]. These tumors are hormonally active and 80% present with androgenic manifestations. Rarely, they can be estrogenic or display mixed features. In addition to testosterone, these tumors may also produce androstenedione and DHEA-S [33].

Average age of diagnosis for Leydig cell tumors is 58 years, with a range from 32 to 78 years [34]. Unlike Sertoli–Leydig cell tumors, the androgenic manifestations of Leydig cell tumors are less severe and can occur very gradually, sometimes over the course of several years [35]. These unilateral tumors are relatively small, with an average diameter of only 2.4 cm. Occasionally, they can be palpable on pelvic exam [35]. Most tumors are situated in the hilum of the ovary and may be difficult to locate. While generally benign, malignant tumors leading to death have been known to occur [33].

The largest group of steroid cell tumors belongs in the category of *steroid cell tumor, not otherwise specified*, accounting for 58% of the total. Approximately half of these tumors are androgenic. As with Leydig cell tumors, the androgenic changes may occur over many years [35]. In addition to testosterone, these tumors have been known to secrete 17-ketosteroids, androstenedione, estrogen, and cortisol. They may occasionally present as Cushing's syndrome. [34]. There is at least one known case of aldosterone production [35].

Average age of occurrence for this group of tumors is 43 years, with a range of 2–80 years. Average tumor size is 8.4 cm, with a range of 1.2–45 cm. Unlike many of the other sex cord-stromal tumors, a significant proportion of these tumors (25–40%) are malignant and extra-ovarian spread is found at time of diagnosis in 20% of cases [34,35]. In younger patients with stage IA tumors, a unilateral salpingo-oophorectomy and staging may be appropriate; otherwise, a TAH BSO with full staging should be performed. Late recurrences after 5 years are common and recurrences may be associated with virilization [34].

Other ovarian tumors associated with androgen production

According to Young and Scully [35], "almost every ovarian tumor has been reported to be associated with steroid hormone production." These tumors may be benign or malignant; malignancy may be primary or metastatic. The elevation in steroid levels may be subclinical; however, some tumors may result in clinical syndromes, such as virilization. While clinically and pathologically fascinating, these cases are very rare. Generally, these tumors may become functional because they contain luteinized stromal cells that resemble lutein or Leydig cells. The tumors can be divided into three broad categories: germ cell tumors containing syncytiotrophoblast cells, tumors with functioning stroma during pregnancy, and idiopathic.

Germ cell tumors containing syncytiotrophoblast cells

Dysgerminomas containing syncytiotrophoblast cells have been associated with luteinization of the ovarian stroma and production of steroid hormones. This includes at least one known case of virilization in a postpubertal woman. In addition to dysgerminomas, other germ cell tumors that produce human chorionic gonadotropin (hCG), such as choriocarcinomas, embryonal carcinomas, mixed primitive germ cell tumors, and polyembryomas, can stimulate the contralateral ovary with subsequent development of luteinized follicles capable of secreting hormones [35].

Tumors with functioning stroma during pregnancy

Hormonally active tumors diagnosed during pregnancy generally will be androgenic. Given the high estrogen levels of pregnancy, an estrogen-secreting tumor would be impossible to detect based on its hormone production [35]. Though extremely rare, luteinized stromal cells may occur with several different types of tumor during pregnancy, the most common in conjunction with Krukenberg tumors, mucinous cystic tumors, and Brenner tumors [35].

In pregnancy, due to the high levels of hCG, all ovaries show a certain degree of hyperplasia and luteinization of the theca cells of atretic follicles and focal stromal luteinization. In rare cases, "focal clones grow into expansile nodular masses that appear to emerge from the hyperplastic lutein cells" and develop into what are termed "luteomas of pregnancy" [36]. Luteomas are not true neoplasms and it is unclear at what point an exaggerated thecal cell hyperplasia and luteinization becomes a "luteoma," but the cellular characteristics of these masses appears to be different from the normal hyperplasia of pregnancy.

Luteomas are usually bilateral [34], and can grow up to 15 cm in diameter. They are highly dependent on hCG and quickly regress after delivery; even large masses can disappear within weeks [36]. A luteoma of pregnancy may be difficult to distinguish from a heavily luteinized thecoma, which will not spontaneously regress, but a thecoma is usually unilateral [34]. Despite the large amount of androgens luteomas are capable of producing, most patients will not experience virilization. This is due to high levels of sex hormone-binding globulin (SHBG) during pregnancy and the ability of the placenta to aromatize the androgens to estrogen. If for some reason the placenta has a decreased capacity for aromatization, or androgen levels exceed the amount the placenta is capable of clearing, then virilization of both the mother and a female fetus can occur [36]. According to Speroff et al. [9], luteomas cause maternal virilization in 35% of cases.

Idiopathic

There are case reports of almost all ovarian tumors causing steroid production, including androgens. These tend to be in postmenopausal women, although it is unclear whether this is because women in this age group tend to have more tumors overall, or because of the high luteinizing hormone (LH) levels in postmenopausal women [35]. Mucinous tumors are most commonly implicated in this unusual group. Germ cell tumors have already been covered. Brenner tumors may also cause androgen production. Metastatic tumors containing mucinous cells, including metastatic lobular carcinoma of the breast and Krukenberg tumors, have been found to cause virilization [35]. Hayes et al. [37] noted a case of a simple ovarian cyst associated with hilus cell hyperplasia in the cyst wall, leading to a 10-year history of progressive virilization. It was thought that pressure exerted by the cyst led to the hyperplasia. Finally, regardless of the nature of the cyst itself, if it undergoes torsion, the resultant enlargement of the ovary due to the partial torsion of the meso-ovarium can lead to luteinized stromal cells in the edematous areas and virilization in some unusual cases [34].

Treatment

Surgery is the mainstay of treatment for ovarian tumors. Removal of just the ovary, or performance of hysterectomy and bilateral oophorectomy, depends on many factors, including the likelihood of malignancy, the age of the patient, and the desire for fertility. Appropriate consultation is important. Overall, malignancy is unusual; in a series of 194 masculinizing tumors, only 5% were malignant [38]. In cases where a hysterectomy is not performed, consideration should be given to the performance of hysteroscopy or endometrial sampling, given the high rate of concomitant endometrial cancer with hormonally active tumors [28]. Other than surgical removal and other procedures as warranted by the tumor type, treatment of the symptoms of androgen excess due to a tumor is no different than from other causes. These treatments are discussed elsewhere. The advantage is that the source of the androgen production is removed and additional symptoms will not continue to develop. As previously noted, while hirsutism may resolve to some extent, if virilization has been severe and includes clitoromegaly and/or voice change, these changes may be permanent.

Occasionally, a situation arises where surgery is refused or is not feasible (e.g., due to the underlying health of the patient). Several documented cases demonstrate successful suppression of testosterone secretion by tumors and presumed tumors [28,39,40]. Gonadotropin-releasing hormone (GnRH) suppression has also been used in cases of persistently elevated testosterone despite tumor removal [41]. The mechanism of action may be due to suppression of a gonadotropin-dependent secretion of testosterone by the tumor or through direct effect of GnRH analog on tumor cells [28,40]. GnRH binding sites and LH receptors have both been identified on Leydig tumors [28].

Conclusion

Getting back to the resident or student in the clinic, my next question is: what characteristics lead you to believe this woman's symptoms may be caused by an androgen-secreting tumor? It is important to include tumor in the initial differential diagnosis. Careful review of all of the data will usually quickly eliminate this as a possibility and eventually lead to a functional cause. In those rare cases it does not, a fascinating array of ovarian tumors can be responsible for elevated androgens. In most cases, removal of the tumor can quickly resolve many of the symptoms that brought the woman to the clinic in the first place.

References

1. Bailey AP, Schutt AK, Carey RM, et al. Hyperandrogenism of ovarian etiology: utilizing differential venous sampling for diagnosis. *Obstet Gynecol* 2012; 120: 476–479.

2. Azziz R, Sanchez IA, Knochenhauer ES, et al. Androgen excess in women: experience with over 1000 consecutive patients. *J Clin Endocrinol Metab* 2004; 89: 453–462.

3. Luton J-P, Cerdas S, Billaud L, et al. Clinical features of adrenocortical carcinoma, prognostic factors, and the effect of Mitotane therapy. *N Engl J Med* 1990; 322: 1195–1201.

4. Cook DM, Loriaux DL. The incidental adrenal mass. *Am J Med* 1996; 101: 88–94.

5. Tekin O, Bünyamin I, Avci Z, et al. Hirsutism: common clinical problem or index of serious disease? *Med Gen Med* 2004; 6: 56.

6. Loriaux DL. An approach to the patient with hirsutism. *J Clin Endocrinol Metab* 2012; 97: 2957–2968.

7. Practice Committee of the American Society for Reproductive Medicine. The evaluation and treatment of androgen excess. *Fertil Steril* 2006; 86: S241–S247.

8. Escobar-Morreale HF, Carmina E, Dewailly D, et al. Epidemiology, diagnosis and management of hirsutism: a consensus statement of the Androgen Excess and Polycystic Ovary Syndrome Society. *Hum Reprod Update* 2012; 18: 146–170.

9. Speroff L, Glass RH, Kase NG. *Clinical Gynecologic Endocrinology and Infertility*, Sixth Edition. Philadelphia: Lippincott Williams & Wilkins, 1999.

10. Rosenfield RL. Clinical practice. Hirsutism. *N Engl J Med* 2005; 353: 2578–2588.

11. Escobar-Morreale, HF. Diagnosis and management of hirsutism. *Ann NY Acad Sci* 2010; 1205: 166–174.

12. Surrey ES, de Ziegler D, Gambone JC, Judd HL. Preoperative localization of androgen-secreting tumors: clinical, endocrinologic, and radiologic evaluation of ten patients. *Am J Obstet Gynecol* 1988; 158: 1313–1322.

13. Derksen J, Nagesser SK, Meinders AE, et al. Identification of virilizing adrenal tumors in hirsute women. *N Engl J Med* 1994; 331: 968–973.

14. Lobo RA. Ovarian hyperandrogenism and androgen-producing tumors. *Endocrinol Metab Clin North Am* 1991; 20: 773–805.

15. Levens ED, Whitcomb BW, Csokmay JM, Nieman LK. Selective venous sampling for androgen-producing ovarian pathology. *Clin Endocrinol* 2009; 70: 606–614.

16. Nishiyama S, Hirota Y, Udagawa Y, et al. Efficacy of selective venous catheterization in localizing a small

androgen-producing tumor in ovary. *Med Sci Monit* 2008; 14: CS9–12.

17. Outwater EK, Marchetto B, Wagner BJ. Virilizing tumors of the ovary: imaging features. *Ultrasound Obstet Gynecol* 2000; 15: 365–371.

18. Ozgun MT, Batukan C, Turkyilmaz C, et al. Selective ovarian vein sampling can be crucial to localize a Leydig cell tumor: an unusual case in a postmenopausal woman. *Maturitas* 2008; 61: 278–280.

19. Demidov VN, Lipatenkova J, Vikhareva O, et al. Imaging of gynecological disease (2): clinical and ultrasound characteristics of Sertoli cell tumors, Sertoli-Leydig cell tumors and Leydig tumors. *Ultrasound Obstet Gynecol* 2008; 31: 85–91.

20. Annaiah TK, Webb B, Buckingham S. Magnetic resonance imaging: a valuable aid to the diagnosis of a rare ovarian tumour – steroid secreting tumour of the ovary not otherwise specified. *J Obstet Gynaecol* 2010; 30: 77–78.

21. Prassopoulos V, Laspas F, Vlachou F, et al. Leydig cell tumour of the ovary localized with Positron Emission Tomography/Computed Tomography. *Gynecol Endocrinol* 2011; 27: 837–839.

22. Hiromura T, Nishioka T, Nishioka S, et al. Reflux in the left ovarian vein: analysis of MDCT findings in asymptomatic women. *Am J Roentgenol* 2004; 183: 1411–1415.

23. Sorensen R, Moltz L, Schwartz U. Technical difficulties of selective blood sampling in the differential diagnosis of female hyperandrogenism. *Cardiovasc Intervent Radiol* 1986; 9: 75–82.

24. Dickerson RD, Putman MJ, Black ME, et al. Selective ovarian vein sampling to localize a Leydig cell tumor. *Fertil Steril* 2006; 84: 19–22.

25. Bohlmann MK, Rabe T, Sinn H-P, et al. Intraoperative venous blood sampling to localize a small androgen-producing ovarian tumor. *Gynecol Endocrinol* 2005; 21: 138–141.

26. Regnier C, Bennet A, Malet D, et al. Intraoperative testosterone assay for virilizing ovarian tumor topographic assessment: report of a Leydig cell tumor of the ovary in a premenopausal woman with an adrenal incidentaloma. *J Clin Endocrinol Metab* 2002; 87: 3074–3077.

27. Petersons CJ, Burt MG. The utility of adrenal and ovarian venous sampling in the investigation of androgen-secreting tumours. *Intern Med J* 2011; 41: 69–70.

28. Klotz RK, Müller-Holzner E, Fressler S, et al. Leydig-cell tumor of the ovary that responded to GnRH-analogue administration: case report and review of the literature. *Exp Clin Endocrinol Diabetes* 2010; 118: 291–297.

29. Chen VW, Ruiz BR, Killeen JL, et al. Pathology and classification of ovarian tumors. *Cancer (supplement)* 2003; 97: 2631–2642.

30. Sadler TW. *Langman's Medical Embryology*, Fifth Edition. Baltimore: Williams & Wilkins, 2005.

31. Scully RE. Classification of human ovarian tumors. *Environ Health Perspect* 1987; 73: 15–24.

32. Lee-Jones L. Ovarian tumours: an overview. *Atlas Genet Cytogenet Oncol Haematol* 2003. http://AtlasGeneticsOncology.org/Tumors/OvarianTumOverviewID5231.htm (accessed January 25, 2013).

33. Fox H. Sex cord-stromal tumours of the ovary. *J Pathol* 1985; 145: 127–148.

34. Deavers MT, Oliva E, Nucci M. Sex cord-stromal tumors of the ovary. In: Nucci MR, Oliva E, Eds. *Gynecologic Pathology: A Volume in Foundations in Diagnostic Pathology*. China: Elsevier Churchill Livingstone. 2007; 445–500.

35. Young RH, Scully RE. Sex cord-stromal, steroid cell, and other ovarian tumors with endocrine, paraendocrine, and paraneoplastic manifestations. In: Kurman RJ, Ed. *Blaustein's Pathology of the Female Genital Tract*, Fifth Edition. New York: Springer-Verlag, Inc., 2002: 905–966.

36. Sternberg WH, Dhurandhar HN. Functional ovarian tumors of stromal and sex cord origin. *Hum Pathol* 1977; 8: 565–582.

37. Hayes FJ, Sheahan K, Rajendiran MB, McKenna TJ. Virilization in a postmenopausal woman as a result of hilus cell hyperplasia associated with a simple ovarian cyst. *Am J Obstet Gynecol* 1997; 176: 719–720.

38. Ireland K, Woodruff JD. Masculinizing ovarian tumors. *Obstet Gynecol Surv* 1976; 31: 83–111.

39. Cheng V, Doshi KB, Falcone T, Faiman C. Hyperandrogenism in a postmenopausal woman: diagnostic and therapeutic challenges. *Endocr Pract* 2011; 17: e21–e25.

40. Picón, MJ, Lara JI, Sarasa JL, et al. Use of a long-acting gonadotropin-releasing hormone analogue in a postmenopausal woman with hyperandrogenism due to a hilus cell tumour. *Eur J Endocrinol* 200; 142: 619–622.

41. Wang P-H, Chao H-T, Lee W-L. Use of a long-acting gonadotropin-releasing hormone agonist for treatment of steroid cell tumors of the ovary. *Fertil Steril* 1998; 69: 353–355.

Chapter

14 Cognitive and behavioral impact of androgen disorders in females: learning from complete androgen insensitivity syndrome and congenital adrenal hyperplasia

Amy B. Wisniewski

Introduction

Many behaviors differ between girls and boys, and women and men [1]; however, it is often difficult or even impossible to determine the causes of these differences. For example, if a girl prefers to play with dolls instead of rough-and-tumble wrestling, is this because she was taught by her family and friends to prefer dolls? Perhaps her proclivity for dolls results from actions of her genes and/or hormonal environment during critical periods of her development. Among the vast majority of children and adults who exhibit sexually dimorphic behaviors, influences such as learning/socialization, sex chromosomes, and hormone exposure cannot be teased apart for independent consideration. For this reason, the study of sexually dimorphic behaviors in people affected by disorders of sex development (DSD) is useful. Two categories of DSD that are particularly instructive about the origins of sexually dimorphic behaviors in females are complete androgen insensitivity syndrome (CAIS) and congenital adrenal hyperplasia (CAH) due to 21-hydroxylase deficiency. Therefore, these specific types of DSD are the focus here.

Investigating genetic and hormonal effects on sex differences in cognition

DSD collectively refers to a group of conditions in which discordance between genetic, hormonal, and anatomic sex exists in a person at birth [2]. For example, a person with 46,XY DSD may possess a female phenotype throughout life despite the presence of a Y chromosome in every cell of her body. A specific type of 46,XY DSD, CAIS, results when a person who possesses a male-typical chromosome complement (46,XY) and testes develops a female phenotype due to her end-organ insensitivity to androgens [3]. In contrast, a person with 46,XX DSD may possess an ambiguous or male phenotype despite the possession of a female chromosome complement. One type of 46,XX DSD, CAH due to 21-hydroxylase deficiency, results when a person with ovaries develops a masculinized phenotype caused by excessive androgen production due to an inability to produce the enzyme 21-hydroxylase [4].

Thus, females with CAIS and CAH allow for the independent assessment of male-typic genetic or hormonal influences on sexually dimorphic behaviors in affected girls and women. In other words, if women with CAIS behave similarly to unaffected women on measures of sexually dimorphic behaviors, then the possession of a Y chromosome alone is insufficient to support the expression of male-typic behaviors in humans. In contrast, if women with CAH behave similarly to men, then early androgen exposure is sufficient to support male-typic behavior. Importantly, by comparing women with CAIS to those with CAH, it is possible to determine the impact of possessing a Y chromosome versus early androgen exposure independent of socialization and learning on the expression of sexually dimorphic behaviors in humans (Table 14.1).

Sexually dimorphic cognitive behaviors

Cognitive behaviors that show strong sex differences in humans include gender identity (GI), gender role (GR), and performance on some visuo-spatial tasks

---

Androgens in Gynecological Practice, ed. Leo Plouffe and Botros Rizk. Published by Cambridge University Press. © Cambridge University Press 2015.

Table 14.1 Comparisons of women with CAIS or CAH to unaffected women and men allow for study of the impact of sex chromosomes and early androgen exposure on sexually dimorphic behaviors in people.

	Unaffected women	Women with CAIS	Women with CAH	Unaffected men
Socialization/Learning	Female	Female	Female	Male
Sex chromosomes	Female	Male	Female	Male
Early androgen exposure	Female	Female	Male	Male

[1]. We will first consider GI and GR, as these reflect some of the largest sex differences observed in human behaviors. GI is the internal experience a person has of being female or male, while GR refers to overt behaviors expressed by people, such as toy preferences, that society categorizes as feminine or masculine. The vast majority of women studied report a female GI [5]; however, GR is more variable among girls and women and standards for feminine and masculine behaviors are continuously evolving over time and across cultures [6]. Most girls and women with a female GI and GR have been socialized as female, possess a 46,XX chromosome complement, and experience minimal early androgen exposure and thus, no phenotypic masculinization. Thus, studies of GI and GR in women with CAIS and CAH are needed to understand if, and how, sex chromosomes and early androgen exposure influence these behaviors apart from learning.

Girls and women with CAIS overwhelmingly report female-typical GI and GR [7–9]. Therefore, female patterns of gender can develop in people despite their possession of a Y chromosome. The majority of girls and women with CAH also report a female GI, but the incidence of male GI development in this group is higher than in unaffected women or women with CAIS. Furthermore, girls and women with CAH are frequently labeled as tomboys because they often prefer male-typic toys and play activities, as well as hobbies and professional interests [10]. Thus, there is a greater impact of early androgen exposure, compared with possessing a Y chromosome in the absence of androgenic effects, on GI. Studies of girls and women with CAH show that the impact of androgens on GR is even greater than the relationship between androgens and GI [11]. However, it is not completely clear whether androgen exposure during critical periods associated with CAH directly masculinizes GR via steroid action at the central nervous system, or whether androgens indirectly impact behavior of affected girls and women by influencing how society responds to their masculinized appearance (i.e., hirsutism and masculinized external genitalia) [12].

Another type of cognition that differs significantly between females and males, and also is thought to be influenced by early androgen exposure, is visuo-spatial processing [13,14]. Examples of visuo-spatial processing that elicit large sex differences in favor of males include judgment of line angle orientation and mental rotation of objects. In a study of visuo-spatial cognition in females affected by CAH, the severity of the condition (i.e., a surrogate marker for excessive prenatal androgen exposure) was positively associated with performance on these male-biased, visuo-spatial tasks. Additionally, excessive postnatal androgen exposure also predicted better visuo-spatial performance of women with CAH [15]. Thus, the influence of androgens on cognition in humans is not restricted to critical periods *prior to* birth, but can also extend throughout childhood. In contrast to GI and GR, visuo-spatial abilities have largely been understudied in girls and women with CAIS. Experiments that employ animal models of XY females, although also few in number, indicate that XY female rodents perform overwhelmingly like their XX counterparts [16; but see 17]. Therefore, similar to GI and GR, androgen exposure during critical periods of early development exerts a greater influence on male-typic, visuo-spatial performance than possession of a sex chromosome complement.

Although clear evidence exists from nonhuman animal studies and clinical investigations of people with DSD for a role of early androgen exposure to influence performance on visuo-spatial tasks, there is also evidence for environmental and learning influences as well. For example, access to education for girls and women ameliorates the male-biased sex difference in a type of visuo-spatial processing known as line angle judgment [18; but see 19 for contrasting results]. Additionally, the male advantage for mental rotation is eliminated following a series of eight 45-minute training sessions for children in first grade [20], indicating that modest training is sufficient to override visuo-spatial advantages associated with early androgen exposure. To what extent education and training might modulate early androgen influences on GI and GR is less amenable to study; however, the differences observed between masculine and feminine GR across history and cultures imply that they too are impacted by learning as well as hormonal influences.

Sex differences in mental health status

Similar to cognition, aspects of mental health differ in prevalence, incidence, and expression between males and females [21–23]. For example, women are more likely to report and seek treatment for depression [24], whereas men have a greater likelihood of being treated for addiction [25]. Because the female preponderance for depressive symptoms is revealed during adolescence, a role for sex steroids is thought to exist [26]. In contrast, the greater tobacco, alcohol, and illicit drug use evident in men does not occur until later adulthood [27], suggesting that perhaps societal influences over time differentially influence addictive behaviors in men and women.

In our long-term outcomes study of 14 women with CAIS, only 1 participant reported problems with depression and addiction [7]. Additional studies conducted by others also report no increase in depressive symptoms or suicidal ideation associated with having CAIS [28–30]; however other investigations of mental health in this group reveal increased suicidal tendencies [8,31,32]. Thus, while it is inconclusive at this time whether women who possess a 46,XY chromosomal complement as a result of CAIS do, or do not, experience increased depressive symptoms, it seems that a male karyotype does not by itself protect women from depression. The variability in reported depressive symptoms across studies is likely due to methodological problems, such as small sample sizes and participation bias [33]. Studies of addictive behavior in women with 46,XY DSD are needed, but to date have not been performed, to understand what role genes encoded on the Y chromosome may play in addiction.

Mental health studies of women with CAH can reveal whether or not prenatal androgen exposure offers protection against depressive symptoms or predisposes a person to addictive behaviors (the male-typical pattern for each). Similar to CAIS, no studies have been conducted to date on tobacco, alcohol, or illicit drug use in women with CAH. Regarding depressive symptoms, women with CAH are more likely to report a history of suicide attempts and psychopharmacologic

treatment for depression than unaffected women [29, 30]. Greater virilization in women with CAH at the time of their diagnosis is positively associated with depressive symptoms, suicide attempts, and psychopharmacologic treatment. Of interest, a similar association between depressive symptoms and polycystic ovary syndrome has also been demonstrated [34], suggesting that signs of virilization such as hirsutism (commonly evident in women with CAH or polycystic ovary syndrome) predispose women to experience depression. Furthermore, surgical treatment and the functional outcomes associated with such procedures are also associated with self-reported depressive symptoms in women with CAH [30]. Thus, it remains to be seen how androgen exposure, either directly or indirectly, exerts an effect on depression in women with CAH. Additionally, surgical treatment and subsequent functional outcomes for women with CAH who are born with ambiguous or even male-typic external genitalia can also impact mental health [30]. For example, a recent study of women with CAH reported distress associated with the timing and type of surgical treatment received for CAH, as well as sexual function following those surgeries [30]. Thus, this is an important reminder that differences noted in cognition or mental health among women with CAH, CAIS, or their unaffected counterparts may be due to medical and/or surgical therapies and not whether a Y chromosome or early androgen exposure is present.

Summary and conclusions

Significant sex differences exist between girls and boys, and women and men, for gender identity and role, certain visuo-spatial abilities, and particular aspects of mental health, such as depression and addiction. Some of these sex differences are evident early, while others emerge at puberty or later. For most individuals, it is impossible to tease apart the influence of sex chromosomes, sex hormones, the environment, and socialization on the development and expression of sexually dimorphic behaviors. Thus, the nature/nurture debate persists in the context of female/male differences in behavior.

People affected by DSD allow for the study of sexually dimorphic behaviors, while accounting for genes, hormones, environment, and learning. For example, girls and women with CAIS provide an opportunity to study sexually dimorphic behaviors in females who possess a 46,XY chromosomal complement.

Conversely, girls and women with CAH provide an opportunity to study behaviors in females who have been exposed to male hormones pre- and post-natally. Therefore, studies of GI and GR, spatial cognition, and mental health in these groups of women can reveal important genetic and hormonal mechanisms underlying higher level brain function in humans.

Concerning cognition, women with CAIS overwhelmingly develop a female GI and GR. XY female rodents perform similarly to their XX counterparts on visuo-spatial tasks. Therefore, there is converging evidence across species as well as cognitive tasks for little or no influence of the Y chromosome on these sexually dimorphic behaviors. Too few data exist to conclude at this point that the presence of a Y chromosome exerts no influence on depression or addictive behaviors, and these are areas that warrant greater study in people with 46,XY DSD. A greater role for early androgen exposure exists for GI, GR, and performance on visuo-spatial tasks as revealed by studies of women with CAH. In contrast, androgens do not protect women with CAH from depression. Again, too little information about addiction and androgens is available to make conclusions at this point. In conclusion, studies of women with DSD inform us that the early hormonal milieu, but not genetic sex, contributes to sexually dimorphic behaviors of importance to humans.

References

1. Hines M. Sex-related variation in human behavior and the brain. *Trends Cogn Sci* 2010; 14: 448–56.

2. Hughes IA, Houk C, Ahmed SF, Lee PA. Consensus statement on management of intersex disorders. *Arch Dis Child* 2006; 91(7): 554–63.

3. Hughes IA, Werner R, Bunch T, Hiort O. Androgen insensitivity syndrome. *Semin Reprod Med* 2012; 30: 432–42.

4. Witchel SF, Azziz R. Congenital adrenal hyperplasia. *J Pediatr Adolesc Gynecol* 2011; 24: 116–26.

5. Bao A, Swaab DF. Sexual differentiation of the human brain: relation to gender identity, sexual orientation and neuropsychiatric disorders. *Front Neuroendocrinol* 2011; 32: 214–26.

6. Bussey K, Bandura A. Social cognitive theory of gender development and differentiation. *Psychol Rev* 1999; 106: 676–713.

7. Wisniewski AB, Migeon CJ, Meyer-Bahlburg HFL, et al. Complete androgen insensitivity syndrome: long-term medical, surgical and psychosexual outcome. *JCEM* 2000; 85: 2664–69.

8. Hines M, Ahmed SF, Hughes IA. Psychological outcomes and gender-related development in complete androgen insensitivity syndrome. *Arch Sex Behav* 2003; 32: 93–101.

9. Hooper HT, Figueiredo BC, Pavan-Senn CC, et al. Concordance of phenotypic expression and gender identity in a large kindred with a mutation in the androgen receptor. *Clin Genet* 2004; 65: 183–90.

10. Meyer-Bahlburg HF. Brain development and cognitive, psychosocial, and psychiatric functioning in classical 21-hydroxylase deficiency. *Endocr Dev* 2011; 20: 88–95.

11. Berenbaum SA, Beltz AM. Sexual differentiation of human behavior: effects of prenatal and pubertal organizational hormones. *Front Neuroendocrinol* 2011; 32: 183–200.

12. Jordan-Young RM. Hormones, context, and "Brain Gender": a review of evidence from congenital adrenal hyperplasia. *Soc Sci Med* 2012; 74: 1738–44.

13. Kimura D. Sex hormones influence human cognitive pattern. *Neuro Endocrinol Lett* 2002; Suppl 4: 67–77.

14. Puts DA, McDaniel MA, Jordan CL, Breedlove SM. Spatial ability and prenatal androgens: meta-analyses of congenital adrenal hyperplasia and digit ratio (2D:4D) studies. *Arch Sex Behav* 2008; 37: 100–11.

15. Mueller SC, Temple V, Oh E, et al. Early androgen exposure modulates spatial cognition in congenital adrenal hyperplasia (CAH). *Psychoneuroendocrinology* 2009; 33: 973–80.

16. Stavnezer AJ, McDowekk CS, Hyde LA, et al. Spatial ability of XY sex-reversed female mice. *Behav Brain Res* 2000; 112: 135–43.

17. Rizk A, Robertson J, Raber J. Behavioral performance of tfm mice supports the beneficial role of androgen receptors in spatial learning and memory. *Brain Res* 2005; 1034: 132–38.

18. Caparelli-Daquer EM, Oliveira-Souza R, Moreira Filho PF. Judgment of line orientation depends on gender, education and type of error. *Brain Cogn* 2009; 69: 116–20.

19. Lippa RA, Collaer ML, Peters M. Sex differences in mental rotation and line angle judgments are positively associated with gender equality and economic development across 53 nations. *Arch Sex Behav* 2010; 39: 990–97.

20. Tzuriel D, Egozi G. Gender differences in spatial ability of young children: the effects of teaching and processing strategies. *Child Dev* 2010; 81: 1417–30.

21. Piccinelli M, Wilkinson G. Gender differences in depression: critical review. *Br J Psychiatry* 2000; 177: 486–92.

22. Becker JB, Hu M. Sex differences in drug abuse. *Front Neuroendocrinol* 2008; 29: 36–47.

23. Eisenberg D, Hunt J, Speer N. Mental health in American colleges and universities: variation across student subgroups and across campuses. *J Nerv Ment Dis* 2013; 201: 60–7.

24. Vasiliadis HM, Gagne S, Jozwiak N, Preville M. Gender differences in health service use for mental health reasons in community dwelling older adults with suicidal ideation. *Int Psychogeriatr* 2012; 5: 1–8.

25. Kelly JF, Hoeppner BB. Does Alcoholics Anonymous work differently for men and women? A moderated multiple-mediation analysis in a large clinical sample. *Drug Alcohol Depend* 2013; 130(1–3): 186–93.

26. Nolen-Hoeksema S, Girgus JS. The emergence of gender differences in depression during adolescence. *Psychol Bull* 1994; 115: 424–43.

27. Brady KT, Randall C. Gender differences in substance use disorders. *Psychiatr Clin North Am* 1999; 22: 241–52.

28. Warne G, Grover S, Hutson J, et al. A long-term outcome study of intersex conditions. *J Pediatr Endocrinol Metab* 2005; 18: 555–67.

29. Johannsen TH, Ripa CPL, Mortensen E, Main KM. Quality of life in 70 women with disorders of sex development. *Eur J Endocrinol* 2006; 155: 877–85.

30. Fagerholm R, Mattila AK, Roine RP, et al. Mental health and quality of life after feminizing genitoplasty. *J Pediatr Surg* 2012; 47: 747–51.

31. Diamond M, Watson LA. Androgen insensitivity syndrome and Klinefelter syndrome: sex and gender considerations. *Child Adolesc Psychiatr Clin N Am* 2004; 32: 93–101.

32. Schutzmann K, Brinkmann L, Schacht M, Richter-Appelt H. Psychological distress, self-harming behavior, and suicidal tendencies in adults with disorders of sex development. *Arch Sex Behav* 2009; 38: 16–33.

33. Wisniewski AB, Mazur T. 46,XY DSD with female or ambiguous external genitalia at birth due to androgen insensitivity syndrome, 5α-reductase-2 deficiency, or 17β-hydroxysteroid dehydrogenase deficiency: a review of quality of life outcomes. *Int J Pediatr Endocrinol* 2009;2009:567430.

34. Rodrigues CE, Ferreira Lde L, Jansen K, et al. Evaluation of common mental disorders in women with polycystic ovary syndrome and its relationship with body mass index. *Rev Bras Ginecol Obstet* 2012; 34: 442–6.

Testosterone replacement in the aging male: lessons learned from the Women's Health Initiative

David Muram and Craig F. Donatucci

Introduction

Aging in men leads to decline in multiple physiologic and psychological parameters. The physical effects of aging broadly recognized include weight gain, loss of muscle mass, and skin changes. The decline is a universal event for men as they age and is due in part to decreasing androgen production. The reduction of serum levels of total, free, and bioavailable testosterone throughout aging has been well documented [1–3]. While the majority of men remain asymptomatic, some complain of symptoms that are attributed to this drop in androgens. The symptoms are often vague and nonspecific like depression, loss of energy, and sexual dysfunction. The symptoms secondary to decreased androgen production are so common that some would say that they are not reflective of a medical condition, but rather the normal consequence of aging. This perspective may cause clinicians to question the wisdom of androgen replacement therapy in aging men, even in men suffering a significant decrease in quality of life [2]. Occasionally, skeptics of androgen replacement therapy point to the surprising results of the Women's Health Initiative (WHI) studies to warn of similar, possibly unrecognized risks associated with hormone replacement therapy (HRT) in aging men. One may ask whether such comparison is appropriate. To address this question, a brief review of HRT in women and the WHI studies is necessary [4].

Menopause, women, and hormone replacement therapy

Menopause is the permanent cessation of the menstrual cycle. Menopause is tied to a specific date, defined as the day after the final menstrual period, and determined retrospectively once 12 months have gone by with no menstrual flow at all. In the United States, the average age of a woman having her last period, menopause, is 51; however, some women have their last period in their forties, though most have it later in their fifties. This transition is typically neither sudden nor abrupt; rather it tends to occur over a period of years, and is a natural consequence of aging. For some women, the accompanying signs and symptoms which may occur during the transition period can significantly disrupt daily activities and sense of well-being [5]. In most women, hormone replacement therapy (HRT) can alleviate the symptoms of menopause.

The issue with HRT is not with its use for the relief of menopausal symptoms, but with the contention that it may have additional health benefits for women. For many decades, HRT was widely recommended not only for the treatment of menopause and menopausal symptoms, but also for the primary prevention of osteoporosis and heart disease. The wide use of HRT in postmenopausal women was bolstered by observational studies that demonstrated that postmenopausal users of HRT had substantially lower rates of cardiovascular events than non-users [6]. HRT became the standard of care, as a form of "preventive medicine," with many physicians believing it reduced risk of cardiovascular disease and improved women's well-being and quality of life. This "primary prevention" aspect was considered especially beneficial, given the longer life expectancy and greater risk of disabling disease after menopause. Despite concerns about possible bias in the observational studies toward healthy women and reports of increased breast cancer risk among women taking HRT, many physicians nevertheless accepted the argument that women should take HRT because of

Androgens in Gynecological Practice, ed. Leo Plouffe and Botros Rizk. Published by Cambridge University Press. © Cambridge University Press 2015.

the much higher prevalence of cardiovascular disease compared with the risk for breast cancer [7].

The Women's Health Initiative

The Women's Health Initiative (WHI) was a long-term national health study that has focused on strategies for preventing heart disease, breast and colorectal cancer, and osteoporotic fractures in postmenopausal women [4]. This 15-year project involved 161 808 women aged 50–79, and is still ongoing. Randomized placebo-controlled clinical trials evaluating hormone therapy were a major part of the WHI. There were two such trials. The "WHI-E+P" compared conjugated equine estrogen, plus progestin (Prempro, Wyeth) to placebo in healthy postmenopausal women. The second, the "WHI-CEE" trial, enrolled menopausal women without a uterus and compared conjugated equine estrogen versus placebo.

The estrogen plus progestin trial was stopped early in July 2002 after investigators concluded that the risk–benefit profile was not consistent with the requirements for a viable intervention for primary prevention of chronic diseases. They reported an increased risk of myocardial infarction, breast cancer, stroke, deep venous thrombosis (DVT), and pulmonary embolism (PE). There was a decreased risk of colorectal cancer and fewer fractures. In announcing the decision, Dr. Claude Lenfant added that "the cardiovascular and cancer risks of estrogen plus progestin outweigh any benefits – and a 26 percent increase in breast cancer risk is too high a price to pay, even if there were a heart benefit. Similarly, the risks outweigh the benefits of fewer hip fractures" [4].

The estrogen only trial was discontinued in March 2004. In announcing the decision, Dr. Barbara Alving said that "after careful consideration of the data, NIH has concluded that with an average of nearly 7 years of follow-up completed, estrogen alone does not appear to affect (either increase or decrease) heart disease, a key question of the study. At the same time, estrogen alone appears to increase the risk of stroke and decrease the risk of hip fracture. It has not increased the risk of breast cancer during the time period of the study" [4].

A similar British study evaluated the risk–benefit ratio of HRT, which was used by around one-third of women aged 50–64 in Britain [8]. Data from the Million Women Study showed that current use of HRT was associated with an increased risk of incident and fatal breast cancer, and that the effect was substantially greater for estrogen–progestin combinations than for other types of HRT [9]. These findings were similar to those observed in the WHI study. Following publication of the Million Women Study results, the use of HRT in Britain declined significantly from 107/1000 (95% confidence interval [CI] 104–110) in 2000 to 87/1000 (95% CI 84–89) in 2003 [10].

Following publication of the WHI data, physicians were urged to use HRT with caution. The investigators recommended that women with normal rather than surgical menopause should take the lowest feasible dose of HRT for the shortest possible time to minimize their risk. The United States Preventive Task Force concluded that the harms of long-term therapy are greater than the potential benefits [11].

Current expert opinion suggests that systemic hormone therapy is the most effective treatment for most menopausal symptoms, including vasomotor symptoms and vaginal atrophy. Although the risks of venous thromboembolic events and ischemic stroke increase with either estrogen therapy or estrogen and progestogen therapy, the risk is rare in the 50- to 59-year-old age group. An increased risk of breast cancer is seen with 5 years or more of continuous estrogen with progestogen therapy, possibly earlier with continuous use since menopause. While treatment duration may vary based on an individual's risk and benefit assessment, in general, the lowest dose of hormone therapy should be used for the shortest amount of time to manage menopausal symptoms [12].

Androgen deficiency in the aging male

As in women, there is a decline in the production of testosterone as men age, though this decline differs from estrogen both in abruptness of onset and totality. The decline of serum testosterone levels is a gradual, age-related process resulting in an approximate 1% annual decline after age 30. In cross-sectional and longitudinal studies of men aged 30 or 40 years and above, total, bioavailable, and free testosterone concentrations fall with increasing age. The observed decline in bioavailable and free testosterone was even more pronounced than the decline in total testosterone levels [13–15].

Multiple mechanisms likely influence the decline of testosterone levels in aging men. Lower testosterone levels may result from reduced testicular responses to gonadotropin stimuli with aging, coupled with a progressive decline in the sensitivity of the hypothalamic–pituitary feedback mechanism, resulting in incomplete compensation for the fall in total and free testosterone levels. Whether the age-dependent decline in androgen levels leads to health problems in older men is being debated vigorously [16–19].

Three large population-based studies document the prevalence of hypogonadism in men. The Boston Area Community Health (BACH) Survey is a random community sample of men age 30–79, which showed that 24% had total testosterone (TT) <300 ng/dL, but only 5.6% of men were symptomatic [3]. The Hypogonadism in Males (HIM) study evaluated men aged 45 and older who came for care to participating primary care clinics. In this population, 38.7% of men had TT <300 ng/dL. Once again, many of these men were asymptomatic and the clinic visit was for other medical conditions [20]. The European Male Aging Study (EMAS) is a random sample of European men aged 40–79 that showed that almost 25% had some form of hypogonadism [21].

Testosterone replacement therapy (TRT) is available and was first approved in 1972 by the Food and Drug Administration (FDA) as a treatment for the signs and symptoms of male hypogonadism. Since the introduction of transdermal testosterone gel in 2000, the market for TRT has increased dramatically for men with age-related decline of testosterone. Is testosterone replacement therapy in the offing for all aging males with low testosterone? Is such a broad androgen replacement strategy based on a positive risk–benefit assessment? What information is available on the effects of TRT in men?

Unlike in women where estrogen replacement was widely believed to carry health benefits (until disproven by the WHI project), testosterone replacement has long been suspected of carrying a significant risk in men. The landmark paper of Huggins and Hodges associating testosterone and prostate cancer outcome in men with metastatic disease, for which they were awarded the Nobel Prize in Medicine, gave rise to fears that testosterone could induce or accelerate prostate cancer in hypogonadal men receiving replacement [22]. Despite some mitigating evidence from small observational series and clinical trials, continuing concern about risk, along with a rapidly expanding number of men receiving replacement therapy led to efforts to clarify the true risk/reward relationship for TRT in aging men.

The Institute of Medicine (IOM) in 2002 conducted an independent assessment of clinical research on testosterone therapy and made recommendations for future direction for the field of TRT for the aging male [1]. In its review of the literature the committee identified "only" 31 placebo-controlled trials of testosterone therapy in older men. The placebo-controlled trial with the largest sample size involved 108 participants and the duration of therapy in 25 of the 31 trials was 6 months or less. Only one placebo-controlled trial lasted longer than a year. Therefore, assessments of risks and benefits have been limited, and uncertainties remain about the value of this therapy for older men.

The paucity of credible information is reflected in current practice guidelines informing physicians on hypogonadism and testosterone replacement therapy. Unlike HRT (that prior to WHI was recommended for all menopausal women), the Endocrine Society recommends against screening for androgen deficiency in the general population, and clinicians should measure serum testosterone levels in patients with clinical manifestations or symptoms of low testosterone.

Hypogonadism is a clinical condition characterized by consistent symptoms and signs and unequivocally low serum testosterone levels. Patients with hypogonadism are treated with testosterone to reverse the signs and symptoms of androgen deficiency; there is no indication for testosterone to be used as primary prevention of other medical conditions. These recommendations are similar to those of HRT for menopausal women issued after the WHI results became public [23].

As with other therapies, careful evaluation of patients is necessary before initiating a TRT regimen. Patients with breast or prostate cancer should not be treated with TRT. It is recommended that TRT not be given to patients with hematocrit above 50%, untreated severe obstructive sleep apnea, severe lower urinary tract symptoms (International Prostate Symptom Score [IPSS] score of ≥19), uncontrolled or poorly controlled heart failure, or in those desiring fertility. Urologic consultation is recommended in patients with palpable prostate nodule or induration, prostate-specific antigen (PSA) ≥4 ng/mL, or PSA ≥3 ng/mL in men at high risk of prostate cancer [23].

Symptoms and signs of hypogonadism

Testosterone has different functional effects; the exact expression depends on the target tissue and the point in the male life cycle that testosterone exposure occurs. In fetal life, testosterone is responsible for male sexual development, while during adolescence testosterone plays a role in male pubertal development. In contrast, symptoms of androgen deficiency in adult men are nonspecific and are often modified by age, comorbid illnesses, severity and duration of androgen deficiency, and variations in androgen sensitivity. The commonly described signs and symptoms of testosterone deficiency may represent the more severe end of the androgen deficiency spectrum since they are based on the clinician's experience and derived from men who have sought medical assistance. In population-based surveys of community-dwelling, middle-aged and older men, low libido, erectile dysfunction (ED), and hot flushes, as well as less specific symptoms such as fatigue or lack of energy, mood changes, and reduced physical performance, were associated with low testosterone levels. In these surveys the prevalence of symptomatic androgen deficiency was around 6% [3,23–25].

Androgens exert their biological effect through a single intracellular receptor that is present in the reproductive tract as well as in many non-reproductive tissues, including bone, skeletal muscle, brain, liver, kidney, and adipocytes. Some actions of androgens are mediated by local enzymes such as 5α-reductase and aromatase [26].

An association between low testosterone (low T) and impaired fasting glucose, insulin resistance, type 2 diabetes, and the metabolic syndrome was seen in patients both with and without ED. Low T is associated not only with the metabolic syndrome, but also with each of its individual components, i.e., type 2 diabetes, visceral obesity, insulin resistance, dyslipidemia, and high blood pressure. The trinucleotide repeat sequence (CAG repeat) length could modulate the impact of testosterone on the metabolic risk factors, and men with longer CAG length and low testosterone concentrations show the highest risk of incident metabolic syndrome. Prospective studies demonstrate that a low T at baseline predicts the development of type 2 diabetes. Similarly, diabetes or metabolic syndrome at baseline may herald the onset of hypogonadism [27].

Bone mass in men declines linearly with age starting at around age 30 and the decrease in trabecular mass is greater than that in cortical mass. Hypogonadism is an important risk factor for osteoporosis in men. Androgens stimulate the proliferation of bone cells in vitro [27]. The incidence of rapid hip-bone loss was higher in men with low testosterone level [27]. The role of aromatization of testosterone to estradiol (E_2) is predominant in the protective effect of testosterone

on bone loss. The association of bone mineral density (BMD) with E_2 is much stronger than that of BMD with testosterone, probably due to the fact that bone has both androgen and estrogen receptors[27]. While BMD in the hip increased significantly in patients receiving testosterone, little is known about the amount of testosterone required to maintain bone mass in men. Furthermore, it is not known whether the beneficial effect of testosterone on bone metabolism is due to the androgen itself or to the estrogen produced from it [28].

Testosterone stimulates erythropoiesis. Administration of testosterone increases reticulocyte counts, hemoglobin concentrations, and bone marrow erythropoietic activity through direct and indirect mechanisms. Administration of androgens results in a significant increase in the levels of endogenous erythropoietin [26]. In addition, experimental data suggest that androgens may also directly enhance erythropoiesis by stimulating erythropoietic stem cells [26].

Frailty is a syndrome of aging characterized by loss of physiologic reserve and generalized vulnerability to disability. Testosterone retains nitrogen, which accounts for its anabolic properties. Prior to puberty, testosterone promotes muscle growth. After puberty, the androgen receptor in striated muscle is downregulated and is saturated with physiologic concentrations of circulating testosterone. Low T was associated with incident frailty as well as worsening frailty, suggesting that age-related changes in blood androgens may contribute to development and progression of frailty [27]. Another study reported that patients with low T demonstrated lower scores on vitality than eugonadal men, but no differences were detected in psychological well-being between hypogonadal and eugonadal men [2].

While androgens have a key role in stimulating and maintaining sexual function in men [29–30], the role of androgens in increasing the frequency and quality of erections is unclear [31–33]. In normal young men, suppression of serum testosterone concentrations to the range associated with castration reduces sexual desire, sexual fantasies, and spontaneous erections [34]. In a random sample of 414 community-dwelling men, bioavailable testosterone was significantly correlated with both erectile function and libido [2].

The threshold testosterone level below which symptoms of androgen deficiency and adverse health outcomes occur is not known. Individual symptoms may have different thresholds (i.e., low libido versus low energy), and individual men may not exhibit the same degree of symptomatology at similar testosterone levels. Such individual thresholds may be the result of variable androgen receptor sensitivity related to CAG repeats. The accepted threshold below which symptoms are more likely to occur is approximately 300 ng/dL and more men may become symptomatic as testosterone levels decline even further [23]. The existence of a measurable threshold at which symptoms appear was further supported by Kelleher et al. [31]. Men may reach a highly reproducible specific serum concentration of testosterone that triggers symptoms of androgen deficiency, though this level differs widely among individuals. Although this threshold varied from very low to values above the lower limit of the eugonadal reference range, on average it approximated the lower limit of the eugonadal reference range for young men, 300 ng/dL [23].

While most clinicians accept the existence of a threshold value for serum testosterone symptom improvement, it may not be 300 ng/dL for everyone. A recent study that investigated the relationship between symptoms and testosterone serum levels offered additional evidence; some improvement was seen in individuals receiving TRT who did not reach testosterone ≥300 ng/dL, suggesting that their individual threshold was reached even if they did not reach the prespecified threshold [27].

While no clear relationship between decreasing testosterone and symptoms has been determined, Zitzmann et al. did observe a general trend between decreasing testosterone and increasing prevalence of groups of symptoms [24]. For example: below testosterone concentrations of 432 ng/dL (15 nmol/L), loss of libido and loss of vigor increased; below testosterone concentrations of 288 ng/dL (10 nmol/L), depression, disturbed sleep, lack of concentration, and diabetes mellitus type 2 occurred in significantly more men; and only at testosterone concentrations below 230 ng/dL (8 nmol/L) did ED occur. These findings suggest that symptom-specific testosterone thresholds occur and that the spectrum of complaints of testosterone deficiency cannot be related to a uniform threshold of testosterone concentration, but that thresholds vary with the various symptoms of testosterone deficiency [24]. As previously mentioned, CAG repeats modulate the androgen receptor activity. Because there is a genetic heterogeneity, individual patients may have different thresholds at which the various symptoms of hypogonadism appear [27].

Patient-reported outcome tools

There are several tools available to collect information from patients about presence and severity of symptoms. However, there is no universally accepted patient-reported outcome (PRO) questionnaire to help assess all the symptoms associated with hypogonadism, and different questionnaires were used in studies of hypogonadism. Thus, it is difficult to compare data across studies.

The Androgen Deficiency of Aging Men (ADAM) questionnaire was evaluated for efficacy in diagnosing hypogonadism. In one trial, 78 men were identified as androgen deficient by the ADAM questionnaire; however, testosterone levels confirmed the diagnosis in only 27 cases (28.1%). Another study evaluated the Aging Male Symptoms (AMS) score in 348 men. Serum testosterone levels correlated with the andrologic symptoms of AMS; the only correlation of the total AMS score was with age [2].

Sexual function and mood changes can be assessed by the Psychosexual Daily Questionnaire (PDQ) [28]. The PDQ is a patient self-reporting instrument designed for the assessment of sexual function and mood on a daily basis. It covers three domains of sexual function including sexual desire, enjoyment and performance; sexual activity; and mood. Data are collected for a period of 7 days prior to and following TRT.

Sexual desire and sexual enjoyment are rated on a 7-point Likert-type scale with 0 indicating none and 7 indicating very high. Sexual performance includes self-assessment of satisfaction of erection, also rated using the 7-point Likert-type scale described above, and percent full erection, which varied from 10% to 100%. The latter two items are left blank if the subject did not have an erection.

Sexual activity is assessed using a checklist format. Subjects record whether they had any of the following on each of the 7 days: sexual daydreams, anticipation of sex, flirting, sexual interactions with partner, erection, masturbation, intercourse, orgasm, and ejaculation. The value is recorded as 0 (none) or 1 (any) for analysis. The weekly value for the sexual activity items is the sum of the number of "any" responses for the week. The sexual activity score is then calculated as the average of the weekly values for all of the items with a possible score of 0 to 7 for each sexual activity.

Positive mood and negative mood are recorded using a 7-point Likert scale with 0 indicating lowest and 7 indicating highest. Positive mood parameters include: alert, full of pep, friendly, and well. Negative mood parameters include: angry, irritable, sad or blue, tired, and nervous.

Pre-treatment evaluation

All TRTs carry an absolute contraindication for use in men with a history of breast or prostate cancer. Thus, prior to initiation of TRT, the clinician should assess patients for the risk of each, in particular for the risk of prostate cancer as it is markedly more common in older men suffering from androgen deficiency. Urologic consultation is recommended for patients with a palpable prostate nodule or prostatic induration. In men at low risk of prostate cancer, a PSA greater than 4 ng/mL should prompt the need for consultation; for men at high risk of prostate cancer (family history in blood relative), the PSA threshold is lowered to ≥3 ng/mL. Because testosterone raises hematocrit, TRT should not be prescribed for men whose baseline hematocrit is ≥50%.

Other conditions where TRT may aggravate pre-existing symptoms include: untreated severe obstructive sleep apnea, severe lower urinary tract symptoms (IPSS >19), and uncontrolled or poorly controlled heart failure. TRT may suppress spermatogenesis through feedback inhibition of pituitary follicle-stimulating hormone (FSH), which could possibly lead to adverse effects on semen parameters, including sperm count.

Type of therapy

Various preparations are available for men who require TRT, from parenteral to topical; no oral agent is approved for testosterone replacement in the USA. Testosterone therapy can be initiated with any of the available products, taking into consideration the patient's preference, pharmacokinetics of testosterone formulation, treatment burden, and cost. Current commercially available preparations of testosterone (with the exception of the 17α-alkylated ones) are well tolerated and effective [27]. Some preparations, such as oral and injectable testosterone undecanoate, are not available in the USA. Patients should be informed about the advantages and disadvantages of each preparation prior to selection of a treatment modality.

The goal of therapy is to alleviate the symptoms of hypogonadism, and in most patients, symptom relief is achieved when serum testosterone levels are within the mid-normal range [23]. To ensure proper dosing, serum testosterone levels should be measured periodically and dose can be adjusted to achieve serum testosterone levels within the normal range. If the serum testosterone concentration is below the normal range, the daily dose should be increased, and when the serum testosterone concentration exceeds the normal range, the daily dose should be decreased. If the serum testosterone concentration consistently exceeds the normal range at the lowest dose, therapy should be discontinued or the patient should be treated with a different preparation [23].

Once therapy is instituted, some of the signs and symptoms can improve starting as soon as 3 weeks after initiation of therapy. Other signs and symptoms require longer treatment duration. Wang et al. [34] reported that sexual function and mood improved by day 30 of TRT. This improvement corresponded to serum testosterone reaching eugonadal levels, but was not related to further increases, suggesting that once a serum testosterone level threshold was achieved, normalization of sexual function occurred. Increasing serum testosterone levels to the upper normal range did not further improve sexual motivation or performance. Administration of androgens in men with ED, but with normal gonadal function, is usually not beneficial [35].

Testosterone improves BMD, mediated through estrogen action on the bone. BMD at the lumbar spine increases, but effects at the femoral neck are less certain. Effects on bone were detectable after 6 months of therapy, and continued for up to 3 years [36].

Zitzmann et al. evaluated effects of TRT in an observational study of over 1400 men who received therapy for up to 12 months. The investigators reported improvements of the overall levels of sexual desire/libido, improvements in the overall level of vigor/vitality, as well as quality of sleep. Moderate hot flushes declined from 17.5% at baseline to 4% at the time of injection 5, and severe hot flushes from 7.6% at baseline to 0.5% at endpoint. Complaints of excessive sweating showed a similar pattern of improvement. Patients reported improvement in erectile function. The percentage of patients reporting moderate, severe, or extremely severe ED declined from 65% at baseline to 19% after therapy [37].

Testosterone may improve cognition, sense of well-being, and mood in young males with organic testosterone deficiency. Clinical trials have shown that normalizing testosterone levels decreases anger, irritability, sadness, tiredness, and nervousness, and improves energy level, friendliness, and sense of well-being. In addition, some trials of TRT, which assessed fatigue or energy in men with androgen deficiency, reported a significant improvement of these symptoms compared with placebo. The improvement of such psychological symptoms and of fatigue seems to occur as early as after 3 to 4 weeks [27].

Randomized clinical trials studying the effects of testosterone on cognition in non-cognitively impaired aging males have shown varying results: approximately half the subjects demonstrated improvement in spatial cognition and memory [27]. Beneficial results have also been observed in two of the randomized clinical trials of testosterone in men with Alzheimer's disease or cognitive decline, but most of these studies were of short duration and enrolled only a small number of patients [2]. Conversely, the majority of the trials that addressed the effect of testosterone on mood of aging males yielded negative results.

Sometimes it may be difficult to differentiate depression from testosterone deficiency based only on the symptoms. In population-based studies, lower serum testosterone levels have been associated with mild symptoms of depression. In one small trial of men with depression, testosterone seemed to demonstrate short-term efficacy in augmenting the effects of antidepressants. However, when the same group of investigators repeated the trial in a larger group of depressed men, testosterone therapy showed no benefit. It thus appears that testosterone supplementation is not effective as a treatment for men with major depression [2,27].

Studies in patients using high doses of anabolic steroids (testosterone abuse) have confirmed a correlation between hostility and aggressiveness after administration of such substances. However, similar correlation between testosterone and aggressiveness was not found in controlled trials in hypogonadal men receiving TRT. In a recent trial in 28 young eugonadal men, one injection of 1000 mg of the long-acting testosterone undecanoate increased serum testosterone concentration above 30 nmol/L without increasing aggressive behavior or inducing any changes in nonaggressive or sexual behavior [38].

Changes in body composition were seen in a 3-year study of 96 hypogonadal men aged 65 and older who were randomized to receive either testosterone or placebo. Patients receiving testosterone showed a significant increase in lean body mass and a reduction in fat mass when compared with patients receiving placebo

[39]. Similar observations were reported by other investigators [28,40].

Isidori et al. performed a systematic review and meta-analysis of placebo-controlled studies of patients receiving TRT published between 1975 and 2004. Evaluable patients (N = 656) from 17 studies were included in the analysis [41]. Two hundred and eighty-four men were randomized to testosterone, 284 to placebo, and 88 were treated in a crossover manner. The average length of therapy across the studies was 3 months (range 1–36). Results of the meta-analysis revealed that in hypogonadal men (baseline testosterone level <12 nmol/L; 346 ng/dL), TRT led to an improvement in nocturnal erections, sexual thoughts and motivation, number of successful intercourses, scores for erectile function, and overall sexual satisfaction. TRT had no effect on sexual function in eugonadal men and was equivalent to placebo. Investigators also observed changes in metabolic parameters. Blood pressure declined slightly, lipid profile improved with reductions in total cholesterol, low-density lipoprotein, and triglycerides and no change in high-density lipoprotein (HDL) levels. While weight reduction was small, waist circumference decreased significantly. Similarly, in a subset of men with impaired glucose tolerance, HgA_{1c} declined. The most common adverse drug reactions were increases of hematocrit and PSA [37].

Monitoring patients following initiation of therapy

To ensure proper dosing, serum testosterone concentrations should be measured after initiation of therapy to ensure that the desired concentrations (300–1050 ng/dL) are achieved [23]. The dose can be adjusted based on the serum testosterone concentration from a blood draw 2–8 hours after application and at least 14 days after starting treatment or following dose adjustment.

Once appropriate dosing is achieved, the Endocrine Society Guidelines recommend that patients be evaluated 3–6 months after treatment initiation and then annually to assess whether symptoms have adequately improved, and to assess for adverse effects. PSA should be measured 3–6 months after initiation of therapy to establish a baseline from which PSA velocity can be calculated. Similarly, hematocrit should be checked at 3–6 months, and then annually. Therapy should be discontinued if the hematocrit level is >54% [23].

Due to lingering concerns about the induction or acceleration of prostate cancer in men receiving

TRT, the Endocrine Society Guidelines recommend that clinicians obtain urological consultation if there is: an increase in serum or plasma PSA concentration greater than 1.4 ng/mL within any 12-month period of testosterone treatment, if a prostatic abnormality is found on digital rectal examination, or if PSA velocity is more than 0.4 ng/mL per year using the PSA level after 6 months of testosterone administration as the reference [23].

Scant data exist on the long-term risks and benefits of TRT in the aging male [1]. A recent systematic review and meta-analyses of testosterone trials evaluated the adverse effects of testosterone treatment in men. The reviewers assessed the methodological quality of the 51 studies from low to medium, and follow-up duration ranged from 3 months to 3 years. They concluded that TRT was associated with an increase in hemoglobin and a decrease in HDL cholesterol. However, there were no significant effects on mortality rates, incidence of prostate cancer, or cardiovascular events [42].

Conclusion

As an FDA-approved treatment for male hypogonadism, TRT has been found to be effective in ameliorating a number of symptoms in markedly hypogonadal males. Despite the increasing use of testosterone, there is only a limited body of data on the long-term risks and benefits associated with TRT in adult male patients with hypogonadism [1].

The current Endocrine Society Guidelines for the diagnosis and treatment of hypogonadism follow to a large extent the current practice for HRT in women, which was established based upon the findings of WHI. Additional information may be available once "The Testosterone Trial in Older Men" is complete. This is a randomized, placebo-controlled, double-blind treatment efficacy and safety study that will determine whether 1 year of active testosterone replacement will lead to improvement in five primary outcome measures: walking speed, sexual activity, vitality scale, verbal memory test, and correction of anemia [43].

Note from the Editors

The present chapter may seem totally out of place in a book dedicated to women's health. In the spirit of "learning from the extremes," we thought it would be important to grasp the place that testosterone therapy plays in men's health, where the notion of testosterone replacement therapy has been long established. In

addition to the excellent review provided in the current chapter, two major regulatory authorities have recently assessed the cardiovascular effects of testosterone in men. These opinions can be found at the following sites. For the US Food and Drug Administration:

www.fda.gov/downloads/advisorycommittees/committeesmeetingmaterials/drugs/reproductive-healthdrugsadvisorycommittee/ucm412536.pdf.

For the European Medicines Agency:

www.ema.europa.eu/docs/en_GB/document_library/Referrals_document/Testosterone_31/Recommendation_provided_by_Pharmacovigilance_Risk_Assessment_Committee/WC500175213.pdf.

References

1. Committee on Assessing the Need for Clinical Trials of Testosterone Replacement Therapy. In: Liverman CT and Blazer DG (Eds.) *Testosterone and Aging: Clinical Research Directions*. Washington, DC: National Academy of Sciences,; 2004.

2. Donatucci CF. The Institute of Medicine white paper on testosterone: current perspective. In: Helstrom WJG (Ed.) *Androgen Deficiency and Testosterone Replacement*. New York: Humana Press; 2013: 1–14.

3. Araujo AB, Esche GR, Kupelian V, et al. Prevalence of symptomatic androgen deficiency in men. *J Clin Endocrinol Metab* 2007; 92:4241–7.

4. Women's Health Initiative. www.nhlbi.nih.gov/whi/index.html (accessed February 2015).

5. National Institute of Health, Age page. www.nia.nih.gov/health/publication/menopause (accessed October 2012).

6. Grady D, Rubin SM, Petitti DB, et al. Hormone therapy to prevent disease and prolong life in postmenopausal women. *Ann Intern Med* 1992;117:1016–37.

7. Krieger N, Löwy I, Aronowitz R, et al. Hormone replacement therapy, cancer, controversies, and women's health: historical, epidemiological, biological, clinical, and advocacy perspectives. *J Epidemiol Community Health* 2005;59:740–8.

8. Million Women Study Collaborators. Patterns of use of hormone replacement therapy in one million women in Britain, 1996–2000. *BJOG* 2002; 109:1319–30.

9. Million Women Study Collaborators. Breast cancer and hormone replacement therapy in the Million Women Study. *Lancet* 2003;362:419–27.

10. Faber A, Bouvy ML, Loskamp L, et al. Dramatic change in prescribing of hormone replacement therapy in The Netherlands after publication of the Million Women Study: a follow-up study. *Br J Clin Pharmacol* 2005;60;641–7.

11. Nelson HD, Walker M, Zakher B, et al. Menopausal hormone therapy for the primary prevention of chronic conditions: a systematic review to update the U.S. Preventive Services Task Force recommendations. *Ann Intern Med* 2012;157:104–13.

12. Stuenkel CA, Gass ML, Manson JE, et al. A decade after the Women's Health Initiative–the experts do agree. *Menopause* 2012;19:846–7.

13. Harman SM, Metter EJ, Tobin JD, et al. Baltimore Longitudinal Study of Aging. Longitudinal effects of aging on serum total and free testosterone levels in healthy men. *J Clin Endocrinol Metab* 2001;86:724–31.

14. Feldman HA, Longcope C, Derby CA, et al. Age trends in the level of serum testosterone and other hormones in middle-aged men: longitudinal results from the Massachusetts Male Aging Study. *J Clin Endocrinol Metab* 2002;87:589–98.

15. Liu PY, Beilin J, Meier C, et al. Age-related changes in serum testosterone and sex hormone binding globulin in Australian men: longitudinal analyses of two geographically separate regional cohorts. *J Clin Endocrinol Metab* 2007;92:3599–603.

16. Wu FCW, Tajar A, Pye SR, et al. Hypothalamic–pituitary–testicular axis disruptions in older men are differentially linked to age and modifiable risk factors: the European Male Aging Study. *J Clin Endocrinol Metab* 2008;93:2737–45.

17. Veldhuis JD. Aging and hormones of the hypothalamo–pituitary axis: gonadotropic axis in men and somatotropic axes in men and women. *Aging Res Rev* 2008;7:189–208.

18. Kaufman JM, Vermeulen A. The decline of androgen levels in elderly men and its clinical and therapeutic implications. *Endocr Rev* 2005;26:833–76.

19. Morales A, Spevack M, Emerson L, et al. Adding to the controversy: pitfalls in the diagnosis of testosterone deficiency syndromes with questionnaires and biochemistry. *Aging Male* 2007;10:57–65.

20. Mulligan T, Frick MF, Zuraw QC, et al. Prevalence of hypogonadism in males aged at least 45 years: the HIM study. *Int J Clin Pract* 2006;60:762–9.

21. Tajar A, Forti G, O'Neill TW, et al. Characteristics of secondary, primary, and compensated hypogonadism in aging men: evidence from the European Male Ageing Study. *J Clin Endocrinol Metab* 2010;95:1810–18.

22. Huggins C, Hodges CV. Studies on prostate cancer. I. The effect of castration, of estrogen and of androgen injection on serum phosphatases in metastatic carcinoma of the prostate. *Cancer Res* 1941;1:293–7.

23. Bhasin S, Cunningham GR, Hayes FJ, et al. Testosterone therapy in men with androgen deficiency syndromes: an Endocrine Society clinical practice guideline. *J Clin Endocrinol Metab* 2010: 95: 2536–59.

24. Zitzmann M, Faber S, Nieschlag E. Association of specific symptoms and metabolic risks with serum testosterone in older men. *J Clin Endocrinol Metab* 2006;91:4335–43.

25. Hall SA, Esche GR, Araujo AB, at al. Correlates of low testosterone and symptomatic androgen deficiency in a population-based sample. *J Clin Endocrinol Metab* 2008;93:3870–7.

26. Shahidi NT. A review of the chemistry, biological action, and clinical applications of anabolic-androgenic steroids. *Clin Ther* 2001;23(9):1355–90.

27. Buvat J, Maggi M, Guay A, et al. Testosterone deficiency in men: systematic review and standard operating procedures for diagnosis and treatment. *J Sex Med* 2013;10:245–84.

28. Svartberg J, Agledahl I, Figenschau Y, et al. Testosterone treatment in elderly men with subnormal testosterone levels improves body composition and BMD in the hip. *Int J Impot Res* 2008;20:378–87.

29. Kwan M, Greenleaf WJ, Mann J, et al. The nature of androgen action on male sexuality: a combined laboratory-self-report study on hypogonadal men. *J Clin Endocrinol Metab* 1983;57:557–62.

30. Bagatell CJ, Heiman JR, Rivier JE, et al. Effects of endogenous testosterone and estradiol on sexual behavior in normal young men. *J Clin Endocrinol Metab* 1994; 78:711–16.

31. Kelleher S, Conway AJ, Handelsman DJ. Blood testosterone threshold for androgen deficiency symptoms. *J Clin Endocrinol Metab* 2004;89:3813–17.

32. Ni X, Muram D. Symptomatic improvement in hypogonadal men receiving testosterone replacement therapy as a function of testosterone level: an open-label subgroup analysis. *Endocr Rev* 33, SUN -75, 2012. Manuscript submitted for publication.

33. Lee KK, Berman N, Alexander GM, et al. A simple self-report diary for assessing psychosexual function in hypogonadal men. *J Androl* 2003;24:688–98.

34. Wang C, Cunningham G, Dobs A, et al. Long-term testosterone gel (AndroGel) treatment maintains beneficial effects on sexual function and mood, lean and fat mass, and bone mineral density in hypogonadal men. *J Clin Endocrinol Metab* 2004;89:2085–98.

35. Hubert W. Psychotropic effects of testosterone. In: Nieschlag EB, Behre HM (Eds.) *Testosterone: Action, Deficiency, Substitution.* Berlin, Germany: Springer-Verlag; 1990: 51–71.

36. Saad F, Aversa A, Isidori AM, et al. Onset of effects of testosterone treatment and time span until maximum effects are achieved. *Eur J Endocrinol* 2011;165:675–85.

37. Zitzmann M, Mattern A, Hanisch J, et al. IPASS: a study on the tolerability and effectiveness of injectable

testosterone undecanoate for the treatment of male hypogonadism in a worldwide sample of 1,438 men. *J Sex Med*. 2013; 10:579–88.

38. Buvat J, Maggi M, Gooren L, et al. Endocrine aspects of male sexual dysfunctions. *J Sex Med* 2010;7:1627–56.

39. Snyder PJ, Peachey H, Hannoush P, et al. Effect of testosterone treatment on body composition and muscle strength in men over 65 years of age. *J Clin Endocrinol Metab* 1999;84:2647–53.

40. Isidori AM, Giannetta E, Greco EA, et al. Effects of testosterone on body composition, bone metabolism and serum lipid profile in middle-aged men: a meta-analysis. *Clin Endocrinol (Oxf)* 2005;63:280–93.

41. Isidori AM, Giannetta E, Gianfrilli D, et al. Effects of testosterone on sexual function in men: results of a meta-analysis. *Clin Endocrinol* 2005;63:381–94.

42. Fernández-Balsells MM, Murad MH, Lane M, et al. Clinical review 1: adverse effects of testosterone therapy in adult men: a systematic review and meta-analysis. *J Clin Endocrinol Metab* 2010;95:2560–75.

43. Randomized, Placebo-controlled, Double-blind Study of Five Coordinated Testosterone Treatment Trials in Older Men (ClinicalTrials.gov Identifier: NCT00799617).

Chapter

16

History and physical examination of polycystic ovary syndrome: detecting too much or too little

John J. Kohorst, Andrew S. Fischer, and Christopher B. Rizk

Introduction

Polycystic ovary syndrome (PCOS) is a syndrome of unsettled features. The National Institutes of Health, Rotterdam, Androgen Excess and PCOS (AE-PCOS) Society (formerly the Androgen Excess Society [AES]) have all previously defined the syndrome with different criteria, presented in Table 16.1 [1– 3]. This has led to the emergence of four recognized phenotypes, presented in Table 16.2. At its core, PCOS is characterized by androgen excess and ovarian dysfunction (whether ovulatory or polycystic), but PCOS is also associated to differing degrees with other features, such as pregnancy complications, metabolic syndrome, psychiatric diseases, and neoplasms. The features of the syndrome appear to be linked, with one feature influencing the presentation and severity of the other. Healthcare professionals must be aware of the full scope of this broad disease to best diagnose and care for their patients.

Although a full workup is necessary to rule out diseases with similar presentation, PCOS is still a subjective diagnosis rooted in history and physical exam findings suggestive of androgen excess and ovulatory dysfunction. The history of the present illness, past medical history, and physical exam should assess first for hyperandrogenism and the presence of ovulatory dysfunction. Hyperandrogenism can be most readily assessed by the presence of certain cutaneous manifestations. Ovulatory dysfunction can be ascertained through the history. Subsequently, additional features of the syndrome should be evaluated, including: metabolic syndrome, obstructive sleep apnea, cardiovascular disease, nonalcoholic fatty liver disease, psychiatric disease, infertility, pregnancy complications, and risk of neoplasms. The presence of polycystic ovaries is primarily a sonographic finding. The differential diagnosis of hyperandrogenism is also of high importance as several disorders can mimic the clinical features of PCOS, and these should be considered before the diagnosis of PCOS is established. After all, PCOS is a diagnosis of exclusion.

Etiology

The prevalence of PCOS is 5–10% in most studies, making it the most common female reproductive disorder. Disease rates are similar across many races and environments. Variation is likely from differences in the definition of PCOS used, since the NIH definition seems to yield a lower estimate of prevalence [4]. The disease has a high genetic heritability as shown by twin studies, which underscores the importance of a detailed family history [5].

The history and physical examination

Cutaneous manifestations

The cutaneous manifestations of PCOS are detectable on the physical exam, and include hirsutism, acne, alopecia, acanthosis nigricans, and skin tags [6]. Whereas hirsutism, acne, and alopecia are manifestations of hyperandrogenism, acanthosis nigricans and skin tags reflect insulin resistance. A detailed history should document the age of onset, rate of progression, previous long-term treatments (including anabolic agents), any changes with treatment or fluctuations in body weight, and severity relative to family members [6]. Many factors can affect the clinical presentation, such as age, medications, and obesity. Circulating androgens progressively decline with age, meaning

Androgens in Gynecological Practice, ed. Leo Plouffe and Botros Rizk. Published by Cambridge University Press. © Cambridge University Press 2015.

Table 16.1 Comparison of the diagnostic criteria for PCOS [1– 3]

NIH (1990)	Rotterdam (2003)	AES/AE-PCOS Society (2006)
Must have:	*Need 2 of 3:*	*Must have:*
Hyperandrogenism[a]	Hyperandrogenism[a]	Hyperandrogenism[a]
+	or	+
Ovulatory dysfunction	Ovulatory dysfunction	Ovulatory dysfunction
	or	+/or
	Polycystic ovaries[b]	Polycystic ovaries[b]
+	+	+
Exclusion of other androgenic etiologies[c]	Exclusion of other androgenic etiologies[c]	Exclusion of other androgenic etiologies[c]

NIH, National Institutes of Health; AES/AE-PCOS, Androgen Excess and PCOS Society (formerly the Androgen Excess Society).

[a] Hyperandrogenism can be a clinical or biochemical assessment.
[b] Polycystic ovaries are confirmed when at least one ovary has ≥12 follicles with a 2–9 mm diameter or an ovarian volume >10 mL.
[c] Including at least: exogenous androgens, thyroid dysfunction, hyperprolactinemia, androgen-secreting neoplasms, 21-hydroxylase deficient nonclassic adrenal hyperplasia, and Cushing's syndrome.

Table 16.2 Overview of the recognized PCOS phenotypes derived from the differing criteria used for diagnosis

PCOS phenotype	Features
A: Frank	Hyperandrogenism, chronic anovulation, polycystic ovaries
B: Classic	Hyperandrogenism, chronic anovulation
C: Ovulatory	Hyperandrogenism, polycystic ovaries
D: Mild	Chronic anovulation, polycystic ovaries

that older patients typically have lower androgen levels and fewer signs of hyperandrogenism [7]. Medications (discussed below) may cause hyperandrogenism, but may also mask hyperandrogenism. Obesity and insulin resistance also tend to increase androgen levels [8]. Acanthosis nigricans and skin tags are often present in women with PCOS and obesity.

Hirsutism

Hirsutism is excessive hair growth in a woman and which is in a male distribution. It reflects the androgenic transformation of vellus hair to terminal hair [9]. Hirsutism serves as the best clinical sign of hyperandrogenism. PCOS is the predominant cause of hirsutism and is diagnosed in 65–75% of PCOS cases, depending on race and ethnicity [3]. Evaluation for hirsutism in the physical exam is done by searching for thick, dark hair on androgen-sensitive areas of the body, such as the linea alba, chin, upper lip, and around the nipple. Patients with hirsutism are presented in Figure 16.1. The degree of hirsutism can be qualitatively estimated using the modified Ferriman–Gallwey score, shown in Figure 16.2 [10]. The nine most androgen-sensitive areas of the body are assigned a value from 0 to 4 (0 indicates no hair, 4 indicates frank virilization), and the values are totaled to determine a hormonal hirsutism score. A score greater than or equal to 8 is diagnostic of hirsutism [11]. The Ferriman–Gallwey score is limited by its subjective nature and a lack of normative data outside of white and black women. Further, it is often criticized for not being specific enough. A point from each area is considered equivalent, when in fact hair on the face of a woman may be more indicative of androgen excess than hair on the trunk. Also, it does not assess the pattern of hair growth, which should be documented additionally by the physical exam.

Many factors must be considered during the evaluation of hirsutism, including ethnicity, age, obesity, and genetics. There is a great amount of racial variation in hair growth and distribution, which is largely a reflection of varying 5α-reductase activity levels at the hair follicle. Black and white women have similar rates of hirsutism. East Asian women tend to have lower rates, while Southeast Asian women generally have higher rates [3]. As previously stated, hyperandrogenism

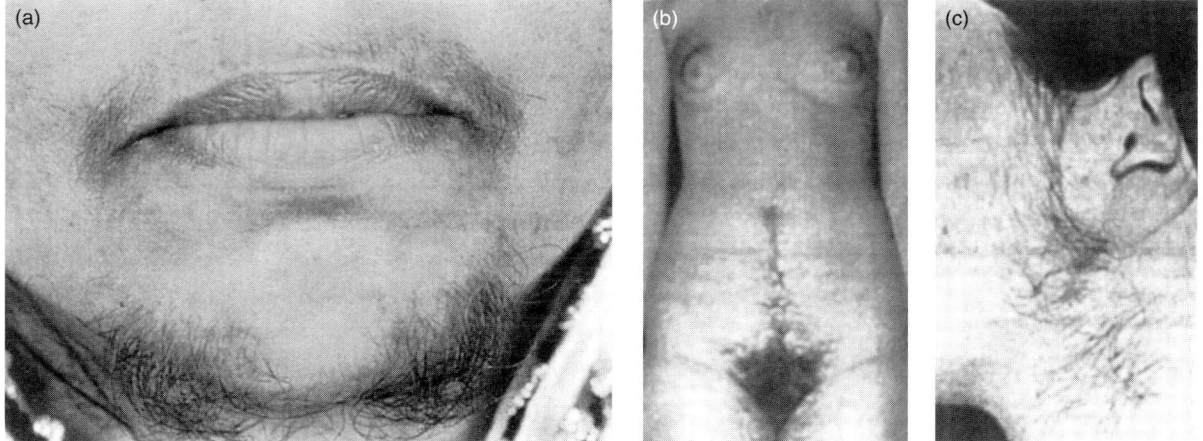

Figure 16.1 Hirsutism is thick, dark terminal hair on androgen-sensitive areas of the body, such as the chin and upper lip (a), pelvis and linea alba (b), and sideburns and neck (c).

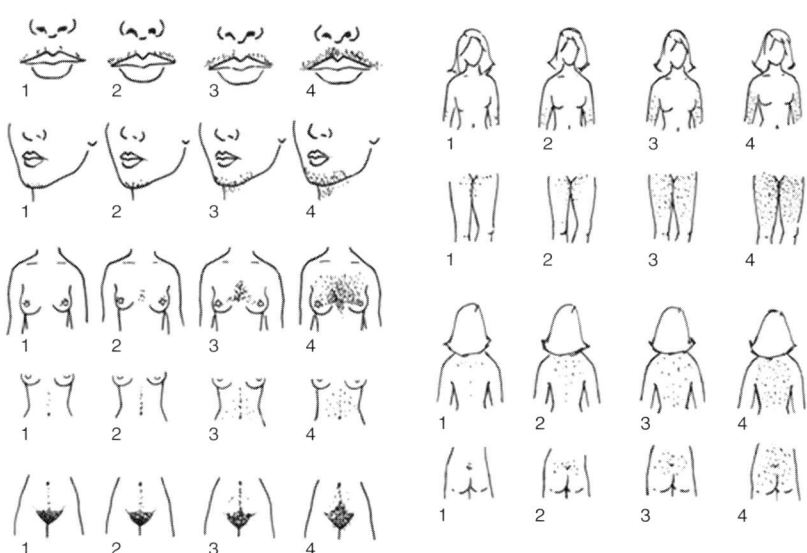

Figure 16.2 Ferriman–Gallwey hirsutism scoring system. Each of the nine body areas most sensitive to androgens is assigned a score from 0 (no hair) to 4 (frankly virile), and these separate scores are summed to provide a hormonal hirsutism score. (Reproduced from Evaluation and treatment of hirsutism in premenopausal women: an Endocrine Society Clinical Practice Guideline) [11].

decreases with age, which may affect phenotype. It is important to evaluate body hair in preadolescents, as pubic hair before age 8 years may be an ominous indication of future PCOS [12]. Further, hirsutism rates seem to correlate directly with body mass index (BMI), and may be a result of increased androgen levels in obesity [13]. Likewise, insulin resistance may increase hirsutism [14]. Family history is also important as mothers and sisters of PCOS patients are often hirsute [15]. Additionally, some hairiness may be genetic. Finally, it is imperative to ask about hirsutism even if there is little evidence on the physical exam as the patient may be controlling hair growth by shaving, waxing, electrolysis, or other cosmetic procedures.

Acne

Both acne vulgaris and alopecia are less common physical manifestations of hyperandrogenism, but if present are still useful to the clinical picture of PCOS. Androgens influence acne by stimulating sebum production, which allows *Propionibacterium acnes* to colonize the follicle; inflammation results. The prevalence of acne in PCOS is similar to the normal population. While some studies have suggested it has a higher prevalence and is a useful finding, the AES determined the prevalence to be 15–25%, which is similar to the normal population [3]. Part of the difficulty is in the assessment. Although systems have been created to grade acne severity, none have been validated as standard. When assessing acne, it may be best to give a global assessment – mild, moderate, or severe – based on the type of lesions (papules, pustules, nodules, cysts) and extent of skin involvement. It is also important to ask about duration of acne and prior treatments. Recalcitrant acne that persists into a woman's late teens or early twenties may be suggestive of a hyperandrogenic etiology [9].

Race, medications, and obesity may affect the presentation of acne. Like hirsutism, the prevalence of acne also varies by race. It is more commonly observed in Asian Indians and less commonly in Pacific Islanders [3]. Topical and oral acne medications can mask the clinical presentation of acne and their use should be evaluated in the history. Obese PCOS women often have less acne, although the reason is unclear [8]. Overall, acne seems to be less helpful in establishing a diagnosis of hyperandrogenism.

Alopecia

Androgenic alopecia, or male pattern balding, reflects endogenous androgen activity on the pilosebaceous unit. It typically begins at the vertex, spreads to involve the entire crown, and eventually leads to diffuse hair loss [9]. Its prevalence in PCOS is variable depending on the study, and there is generally poor correlation between biochemical hyperandrogenism and alopecia [3]. This finding is likely explained by varying sensitivities of the pilosebaceous unit to circulating androgens. Alopecia is relatively underdiagnosed, which may be because hair loss is typically noticed later than hair gain [16]. Generally, a woman becomes aware of thinning hair on her scalp only after a 25% reduction. On the one hand, the presence of alopecia is a strong indicator of PCOS; PCOS may account for 10–40% of all women with alopecia [3]. However, PCOS rarely presents with alopecia. Further, alopecia has low specificity for PCOS, since it has many other causes including genetic, environmental, and nutritional. In general, it seems to be more effective to evaluate hirsutism, since alopecia is rarely present independently from hirsutism [17].

Acanthosis nigricans and skin tags

A key physical exam finding of insulin resistance in PCOS is acanthosis nigricans. The lesion is characterized by thick, velvety, hyperpigmented plaques on the intertriginous surfaces, predominantly the neck and axilla. This is depicted in Figure 16.3. It is often asymptomatic, but may be occasionally pruritic. Its prevalence in PCOS is 15% alone, but 65% with obesity [18]. In fact, it appears to be relatively dependent on weight, and may regress with weight loss. It is important to note the physical exam has a low sensitivity for detecting acanthosis nigricans. Further, acanthosis nigricans is a fairly nonspecific finding as it is common to all conditions with insulin resistance including type 2 diabetes (DM) and Cushing's syndrome.

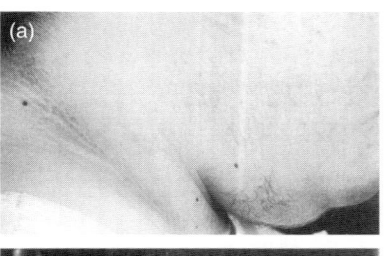
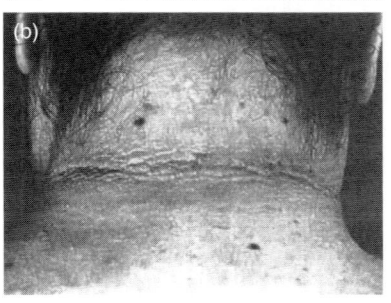
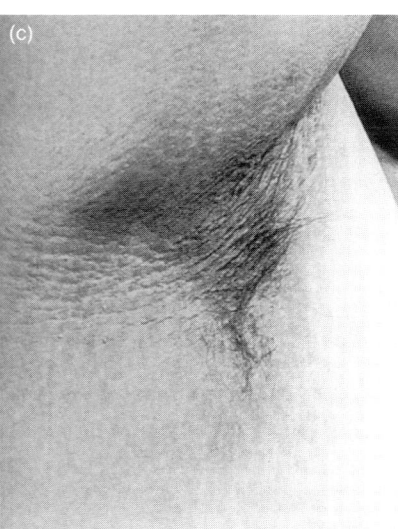

Figure 16.3 Acanthosis nigricans is characterized by thick, velvety, hyperpigmented plaques on the intertriginous surfaces, predominantly the neck (a, b) and axilla (c). Note also the terminal hair growth on the chin in (a).

Acrochordons, or skin tags, are small, soft, flesh-colored, pedunculated, benign neoplasms. Their growth is linked to insulin resistance and they can often be found in and around the lesions of acanthosis nigricans. Their presence is also rather nonspecific.

Ovulatory dysfunction

Ovulatory dysfunction is a cornerstone feature of PCOS and is identifiable from a history of menstrual features and timing. PCOS is the most common cause of ovulatory dysfunction [6]. The root of ovulatory dysfunction in PCOS is decreased ovulation. The level of anovulation in each patient determines her menstrual characteristics. The PCOS phenotype ranges from no menstruation (amenorrhea) to limited menstruation (oligomenorrhea) to normal cycles (eumenorrhea) and even, in rare cases, increased menstruation (polymenorrhea). The average menstrual interval is 28 days with a range of 21–35 days. Oligomenorrhea is menstrual bleeding occurring at equal to or greater than a 35-day interval or fewer than 8 times per year. Polymennorhea is menstrual bleeding occurring more frequently than every 21 days. A history of irregular bleeding without the typical premenstrual symptoms – bloating, mood changes, and breast fullness – is suggestive of anovulation.

Anovulatory cycles in PCOS can present immediately after menstruation or at a later time [18]. The prevalence of anovulation in PCOS is high (75–85%) and four times more prevalent than the regular population [3]. Amenorrhea in PCOS is present in 16%, while oligomenorrhea is present in 68%. [18] Eumenorrhea is present in 20–30% of women with PCOS, although these women often have biochemical evidence of ovulatory dysfunction [3]. Thus a history of regular menstruation does not exclude the diagnosis. An absence of premenstrual symptoms in a eumenorrheic-reporting woman may indicate anovulation. In this case, a midluteal serum progesterone may be a useful consideration [6]. Less than 2% of PCOS patients are estimated to have polymenorrhea [3]. Patients with a more severe PCOS phenotype have more infertility and higher androgen levels.

Several criteria must be considered in evaluating ovulatory dysfunction. PCOS females may have transitory menstrual regularity over time; they may initially have regular cycles, and then develop irregularity. However, most patients state that their cycles never established regularity after menarche.

Adolescents with persistent oligomenorrhea or amenorrhea for more than 2 years after menarche should be evaluated for PCOS [6]. Cycle regularity can also return to PCOS patients, as was shown in a 4- to 7-year follow-up [19]. An accurate history including frequency and duration is critical, as PCOS females may incorrectly believe they have a normal menstrual interval, not oligomenorrhea. Age has a strong impact on menstrual cycles. Early adrenarche and early breast development is associated with PCOS [20]. Menstrual dysfunction commonly decreases closer to menopause [21]. Anovulation in a woman's twenties can disappear by her forties, while ovulatory cycles increase [22]. Additionally, later menopause onset may be associated with PCOS [23]. Thus, the patient's age at thelarche, adrenarche, and menarche should be documented in the history [9]. Higher body weight also commonly worsens ovulatory dysfunction [24]. Drug history must be considered as oral contraceptives can mask ovulatory dysfunction.

A thorough family history of ovulatory dysfunction is important for complete evaluation. Mothers of PCOS patients had increased menstrual dysfunction [15]. Thirty percent of the mothers of PCOS patients also had amenorrhea [25]. Finally, it is also very important to realize that it takes time to establish an individual's menstrual cycle pattern. A provider must determine what is normal for each patient before making the diagnosis.

The differential diagnosis for menstrual dysfunction can be elucidated by the history. Lifestyle changes, such as the start of intense exercise or a rapid weight change, can both lead to transitory menstrual irregularity. Patients should be questioned about eating disorders, including anorexia and bulimia. Further testing will be needed to sort through other potential causes such as pregnancy, premature ovarian failure, hyperprolactinemia, outflow tract obstruction, thyroid dysfunction, and hypothalamic amenorrhea.

Metabolic syndrome

PCOS patients seem to be at a moderately increased risk of developing metabolic syndrome. Metabolic syndrome is defined as a syndrome that predisposes a person to increased risk of type 2 DM and coronary artery disease (CAD). There are multiple definitions of the specific features of the syndrome, which can complicate research. Loosely, metabolic syndrome is a constellation of central obesity, high fasting

Table 16.3 ATP III clinical criteria for the identification of metabolic syndrome in women [26,27]; the diagnosis is made when 3 of the 5 criteria are met

Risk factor	Criteria
Abdominal obesity (defined by waist circumference)	>88 cm (>35 in)
Triglycerides	≥150 mg/dL
HDL cholesterol	<50 mg/dL
Blood pressure	≥130/≥85 mmHg
Fasting glucose	≥100 mg/dL

glucose, hypertension, and dyslipidemia. The National Cholesterol Education Program (NCEP) Adult Treatment Panel III (ATP III) definition for metabolic syndrome is one of the most widely used [26]. It was later updated by the American Heart Association and the National Heart Lung and Blood Institute [27]. The NCEP ATP III definition is presented in Table 16.3.

Metabolic syndrome in PCOS can be evaluated in the initial history and physical exam [6]. BMI and waist circumference should be measured initially and monitored over time. Insulin resistance must be evaluated through patient history as well as physical exam by careful screening for acanthosis nigricans. Blood pressure can be measured in office, and dyslipidemia can be evaluated in subsequent laboratory testing.

Several studies support PCOS as a metabolic syndrome risk, independent of BMI or age. A meta-analysis found PCOS patients to be at increased risk for metabolic syndrome, with and without BMI controlled [28]. Metabolic syndrome was found at a variable prevalence in PCOS patients in recent studies: around 25–35% of PCOS patients compared with approximately 10% in the regular population [29]. The definition used to define PCOS – NIH, Rotterdam, or AES – may affect the prevalence values.

The link between PCOS and metabolic syndrome may be affected by many factors, including BMI, race, and lifestyle. The BMI of PCOS patients can help determine the prevalence of metabolic syndrome [30]. For physical exam purposes, abdominal obesity seems to be the most critical PCOS risk factor to evaluate for metabolic syndrome and insulin resistance [31]. Race, not PCOS, may predispose certain patients to metabolic syndrome. South Asian PCOS patients have higher metabolic syndrome prevalence than Caucasian patients [32]. Culture may also play a large role, as shown by a study comparing Chinese women living in China to American Chinese women [33]. It was concluded that mainland Chinese women with PCOS had a lower risk of metabolic syndrome than Chinese women in Westernized societies. Similarly, diet may be a factor.

Metabolic syndrome screening should be performed in every patient suspected of PCOS. Acanthosis nigricans of insulin resistance, obesity, and hypertension can be easily evaluated in the physician office. Lipid screening should also be done as obese PCOS patients are at risk for dyslipidemia, especially with accompanying hyperandrogenism [34]. As complications of metabolic syndrome, DM and CAD should be fully evaluated in each PCOS patient.

Obesity

Though not part of the common PCOS definitions, obesity is highly prevalent in PCOS patients. It is responsible for excessive morbidity and should be evaluated during the physical exam with BMI calculation and measurement of waist circumference, using the National Health and Nutrition Examination Survey method [6,35]. With the patient standing upright, the waist circumference is measured using a nonfolded measuring tape positioned at the top of the iliac crest and parallel to the floor. The measurement is then taken at the end of the patient's expiration. A waist circumference of greater than 88 cm is considered abnormal in Caucasian American and African American women and greater than 80 cm in Hispanic, Native American, Asian, and European women [35]. The prevalence of obesity in PCOS patients is between 30% to 70%, 50% even in adolescents, and appears across cultures and races [29,24,36].

The relationship between obesity and PCOS is not clearly defined. Some evidence indicates that PCOS is more prevalent in the severely obese, suggesting obesity may influence PCOS development [37]. Others argue that the obesity in PCOS is a greater reflection of environmental factors, noting a parallel increase in the BMI of PCOS patients with the prevalence of obesity in a surrounding population over a 15-year period [38]. Regardless, obesity in PCOS patients causes a multitude of other effects. Obesity increases the severity of other PCOS features such as androgen levels, hirsutism, ovulatory dysfunction, infertility, and pregnancy complications. Interestingly, it tends to ameliorate acne. The influence of obesity on the features of PCOS is likely due to a worsened underlying relative hyperandrogenism [6]. This results when high levels of insulin suppress the hepatic production of sex hormone-binding globulin (SHBG), leading to increased bioavailable androgens.

The presentation of obesity in PCOS patients can occur in childhood before the cornerstone PCOS features of hyperandrogenism and ovulatory dysfunction develop, or it may occur in adulthood. If obesity is present at the time of evaluation, the patient should be questioned about its onset and progression [9]. Body fat distribution can be easily evaluated in the PCOS physical exam. Abdominal fat, specifically, is associated with PCOS obesity. Patients with abdominal obesity, not peripheral, have worse hyperandrogenemia and are at increased risk for developing metabolic syndrome [39,40].

Race, family history, culture, age, and diet should be considered in obesity evaluation in PCOS. Black and Hispanic PCOS women are more likely to be obese than Asian and white PCOS women [41]. Male relatives of PCOS patients are more obese and thus, more prone to metabolic syndrome [42]. First-degree relatives of a PCOS patient have higher body fat levels than the unrelated population [43]. Culture also seems to affect obesity development in PCOS patients. American women with PCOS are more likely to be obese than those from other countries, suggesting a lifestyle effect [3]. Older, postmenopausal PCOS patients also have higher obesity rates [44]. Patient eating habits must also be accounted for. A poor diet may worsen other PCOS features, like insulin resistance; a healthy diet should always be encouraged.

Insulin resistance and glucose dysregulation

Insulin resistance is a metabolic syndrome feature that should be evaluated in the history of any suspected PCOS patient. Patients may complain of lethargy, constant hunger, and difficulty concentrating, which are all nonspecific. Acanthosis nigricans is a clinical manifestation that can be evaluated in the physical exam and is the direct result of a compensatory hyperinsulinemia. In fact, some hypothesize that insulin resistance and the subsequent hyperinsulinemia largely underpin the formation of PCOS in many women. Hyperinsulinemia drives ovarian androgen production and suppresses hepatic SHBG production. The result is increased circulating free androgens. The capacity of otherwise healthy, young women to produce high levels of insulin may mask overt diabetes measured by fasting blood glucose. For this reason, each suspected patient should be screened with a 2-hour 75 g oral glucose tolerance test (OGTT) measuring both insulin and glucose. PCOS patients are at risk for insulin resistance and the prevalence of insulin resistance in PCOS is around 50–70% [3].

Insulin resistance in PCOS can be modulated by many factors. Obesity tends to increase insulin resistance [45]. Likewise, insulin resistance may occur in up to 95% of obese women with PCOS [35]. Insulin resistance worsens with age, while other portions of the disorder may improve [7]. Anovulatory PCOS patients also have higher and more variable insulin resistance than ovulatory patients [46]. Diet also seems to affect insulin resistance in PCOS patients, although the effect may be small [38]. A family history of insulin resistance or type 2 DM can also be supportive as first-degree relatives of PCOS patients have more insulin resistance, and insulin resistance is also more common in individuals with a family history of type 2 DM [36, 43].

PCOS is associated with a 5- to 10-fold increase in the risk of type 2 DM, which should be suspected in any female with the disease. The prevalence of type 2 DM amongst American PCOS patients is 3–10%, and that of glucose intolerance is 30–35% [6]. Over time, many PCOS patients progress from glucose intolerance to type 2 DM, which underscores the importance of continued monitoring [47]. Symptoms that may clue to the onset of diabetes are excessive thirst, frequent urination, polyphagia, and weight loss. Many patients, though, will be detected only on routine screening. The disease should be checked initially using an OGTT, or hemoglobin A_{1c} if the patient is unable to complete an OGTT [6]. Patients should be rescreened every 3–5 years, or more frequently if there is central obesity, substantial weight gain, or clinical symptoms of diabetes [6]. Factors that may predispose a patient with PCOS to type 2 DM development should also be noted, specifically abdominal obesity and a family history of type 2 DM.

Hypertension

The association between PCOS and hypertension is contentious, but blood pressure should be evaluated as part of the complete PCOS physical exam. Ideally, blood pressure should be 120 mmHg systolic and 80 mmHg diastolic or lower. Hypertension, and prehypertension, should be treated to reduce cardiovascular disease. Studies have shown hypertension to be associated with PCOS, even with obesity and insulin resistance controlled, while others have linked the association to particular subgroups such as women aged 35–55 [48,49]. Other research has indicated a hypertension risk only with obesity and high BMI [50,51]. A family history of hypertension may be important to a PCOS workup, as relatives of a woman with PCOS are more likely to have hypertension [15]. Race of the patient should also be considered; black women with PCOS have a higher prevalence of hypertension when corrected for BMI [41].

Obstructive sleep apnea

Obstructive sleep apnea (OSA) is characterized by episodes of pharyngeal obstruction during sleep and may manifest as daytime somnolence. Generally, sleep disordered breathing is seen in older, obese males and the prevalence amongst premenopausal women is less than 1%. However, amongst women with PCOS, the prevalence increases 30- to 40-fold [52]. The link between PCOS and OSA has not been well elucidated, but likely results from the interplay between hyperandrogenism, insulin resistance, and obesity. OSA in PCOS tends to develop after adolescence, later in the course of the disease [53].

Patients should be carefully questioned about sleep disturbances, as OSA is a risk factor for cardiovascular mortality and morbidity and its treatment may help stave off further metabolic sequelae in PCOS. Patients with OSA may mention a history of snoring, night restlessness, morning headache, and subsequent daytime sleepiness. Their partner may be able to attest to interrupted breathing. OSA is difficult to diagnose on history and physical exam, so patients with a suspicious history should be referred to a certified sleep laboratory for polysomnography [6].

Cardiovascular risk

PCOS patients have cardiovascular disease (CVD) risk factors, such as hypertension, obesity, insulin resistance, and type 2 DM. PCOS patients also have increased levels of CVD risk markers [54]. Meta-analysis indicates that PCOS patients are twice as likely to have a myocardial infarction (MI) or stroke as the regular population and 1.5 times as likely when corrected for obesity [55]. Carotid intima thickness and electron beam studies both show increased disease in PCOS patients [56,57]. However, strong evidence also exists against PCOS patients being at increased risk of a cardiovascular attack, aside from PCOS features. A Mayo Clinic population-based study showed no increase in cardiovascular events, with obesity being the only increased CVD risk factor over controls [58].

In all, there is insufficient evidence to determine the rate of cardiovascular disease or its onset. Therefore, attention has turned to identifying CVD risk factors, which can be screened for in the history and physical exam. These include age, family history of early CVD, cigarette smoking, impaired glucose tolerance and type 2 DM, hypertension, dyslipidemia, OSA, and obesity [6]. Age seems to be associated with CVD risk. MI risk is increased in the over-45 cohort, high in the over-65, and worse with smoking and hypertension [59]. Postmenopausal PCOS patients have more CAD by angiography [44]. The AE-PCOS Society conducted a systematic review of the literature to identify PCOS–CVD risk relationships [35]. They identified PCOS women with metabolic syndrome, type 2 DM, and/or overt vascular or renal disease to be at highest risk for CVD, and women with any of obesity (especially abdominal), cigarette smoking, dyslipidemia, hypertension, impaired glucose tolerance, a family history of premature CVD (male <55 years, female <65 years), and subclinical vascular disease to be at risk, albeit lower.

NAFLD/NASH

Nonalcoholic fatty liver disease (NAFLD) has a relationship to PCOS and should be evaluated by history and physical exam. NAFLD is characterized by hepatic steatosis not due to alcohol consumption and is the most common liver disorder in the Western world. Nonalcoholic steatohepatitis (NASH) is an extreme form of NAFLD where hepatitis results from steatosis and may progress to cirrhosis and liver failure. NAFLD and NASH share many risk factors with PCOS, such as metabolic syndrome, type 2 DM, and central obesity. The majority of NAFLD patients have PCOS and PCOS patients have high NAFLD association, even with obesity controlled [60,61]. Around 60% of PCOS patients have been found to have NAFLD, and abnormal liver enzymes have been found in 15% of PCOS adolescents [62,63]. Whether or not NAFLD is related to hyperandrogenism is unclear.

NAFLD/NASH may be asymptomatic or present with fatigue, malaise, or dull right upper quadrant pain. Jaundice is rare. A history of NAFLD/NASH should be evaluated through patient history and record. Hepatomegaly may occasionally be detected on abdominal exam. Past serum liver function tests

Table 16.4 The two-item Patient Health Questionnaire depression module (PHQ-2) for depression screening [64]. The authors set a score of 3 as the optimal cut-off for screening purposes; patients with a score greater than or equal to 3 had a sensitivity of 83% and a specificity of 92% for major depression

Over the past 2 weeks, how often have you been bothered by any of the following problems?	Not at all	Several days	More than half the days	Nearly every day
1. Little interest or pleasure in doing things	0	1	2	3
2. Feeling down, depressed, or hopeless	0	1	2	3

can be evaluated in the patient record or obtained at a later date.

Psychiatric manifestations

PCOS and its clinical manifestations can have a severe psychological impact on patients. Patients with PCOS appear to be at increased risk for depression and anxiety. Therefore, any patient suspected of or confirmed to have PCOS should be questioned about prior psychiatric diagnoses, specifically depression and anxiety. If there is no history, these patients should be screened in office and referred to a psychiatrist if needed [6]. Screening can be done easily with a two-question questionnaire, the PHQ-2, presented in Table 16.4 [64,65]. It is also important to ask about the use of alcohol and drugs, which patients may be using as forms of self-medication.

Women with PCOS may be at a four- to eightfold increased risk of depression [66,67]. Patients with depression most commonly complain of fatigue and sleep disturbances, followed by appetite changes and diminished interest in doing things [67]. The predilection towards depression seems to be independent of PCOS features such as obesity, hyperandrogenism, hirsutism, acne, and infertility [6]. However, PCOS features are certainly implicated as well. Body weight primarily decreases quality of life, but hirsutism, ovulatory dysfunction, infertility, sexual behavior, and negative healthcare experiences with the PCOS diagnosis also contribute [68]. A meta-analysis concluded that PCOS patients have a mild tendency towards depression and anxiety which is influenced by BMI [69]. Body image relates directly to depression severity in PCOS patients [70]. Women with PCOS are less likely to identify with a female gender scheme and more likely to view themselves as androgenous [71]. These alterations in gender identification may partially contribute to psychosocial and emotion problems, including depression. Infertility and the worry of being childless may contribute, although a clear link has yet to be established. Further, sex in PCOS patients may be associated with depression. Some research indicates that women with PCOS have reduced sexual satisfaction with regard to orgasm [72].

Anxiety disorders are the most common psychiatric diagnosis amongst patients with endocrine abnormalities, and depression and anxiety often go hand in hand. PCOS patients have a higher prevalence of anxiety symptoms and generalized anxiety disorder (GAD), which likely additionally contributes to the development of social phobias, specific phobias, and panic disorders [73]. It is hard to clearly define the nature of these disorders in this patient population because anxiety disorders often have their onset during adolescence and may develop slowly over time. They frequently follow a chronic, recurring natural history and become clinically recognizable only when they begin to impair functioning, at a relatively advanced stage. Women with PCOS are more likely to have eating disorders, specifically binge eating [67]. This may result from low self-esteem or a reaction to the appetite-stimulating effects of androgens, impairing impulse control.

It is important to have a prompt diagnosis of PCOS in patients. Some research has shown delayed PCOS diagnosis to be associated with depression and anxiety [74]. Moreover, PCOS patients must be continually monitored over time. A longitudinal study showed a high conversion risk to depression over a 1- to 2-year period [75]. Suicide attempts were seven times more common in PCOS patients in a case-control study [76]. Further, anxiety is a chronic, undulating disease process that may require frequent follow-up.

Encourage physical activity in all suspected patients. Less physically active patients have higher depression rates [77]. Improved diet and exercise can improve depression [78]. Supervised eating may reduce the opportunity for binge eating in bulimic patients. In general, correction of the underlying condition will help improve psychological function [9].

Patients should be reminded that PCOS is a true medical issue with symptoms that can be managed. Support should be provided in a non-judgmental manner and patients should be encouraged to practice self-care and seek help when needed.

Infertility

PCOS and infertility are tightly associated. A history of infertility should be carefully documented in any PCOS evaluation. Likewise, PCOS should always be considered in an infertility presentation, as infertility is often the presenting complaint of PCOS. Infertility is a reliable PCOS feature, and may affect more than half of women with PCOS. PCOS women are far more likely to be using reproductive technology, 13% as compared with 1% [79]. Infertility in PCOS patients is suspected to be a consequence of ovulatory dysfunction and irregularity. Oligomenorrhea decreases the number of conception opportunities and cycle irregularity complicates timing intercourse with ovulation. But, as discussed, anovulation is not necessarily complete or static throughout a woman's life. It is important to remember that PCOS women may conceive a child. Over time, research has shown that a PCOS woman is just as likely to conceive a child as the normal population [80]. Family history is also important to infertility in PCOS. Sisters of PCOS women have more substantial infertility history and have fewer children overall [15]. Obesity should also be noted, as it tends to worsen infertility and decreases effectiveness of fertility treatments [81]. When evaluating women with PCOS for infertility, it is important to keep in mind causes beyond anovulation, such as male factor infertility and tubal occlusion [6].

Pregnancy complications

PCOS is associated with a number of adverse pregnancy and neonatal complications, which have been established by meta-analysis [82]. Evaluation of PCOS should include a thorough history to document complications with previous pregnancies including miscarriage, gestational diabetes, pregnancy-induced hypertensive disorders, cesarean sections, and preterm births. The history should also inquire about neonatal complications, including low birth rate and admission to the neonatal intensive care unit (NICU). The method of conception should be confirmed – natural, ovulation induction, or assisted reproductive technology (ART). Women who require ovulation induction or ART may have a history of multiple births. Because gestational diabetes, preterm delivery, and pre-eclampsias are exacerbated by obesity, all women with PCOS who are planning to get pregnant should have measurements of BMI, OGTT, and blood pressure prior to conception [6].

Recurrent miscarriage

A miscarriage refers to a spontaneous abortion during the first 12 weeks of a pregnancy. There are many reasons for miscarriage and it is currently unresolved whether PCOS carries an elevated risk of miscarriage, and if so, whether this risk is due to PCOS directly, obesity, or use of anovulation medications. Large randomized controlled trials do not support an increased risk of miscarriage [83]. A meta-analysis shows minor support for PCOS leading to increased miscarriage [84]. However, only 8–10% percent of women with recurrent miscarriage have PCOS [85]. With the same infertility treatment, PCOS women have higher miscarriage rates [86]. However, obesity may be the main driver of spontaneous abortion, as obese PCOS women have higher miscarriage rates than non-obese [87]. Clomiphene citrate, a first-line ovulation medication in PCOS, was once thought to contribute to a high miscarriage rate, but this has largely been refuted. Miscarriages in PCOS exhibit fewer chromosomal abnormalities, suggesting a role for other mechanisms such as placental bed thrombosis or endometrial defects.[88,89].

Gestational diabetes mellitus

Pregnancy induces a state of insulin resistance, which may manifest as gestational diabetes mellitus (GDM) if there is overt glucose intolerance. GDM is diagnosed after a 75 g OGTT demonstrates a plasma glucose concentration greater than 92 mg/dL while fasting, 180 mg/dL after 1 hour, and 153 mg/dL after 2 hours. As women with PCOS have an elevated baseline level of insulin resistance, they are at a significantly increased risk for developing GDM [82]. This is independent of obesity [84]. The presence of the disease should be investigated in a history of past pregnancies. GDM presents similarly to type 2 DM with excessive thirst, frequent urination, and fatigue. A large cohort study showed GDM to be two times more prevalent in PCOS than in the regular population [79]. Some studies show women with PCOS to be prone to GDM if they have a past infertility history [90]. Finally, it is important to further evaluate all women with PCOS with an infertility history as they show an increased risk of glucose intolerance after pregnancy [91].

Pregnancy-induced hypertensive disorders

Pregnancy-induced hypertension (PIH) is defined as a blood pressure greater than or equal to 140/90 without proteinuria at more than 20 weeks of gestation. Pre-eclampsia is PIH with proteinuria of greater than 0.3 g/24 hours and its onset is usually accompanied by peripheral edema and occurs during labor, but can be before or even up to 48 hours afterward. Women with PCOS have a higher risk of both hypertensive disorders [82]. The risk of PIH is greater than that of pre-eclampsia. Whether PCOS is directly associated with pregnancy-induced hypertension, or it is the features of obesity or gestational DM that are responsible, is not well established. Insulin resistance may exacerbate the frequency of pre-eclampsia [92].

Preterm birth

Preterm birth is the delivery of a fetus before 37 weeks and its history should also be evaluated in a history of PCOS. PCOS mothers are more prone to preterm births, and this association is maintained even independently of assisted reproductive technology [79,82].

Neonatal complications

Babies born to mothers with PCOS have higher rates of complications and admission to the NICU [82]. Babies born at less than 2500 g are considered to be low birth weight regardless of gestational age. The average birth weight at term is 2500–4200 g. Low birth weight could be due to prematurity or small for gestational age (SGA), which is a fetal index below the 10th percentile after adjustment for gestational age. PCOS mothers tend to give birth to babies that have lower birth weight [82]. Also, there is some evidence indicating that individuals born with a low birth weight are more prone to later PCOS [93]. Therefore, the birth weight of the presenting patient and her offspring is worth considering in a PCOS evaluation. Another consideration is the use of ART in PCOS infertility as ART is associated with low birth weight [94]. Neonatal complications that may contribute to more frequent admission to the NICU include being very preterm at birth, having an increased risk of meconium aspiration, and low APGAR scores at 5 minutes [79].

Neoplasms

The association between PCOS and neoplasms has been contentious, as the link between androgens and carcinogenesis has yet to be fully established. A risk for endometrial neoplasia has been suspected owing to unopposed estrogen during anovulation stimulating endometrial hyperplasia and carcinogenesis. Additionally, PCOS and endometrial carcinoma share the risk factors of elevated BMI, type 2 DM, nulliparity, and late menopause. Systematic review and meta-analysis of observation studies has estimated that the odds of developing endometrial carcinoma are nearly twofold higher in all women with PCOS, and this is increased to threefold in premenopausal women only [95]. However, the evidence is not strong enough to recommend routine assessments of endometrial thickness. Regardless, patients should be asked about unexpected bleeding and spotting.

There is a potential role for breast cancer in PCOS, considering shared risk factors of infertility and obesity. A retrospective study showed breast cancer to be the leading cause of death in PCOS patients [96]. However, the majority of research does not support a clear link between breast cancer and PCOS. Similarly, androgenic stimulation of ovarian epithelial cell proliferation in the pathogenesis of PCOS may contribute to ovarian cancers, although this theory has never been definitely proven. There is evidence for an increased risk of ovarian cancer in premenopausal women, but the data do not support an increased risk overall [95]. Regardless part of the assessment should include a thorough abdominal, pelvic, and breast exam searching for enlargement (uterus) or masses to help exclude neoplasms [9]. The ovaries of most women with PCOS are grossly enlarged, thus bilateral ovarian masses palpated on physical exam may be consistent with PCOS. Ultrasonography may be helpful if the patient's habitus limits the ability to perform an accurate pelvic exam.

Differential diagnosis

Several disorders mimic the clinical manifestations of PCOS and should be considered in the evaluation of PCOS. These include drugs, thyroid disease, hyperprolactinemia, nonclassic congenital adrenal hyperplasia (CAH; 21-hydroxylase deficiency), and – in certain cases – pregnancy, hypothalamic amenorrhea, primary ovarian insufficiency, androgen-secreting neoplasms, Cushing's syndrome, and acromegaly [6]. Disorders included in the differential diagnosis, including common and distinguishing features, are presented in Table 16.5. A list of medications that can cause hyperandrogenism is presented in Table 16.6.

Table 16.5 Differential diagnosis of PCOS including features shared with PCOS and features that distinguish each disorder from PCOS

Disorder	Common clinical manifestation(s)	Distinguishing clinical manifestation(s)
Thyroid disease	Oligomenorrhea, obesity (hypothyroid), psychiatric disturbances	Thyroid nodule or goiter; hyperthyroidism may present with weight loss, peripheral edema, exophthalmos, or pretibial myxedema; hypothyroidism may present with dry skin, cold sensitivity, fatigue, muscle cramps, or constipation
Hyperprolactinemia	Amenorrhea, hirsutism	Galactorrhea
Nonclassic CAH	Hyperandrogenism including hirsutism and acne, oligomenorrhea, obesity	Maybe earlier presentation with premature pubarche and accelerated linear growth velocity yielding short stature in adulthood
Pregnancy	Amenorrhea	Breast fullness, uterine cramps
Hypothalamic amenorrhea	Amenorrhea	Low body weight and BMI, history of stress or excessive exercise
Primary ovarian insufficiency	Amenorrhea	Hot flashes, urogenital symptoms of estrogen deficiency
Androgen-secreting neoplasm	Hyperandrogenism including hirsutism and acne	Rapid onset, frank virilization, clitoromegaly, increased muscle mass, a deepened voice
Cushing's syndrome	Hyperandrogenism, features of metabolic syndrome including obesity and glucose intolerance, and psychiatric disturbances	Proximal myopathy, striae, plethora, thin skin and easy bruising
Acromegaly	Oligomenorrhea, hirsutism, skin tags	Headaches, peripheral vision loss, frontal bossing, macrognathia, macroglossia, increased shoe and glove size

CAH, congenital adrenal hyperplasia.
Differential diagnosis and clinical manifestations from [6,97-99].

Table 16.6 Drugs that can cause hyperandrogenism and hypertrichosis [100]

Hyperandrogenism	Hypertrichosis
ACTH analogs	Acetazolamide
• Cosyntropin	Cyclosporine
• Tetracosactide	Diazoxide
Androgenic anabolic steroids	Glucocorticoids
Androgenic progestins	Interferon-alpha
• Levonorgestrel	Minocycline
• Norethindrone	Minoxidil
• Norgestrel	Penicillamine
Danazol	Phenytoin
Metyrapone	Psoralens
Phenothiazine	Streptomycin
Valproic acid (?)	Zidovudine

These medications either have androgenic properties or stimulate androgen production. Epidemiological studies show that PCOS is more common in women with epilepsy on long-term therapy with valproic acid. There is some evidence to suggest that valproic acid stimulates androgen biosynthesis, but there has not yet been a causal link to clinical hyperandrogenism.

All women should be screened with a thorough medication history and a thyroid-stimulating hormone, prolactin, and 17-hydroxyprogesterone level [6]. Androgenic medications such as danazol, topical testosterone, and anabolic steroids are causes of hyperandrogenism. Other medications associated with hyperandrogenism are the ACTH analogs (which stimulate adrenal androgen production), progesterones (which share a similar molecular structure with androgens), and valproic acid. Thyroid dysfunction, either hypo- or hyperthyroid, is a common cause of menstrual irregularity. At times, hypothyroidism may present similarly to PCOS with chronic anovulation and increased androgens, mediated through reduced SHBG. Thus, the physical exam should include a thorough examination of the thyroid for nodules and gross enlargement. Hyperprolactinemia may present with amenorrhea, or even hirsutism. Galactorrhea may be noted on the physical exam. Although a relatively uncommon cause of androgen excess, nonclassic CAH can present in adolescence or adulthood with

Table 16.7 A summary of the history-taking component in the evaluation of PCOS

History

History of present illness (HPI)

Hyperandrogenism	• Ask about manifestations including hirsutism, acne, alopecia, and obesity.
	• Note the age of onset, rate of progression, previous treatments, any changes with treatment or fluctuation with body weight, and severity relative to family members.
	• A history of the quick onset of frank virilization with increased muscle mass and deepened voice is more suggestive of an androgen-secreting neoplasm than PCOS.
Ovulatory dysfunction	• A thorough history of menstrual features and timing is imperative.
	• This should include the age at thelarche, adrenarche, and menarche, as well as the frequency and duration of menstruation, and presence or absence of premenstrual symptoms.
	• Ask about the possibility of pregnancy and other reasons for menstrual dysfunction including the start of an intense exercise program, rapid weight change, and eating habits.
Psychiatric evaluation	• The HPI should also include a brief psychiatric evaluation, including a screening for depression and anxiety, identification of sources of major stress, and questioning about thoughts of suicide.
	• Remember to provide non-judgmental support.

Pregnancy history

• This should include questioning about past pregnancies, if there have been any.

• Document any complications of past pregnancies including miscarriage, gestational diabetes, pregnancy-induced hypertensive disorders, and preterm births.

• Ask about neonatal complications including low birth weight and admission to the NICU.

Past medical history

• Presence or absence of type 2 DM, hypertension, dyslipidemia, OSA, NAFLD/NASH, and psychiatric disturbances.

Medication history

• A history of current and past medications should survey for the use of drugs with androgenic effects, and medication that may mask the clinical manifestations of the disease, such as oral contraceptives or topical and oral acne medications.

Surgical history

• Specifically, inquire about a past cesarean section and in vitro fertilization (IVF).

Family history

• This should include questioning about a history of PCOS, hirsutism, menstrual irregularity, infertility, early CVD, obesity, type 2 diabetes, hypertension, and dyslipidemia.

Social/Cultural history

• Note ethnicity, which is important for the evaluation of hirsutism.

• Also ask about the use of cigarettes, which is a risk for cardiovascular disease, and the use of drugs and alcohol, which may be a form of self-medication for depression or anxiety.

Review of symptoms (ROS)

• The ROS may be useful to screen for the complications of PCOS and to help work through the differential diagnosis (Table 16.5).

• Excessive thirst, frequent urination, polyphagia, weight loss, or weight gain may suggest overt diabetes.

• Snoring, night restlessness, morning headache, and daytime sleepiness is indicative of OSA.

• Dull right upper quadrant pain and jaundice suggest liver involvement.

• Cold intolerance, muscle cramps, and constipation should alert the clinician to the presence of a thyroid disturbance.

• Headaches and peripheral vision loss are ominous of a pituitary tumor.

hirsutism, oligomenorrhea, and acne. PCOS and non-classic CAH may be virtually identical on history and physical exam, making these disorders very difficult to distinguish without lab work.

The presence of certain features broadens the differential diagnosis to include additional disorders. Frank amenorrhea may be due to pregnancy, hypothalamic amenorrhea, or primary ovarian

Table 16.8 A summary of the pertinent aspects of the physical examination in the evaluation of PCOS

Hair and skin

• Hirsutism is an important clinical feature of hyperandrogenism, and if present should be quantified with the Ferriman–Gallwey score and its distribution noted.

• Alopecia should be noted, with its distribution. Temporal thinning is suggestive of androgen excess.

• Examination of the skin in a patient with PCOS may reveal acne, acanthosis nigricans, and skin tags.

• Dry, cold skin may be indicative of hypothyroidism.

• Striae, plethora, thin skin, or easy bruising may suggest glucocorticoid excess.

Head and neck

• Frontal bossing, macrognathia, or macroglossia may suggest acromegaly.

• Exophthalmos may be present with hyperthyroidism.

• The thyroid should be examined for nodules and gross enlargement.

Chest and abdomen

• Examination of the breasts may reveal galactorrhea in hyperprolactinemia.

• Abdominal exam should note hepatomegaly, and search for a mass to help exclude a neoplasm.

Fat distribution

• Abdominal obesity is common to PCOS.

• The presence of a buffalo hump or supraclavicular fat pad may suggest glucocorticoid excess.

• BMI and waist circumference should be measured and followed over time.

Pelvis

• Examine the pelvis for clitoromegaly, vaginal patency, atrophy, uterine enlargement, and ovarian masses.

• Remember that bilateral ovarian enlargement is suggestive of PCOS, but unilateral should be further evaluated for malignancy.

Extremities

• The presence of peripheral edema or pretibial myxedema may suggest a thyroid dysfunction.

• Proximal myopathy is often a feature of Cushing's syndrome.

Blood pressure should be measured to evaluate for hypertension.

failure. Therefore, a serum or urine pregnancy test and serum luteinizing hormone, follicle-stimulating hormone, and estradiol may be considered in these patients. An androgen-producing neoplasm must also be considered, especially if virilization, clitoromegaly, and a deepened voice are present. Neoplasms tend to be more progressive than PCOS. A serum testosterone and DHEA or imaging of the adrenal and ovaries can be used for screening. Cushing's syndrome shares the signs of hyperandrogenism, metabolic syndrome, and psychiatric disturbances with PCOS if the cause of the hypercortisolemia is endogenous, such as an ACTH-secreting pituitary neoplasm. However, Cushing's syndrome will often present with more specific signs of hypercortisolism such as proximal myopathy, striae, and thin skin. Acromegaly may present with ovulatory dysfunction and similar cutaneous manifestations to PCOS, including hirsutism and skin tags. Typically,

patients will also present with symptoms of a pituitary mass effect (headaches, peripheral vision loss) and unusual growth (frontal bossing, macrognathia, macroglossia). If these are present, screening with serum free insulin-like growth factor-1 or a pituitary MRI should be considered.

Hypertrichosis, or body hair in the absence of androgen excess, may be confused with hirsutism. The hair of hypertrichosis is typically finer, less dark, and in a non-sexual distribution. It reflects growth of the vellus hairs. Hypertrichosis has many etiologies including genetic, hypothyroidism, anorexia, dermatomyositis, and many drugs, which are listed in Table 16.6.

Conclusion

PCOS is characterized by androgen excess and ovarian dysfunction, but it is also associated to varying degrees with other features, such as pregnancy complications, metabolic syndrome, psychiatric diseases, and neoplasms. The evaluation of PCOS is not straightforward owing to its variable clinical features. An evaluation should focus on the time course of symptoms and the presence or absence of the signs of hyperandrogenism and ovulatory dysfunction. A brief psychiatric evaluation is also imperative because of the high rates of psychiatric disturbances amongst women with PCOS. Because there are many conditions that may mimic certain features of the syndrome, the physician must conduct a broad review of systems and have a keen eye for detail on the physical exam. Tables 16.7 and 16.8 provide a succinct summary of the pertinent pieces of the history and physical examination to consider when evaluating a patient for PCOS. Healthcare professionals must be aware of the full scope of this broad disease to best diagnose and care for their patients.

References

1. Zawadzki JK, Dunaif, A. *Diagnostic Criteria for Polycystic Ovary Syndrome: towards a rational approach.* Boston: Blackwell Scientific Publications, 1992.

2. Rotterdam ESHRE/ASRM-Sponsored PCOS Consensus Workshop Group. Revised 2003 consensus on diagnostic criteria and long-term health risks related to polycystic ovary syndrome. *Fertil Steril* 2004;81(1):19–25.

3. Azziz R, Carmina E, Dewailly D, et al. The androgen excess and PCOS society criteria for the polycystic ovary syndrome: the complete task force report. *Fertil Steril* 2009;91(2):456–488.

4. March WA, Moore VM, Willson KJ, Phillips DI, Norman RJ, Davies MJ. The prevalence of polycystic ovary syndrome in a community sample assessed under contrasting diagnostic criteria. *Hum Reprod* 2010;25(2):544–551.

5. Vink JM, Sadrzadeh S, Lambalk CB, Boomsma DI. Heritability of polycystic ovary syndrome in a Dutch twin-family study. *J Clin Endocrinol Metab* 2006;91(6):2100–2104.

6. Legro RS, Arslanian SA, Ehrmann DA, et al. Diagnosis and treatment of polycystic ovary syndrome: an Endocrine Society clinical practice guideline. *J Clin Endocrinol Metab* 2013;98(12):4565–4592.

7. Panidis D, Tziomalos K, Macut D, et al. Cross-sectional analysis of the effects of age on the hormonal, metabolic, and ultrasonographic features and the prevalence of the different phenotypes of polycystic ovary syndrome. *Fertil Steril* 2012;97(2):494–500.

8. Liou TH, Yang JH, Hsieh CH, Lee CY, Hsu CS, Hsu MI. Clinical and biochemical presentations of polycystic ovary syndrome among obese and nonobese women. *Fertil Steril* 2009;92(6):1960–1965.

9. Goodman NF, Bledsoe MB, Cobin RH, et al. American Association of Clinical Endocrinologists medical guidelines for the clinical practice for the diagnosis and treatment of hyperandrogenic disorders. *Endocr Pract* 2001;7(2):120–134.

10. Ferriman D, Gallwey JD. Clinical assessment of body hair growth in women. *J Clin Endocrinol Metab* 1961;21:1440–1447.

11. Martin KA, Chang RJ, Ehrmann DA, et al. Evaluation and treatment of hirsutism in premenopausal women: an Endocrine Society clinical practice guideline. *J Clin Endocrinol Metab*. 2008;93(4):1105–1120.

12. Ibanez L, Potau N, Francois I, de Zegher F. Precocious pubarche, hyperinsulinism, and ovarian hyperandrogenism in girls: Relation to reduced fetal growth. *J Clin Endocrinol Metab* 1998;83(10):3558–3562.

13. Guzel AI, Kuyumcuoglu U, Celik Y. Factors affecting the degree of hirsutism in patients with polycystic ovary syndrome. *Arch Gynecol Obstet* 2012;285(3):767–770.

14. Bhattacharya SM, Ghosh M. Insulin resistance and adolescent girls with polycystic ovary syndrome. *J Pediatr Adolesc Gynecol* 2010;23(3):158–161.

15. Torvinen A, Koivunen R, Pouta A, et al. Metabolic and reproductive characteristics of first-degree relatives of women with self-reported oligo-amenorrhoea and hirsutism. *Gynecol Endocrinol* 2011;27(9):630–635.

16. Cela E, Robertson C, Rush K, et al. Prevalence of polycystic ovaries in women with androgenic alopecia. *Eur J Endocrinol* 2003;149(5):439–442.

17. Vexiau P, Chaspoux C, Boudou P, et al. Role of androgens in female-pattern androgenetic alopecia, either alone or associated with other symptoms of hyperandrogenism. *Arch Dermatol Res* 2000;292(12):598–604.

18. Shi Y, Guo M, Yan J, et al. Analysis of clinical characteristics in large-scale Chinese women with polycystic ovary syndrome. *Neuro Endocrinol Lett* 2007;28(6):807–810.

19. Brown ZA, Louwers YV, Fong SL, et al. The phenotype of polycystic ovary syndrome ameliorates with aging. *Fertil Steril* 2011;96(5):1259–1265.

20. Bronstein J, Tawdekar S, Liu Y, Pawelczak M, David R, Shah B. Age of onset of polycystic ovarian syndrome in girls may be earlier than previously thought. *J Pediatr Adolesc Gynecol* 2011;24(1):15–20.

21. Elting MW, Korsen TJ, Rekers-Mombarg LT, Schoemaker J. Women with polycystic ovary syndrome gain regular menstrual cycles when ageing. *Hum Reprod* 2000;15(1):24–28.

22. Carmina E, Campagna AM, Lobo RA. A 20-year follow-up of young women with polycystic ovary syndrome. *Obstet Gynecol* 2012;119(2 Pt 1):263–269.

23. Dahlgren E, Johansson S, Lindstedt G, et al. Women with polycystic ovary syndrome wedge resected in 1956 to 1965: a long-term follow-up focusing on natural history and circulating hormones. *Fertil Steril* 1992;57(3):505–513.

24. Nazir F, Tasleem H, Tasleem S, Sher Z, Waheed K. Polycystic ovaries in adolescent girls from Rawalpindi. *J Pak Med Assoc* 2011;61(10):961–963.

25. Sam S, Legro RS, Essah PA, Apridonidze T, Dunaif A. Evidence for metabolic and reproductive phenotypes in mothers of women with polycystic ovary syndrome. *Proc Natl Acad Sci USA* 2006;103(18):7030–7035.

26. Third report of the National Cholesterol Education Program (NCEP) expert panel on detection, evaluation, and treatment of high blood cholesterol in adults (Adult Treatment Panel III). Final report. *Circulation* 2002;106(25): 3143–3421.

27. Grundy SM, Cleeman JI, Daniels SR, et al. Diagnosis and management of the metabolic syndrome: an American Heart Association/National Heart, Lung, and Blood Institute scientific statement. *Circulation* 2005;112(17):2735–2752.

28. Moran LJ, Misso ML, Wild RA, Norman RJ. Impaired glucose tolerance, type 2 diabetes and metabolic syndrome in polycystic ovary syndrome: a systematic review and meta-analysis. *Hum Reprod Update* 2010;16(4):347–363.

29. Rahmanpour H, Jamal L, Mousavinasab SN, Esmailzadeh A, Azarkhish K. Association between polycystic ovarian syndrome, overweight, and metabolic syndrome in adolescents. *J Pediatr Adolesc Gynecol* 2012;25(3):208–212.

30. Ehrmann DA, Liljenquist DR, Kasza K, et al. Prevalence and predictors of the metabolic syndrome in women with polycystic ovary syndrome. *J Clin Endocrinol Metab* 2006;91(1):48–53.

31. Lord J, Thomas R, Fox B, Acharya U, Wilkin T. The central issue? Visceral fat mass is a good marker of insulin resistance and metabolic disturbance in women with polycystic ovary syndrome. *BJOG* 2006;113(10):1203–1209.

32. Palep-Singh M, Picton HM, Barth JH, Balen AH. Ethnic variations in the distribution of obesity and biochemical metabolic abnormalities in fertility clinic attendees. *J Reprod Med* 2008;53(2):117–123.

33. Guo M, Chen ZJ, Macklon NS, et al. Cardiovascular and metabolic characteristics of infertile Chinese women with PCOS diagnosed according to the Rotterdam consensus criteria. *Reprod Biomed Online* 2010;21(4):572–580.

34. Castelo-Branco C, Steinvarcel F, Osorio A, Ros C, Balasch J. Atherogenic metabolic profile in PCOS patients: role of obesity and hyperandrogenism. *Gynecol Endocrinol* 2010;26(10):736–742.

35. Wild RA, Carmina E, Diamanti-Kandarakis E, et al. Assessment of cardiovascular risk and prevention of cardiovascular disease in women with the polycystic ovary syndrome: a consensus statement by the Androgen Excess and Polycystic Ovary Syndrome (AE-PCOS) Society. *J Clin Endocrinol Metab* 2010;95(5):2038–2049.

36. Ehrmann DA, Kasza K, Azziz R, Legro RS, Ghazzi MN, PCOS/Troglitazone Study Group. Effects of race and family history of type 2 diabetes on metabolic status of women with polycystic ovary syndrome. *J Clin Endocrinol Metab* 2005;90(1):66–71.

37. Escobar-Morreale HF, Luque-Ramirez M, San Millan JL. The molecular-genetic basis of functional hyperandrogenism and the polycystic ovary syndrome. *Endocr Rev* 2005;26(2):251–282.

38. Yildiz BO, Knochenhauer ES, Azziz R. Impact of obesity on the risk for polycystic ovary syndrome. *J Clin Endocrinol Metab* 2008;93(1):162–168.

39. Gambineri A, Pelusi C, Vicennati V, Pagotto U, Pasquali R. Obesity and the polycystic ovary syndrome. *Int J Obes Relat Metab Disord* 2002;26(7):883–896.

40. Wehr E, Moller R, Horejsi R, et al. Subcutaneous adipose tissue topography and metabolic disturbances in polycystic ovary syndrome. *Wien Klin Wochenschr* 2009;121(7–8):262–269.

41. Lo JC, Feigenbaum SL, Escobar GJ, Yang J, Crites YM, Ferrara A. Increased prevalence of gestational diabetes mellitus among women with diagnosed polycystic ovary syndrome: a population-based study. *Diabetes Care* 2006;29(8):1915–1917.

42. Coviello AD, Sam S, Legro RS, Dunaif A. High prevalence of metabolic syndrome in first-degree male relatives of women with polycystic ovary syndrome is related to high rates of obesity. *J Clin Endocrinol Metab* 2009;94(11):4361–4366.

43. Trottier A, Battista MC, Geller DH, et al. Adipose tissue insulin resistance in peripubertal girls with first-degree family history of polycystic ovary syndrome. *Fertil Steril* 2012;98(6):1627–1634.

44. Shaw LJ, Bairey Merz CN, Azziz R, et al. Postmenopausal women with a history of irregular

menses and elevated androgen measurements at high risk for worsening cardiovascular event-free survival: results from the National Institutes of Health–National Heart, Lung, and Blood Institute sponsored women's ischemia syndrome evaluation. *J Clin Endocrinol Metab* 2008;93(4):1276–1284.

45. Cupisti S, Kajaia N, Dittrich R, Duezenli H, W Beckmann M, Mueller A. Body mass index and ovarian function are associated with endocrine and metabolic abnormalities in women with hyperandrogenic syndrome. *Eur J Endocrinol* 2008;158(5):711–719.

46. Cho LW, Kilpatrick ES, Keevil BG, et al. Insulin resistance variability in women with anovulatory and ovulatory polycystic ovary syndrome, and normal controls. *Horm Metab Res* 2011;43(2):141–145.

47. Legro RS, Gnatuk CL, Kunselman AR, Dunaif A. Changes in glucose tolerance over time in women with polycystic ovary syndrome: a controlled study. *J Clin Endocrinol Metab* 2005;90(6):3236–3242.

48. Holte J, Gennarelli G, Berne C, Bergh T, Lithell H. Elevated ambulatory day-time blood pressure in women with polycystic ovary syndrome: a sign of a pre-hypertensive state? *Hum Reprod* 1996;11(1):23–28.

49. Elting MW, Korsen TJ, Bezemer PD, Schoemaker J. Prevalence of diabetes mellitus, hypertension and cardiac complaints in a follow-up study of a Dutch PCOS population. *Hum Reprod* 2001;16(3):556–560.

50. Luque-Ramirez M, Alvarez-Blasco F, Mendieta-Azcona C, Botella-Carretero JI, Escobar-Morreale HF. Obesity is the major determinant of the abnormalities in blood pressure found in young women with the polycystic ovary syndrome. *J Clin Endocrinol Metab* 2007;92(6):2141–2148.

51. Barcellos CR, Rocha MP, Hayashida SA, Mion Junior D, Lage SG, Marcondes JA. Impact of body mass index on blood pressure levels in patients with polycystic ovary syndrome. *Arq Bras Endocrinol Metabol* 2007;51(7):1104–1109.

52. Tasali E, Van Cauter E, Ehrmann DA. Relationships between sleep disordered breathing and glucose metabolism in polycystic ovary syndrome. *J Clin Endocrinol Metab* 2006;91(1):36–42.

53. de Sousa G, Schluter B, Menke T, Trowitzsch E, Andler W, Reinehr T. A comparison of polysomnographic variables between adolescents with polycystic ovarian syndrome with and without the metabolic syndrome. *Metab Syndr Relat Disord* 2011;9(3):191–196.

54. Toulis KA, Goulis DG, Mintziori G, et al. Meta-analysis of cardiovascular disease risk markers in women with polycystic ovary syndrome. *Hum Reprod Update* 2011;17(6):741–760.

55. de Groot PC, Dekkers OM, Romijn JA, Dieben SW, Helmerhorst FM. PCOS, coronary heart disease,

stroke and the influence of obesity: a systematic review and meta-analysis. *Hum Reprod Update* 2011;17(4):495–500.

56. Luque-Ramirez M, Mendieta-Azcona C, Alvarez-Blasco F, Escobar-Morreale HF. Androgen excess is associated with the increased carotid intima-media thickness observed in young women with polycystic ovary syndrome. *Hum Reprod* 2007;22(12):3197–3203.

57. Christian RC, Dumesic DA, Behrenbeck T, Oberg AL, Sheedy PF, 2nd, Fitzpatrick LA. Prevalence and predictors of coronary artery calcification in women with polycystic ovary syndrome. *J Clin Endocrinol Metab* 2003;88(6):2562–2568.

58. Iftikhar S, Collazo-Clavell ML, Roger VL, et al. Risk of cardiovascular events in patients with polycystic ovary syndrome. *Neth J Med* 2012;70(2):74–80.

59. Mani H, Levy MJ, Davies MJ, et al. Diabetes and cardiovascular events in women with polycystic ovary syndrome: a 20-year retrospective cohort study. *Clin Endocrinol (Oxf)* 2013;78(6):926–934.

60. Brzozowska MM, Ostapowicz G, Weltman MD. An association between non-alcoholic fatty liver disease and polycystic ovarian syndrome. *J Gastroenterol Hepatol* 2009;24(2):243–247.

61. Zueff LF, Martins WP, Vieira CS, Ferriani RA. Ultrasonographic and laboratory markers of metabolic and cardiovascular disease risk in obese women with polycystic ovary syndrome. *Ultrasound Obstet Gynecol* 2012;39(3):341–347.

62. Gutierrez-Grobe Y, Ponciano-Rodriguez G, Ramos MH, Uribe M, Mendez-Sanchez N. Prevalence of non alcoholic fatty liver disease in premenopausal, postmenopausal and polycystic ovary syndrome women the role of estrogens. *Ann Hepatol* 2010;9(4):402–409.

63. Barfield E, Liu YH, Kessler M, Pawelczak M, David R, Shah B. The prevalence of abnormal liver enzymes and metabolic syndrome in obese adolescent females with polycystic ovary syndrome. *J Pediatr Adolesc Gynecol* 2009;22(5):318–322.

64. Kroenke K, Spitzer RL, Williams JB. The patient health questionnaire-2: Validity of a two-item depression screener. *Med Care* 2003;41(11):1284–1292.

65. Gilbody S, Richards D, Brealey S, Hewitt C. Screening for depression in medical settings with the patient health questionnaire (PHQ): a diagnostic meta-analysis. *J Gen Intern Med* 2007;22(11):1596–1602.

66. Cinar N, Kizilarslanoglu MC, Harmanci A, et al. Depression, anxiety and cardiometabolic risk in polycystic ovary syndrome. *Hum Reprod* 2011;26(12):3339–3345.

67. Hollinrake E, Abreu A, Maifeld M, Van Voorhis BJ, Dokras A. Increased risk of depressive disorders in women with polycystic ovary syndrome. *Fertil Steril* 2007;87(6):1369–1376.

68. Jones GL, Hall JM, Lashen HL, Balen AH, Ledger WL. Health-related quality of life among adolescents with polycystic ovary syndrome. *J Obstet Gynecol Neonatal Nurs* 2011;40(5):577–588.

69. Barry JA, Kuczmierczyk AR, Hardiman PJ. Anxiety and depression in polycystic ovary syndrome: a systematic review and meta-analysis. *Hum Reprod* 2011;26(9):2442–2451.

70. Pastore LM, Patrie JT, Morris WL, Dalal P, Bray MJ. Depression symptoms and body dissatisfaction association among polycystic ovary syndrome women. *J Psychosom Res* 2011;71(4):270–276.

71. Kowalczyk R, Skrzypulec V, Lew-Starowicz Z, Nowosielski K, Grabski B, Merk W. Psychological gender of patients with polycystic ovary syndrome. *Acta Obstet Gynecol Scand* 2012;91(6):710–714.

72. Stovall DW, Scriver JL, Clayton AH, Williams CD, Pastore LM. Sexual function in women with polycystic ovary syndrome. *J Sex Med* 2012;9(1):224–230.

73. Dokras A, Clifton S, Futterweit W, Wild R. Increased prevalence of anxiety symptoms in women with polycystic ovary syndrome: systematic review and meta-analysis. *Fertil Steril* 2012;97(1):225–30.e2.

74. Deeks AA, Gibson-Helm ME, Paul E, Teede HJ. Is having polycystic ovary syndrome a predictor of poor psychological function including anxiety and depression? *Hum Reprod* 2011;26(6):1399–1407.

75. Kerchner A, Lester W, Stuart SP, Dokras A. Risk of depression and other mental health disorders in women with polycystic ovary syndrome: a longitudinal study. *Fertil Steril* 2009;91(1):207–212.

76. Mansson M, Holte J, Landin-Wilhelmsen K, Dahlgren E, Johansson A, Landen M. Women with polycystic ovary syndrome are often depressed or anxious–a case control study. *Psychoneuroendocrinology* 2008;33(8):1132–1138.

77. Lamb JD, Johnstone EB, Rousseau JA, et al. Physical activity in women with polycystic ovary syndrome: prevalence, predictors, and positive health associations. *Am J Obstet Gynecol* 2011;204(4):352. e1–352.e6.

78. Thomson RL, Buckley JD, Lim SS, et al. Lifestyle management improves quality of life and depression in overweight and obese women with polycystic ovary syndrome. *Fertil Steril* 2010;94(5):1812–1816.

79. Roos N, Kieler H, Sahlin L, Ekman-Ordeberg G, Falconer H, Stephansson O. Risk of adverse pregnancy outcomes in women with polycystic ovary syndrome: population based cohort study. *BMJ.* 2011;343:d6309.

80. Koivunen R, Pouta A, Franks S, et al. Fecundability and spontaneous abortions in women with self-reported oligo-amenorrhea and/or hirsutism: Northern Finland birth cohort 1966 study. *Hum Reprod* 2008;23(9):2134–2139.

81. Kabiru W, Raynor BD. Obstetric outcomes associated with increase in BMI category during pregnancy. *Am J Obstet Gynecol* 2004;191(3):928–932.

82. Qin JZ, Pang LH, Li MJ, Fan XJ, Huang RD, Chen HY. Obstetric complications in women with polycystic ovary syndrome: a systematic review and meta-analysis. *Reprod Biol Endocrinol* 2013;11:56.

83. Legro RS, Barnhart HX, Schlaff WD, et al. Clomiphene, metformin, or both for infertility in the polycystic ovary syndrome. *N Engl J Med.* 2007;356(6):551–566.

84. Boomsma CM, Eijkemans MJ, Hughes EG, Visser GH, Fauser BC, Macklon NS. A meta-analysis of pregnancy outcomes in women with polycystic ovary syndrome. *Hum Reprod Update* 2006;12(6):673–683.

85. Cocksedge KA, Saravelos SH, Metwally M, Li TC. How common is polycystic ovary syndrome in recurrent miscarriage? *Reprod Biomed Online* 2009;19(4):572–576.

86. Winter E, Wang J, Davies MJ, Norman R. Early pregnancy loss following assisted reproductive technology treatment. *Hum Reprod* 2002;17(12):3220–3223.

87. Ozgun MT, Uludag S, Oner G, Batukan C, Aygen EM, Sahin Y. The influence of obesity on ICSI outcomes in women with polycystic ovary syndrome. *J Obstet Gynaecol* 2011;31(3):245–249.

88. Hasegawa I, Tanaka K, Sanada H, Imai T, Fujimori R. Studies on the cytogenetic and endocrinologic background of spontaneous abortion. *Fertil Steril* 1996;65(1):52–54.

89. Moini A, Tadayon S, Tehranian A, Yeganeh LM, Akhoond MR, Yazdi RS. Association of thrombophilia and polycystic ovarian syndrome in women with history of recurrent pregnancy loss. *Gynecol Endocrinol* 2012;28(8):590–593.

90. Reyes-Munoz E, Castellanos-Barroso G, Ramirez-Eugenio BY, et al. The risk of gestational diabetes mellitus among Mexican women with a history of infertility and polycystic ovary syndrome. *Fertil Steril* 2012;97(6):1467–1471.

91. Palomba S, Falbo A, Russo T, et al. The risk of a persistent glucose metabolism impairment after gestational diabetes mellitus is increased in patients with polycystic ovary syndrome. *Diabetes Care* 2012;35(4):861–867.

92. Bjercke S, Dale PO, Tanbo T, Storeng R, Ertzeid G, Abyholm T. Impact of insulin resistance on pregnancy complications and outcome in women with polycystic ovary syndrome. *Gynecol Obstet Invest* 2002;54(2):94–98.

93. Hizli D, Kosus A, Kosus N, Kamalak Z, Ak D, Turhan NO. The impact of birth weight and maternal history on acne, hirsutism, and menstrual disorder symptoms in Turkish adolescent girls. *Endocrine* 2012;41(3):473–478.

94. McDonald SD, Han Z, Mulla S, et al. Preterm birth and low birth weight among in vitro fertilization singletons: a systematic review and meta-analyses. *Eur J Obstet Gynecol Reprod Biol* 2009;146(2):138–148.

95. Barry JA, Azizia MM, Hardiman PJ. Risk of endometrial, ovarian and breast cancer in women with polycystic ovary syndrome: a systematic review and meta-analysis. *Hum Reprod Update* 2014;20(5):748–758.

96. Pierpoint T, McKeigue PM, Isaacs AJ, Wild SH, Jacobs HS. Mortality of women with polycystic ovary syndrome at long-term follow-up. *J Clin Epidemiol* 1998;51(7):581–586.

97. Witchel SF, Azziz R. Nonclassic congenital adrenal hyperplasia. *Int J Pediatr Endocrinol* 2010;2010:625105.

98. Bahn RS, Burch HB, Cooper DS, et al. Hyperthyroidism and other causes of thyrotoxicosis: management guidelines of the American Thyroid Association and American Association of Clinical Endocrinologists. *Endocr Pract* 2011;17(3):456–520.

99. Garber JR, Cobin RH, Gharib H, et al. Clinical practice guidelines for hypothyroidism in adults: cosponsored by the American Association of Clinical Endocrinologists and the American Thyroid Association. *Endocr Pract* 2012;18(6):988–1028.

100. Neraud B, Dewailly D. Drug-induced hyperandrogenism. In: Azziz R, Nestler JE, Dewailly D, eds. *Contemporary Endocrinology: Androgen Excess Disorders in Women, Polycystic Ovarian Syndrome and Other Disorders*. Second edition. Totowa, NJ: Humana Press Inc, 2006:121–124.

Androgen assays: of mice and men

Keith A. Hansen

Measurement error is one of the most important considerations in clinical medicine, but the effect of measurement error on clinical practice continues to be underestimated. (P. G. McDonough, 2003 [1])

Introduction

Laboratory data are meant to support the clinician's diagnosis based on a thorough history and physical examination. When this laboratory information conflicts with the differential diagnosis, one needs a high index of suspicion for a wrong diagnosis or an error in the laboratory information. Hence, it is important that all clinicians have an understanding of the basics of the assays they order in their clinical practices, and how these assays can be influenced by both exogenous and endogenous factors. A laboratory value not only depends upon analytical factors, but also on pre-analytical handling, in other words how the sample was collected, processed, and stored.

In the human, androgens are primarily made by the gonads, ovary in the female and testicle in the male, as well as the adrenal gland. In the ovary, androgens are necessary precursors for the formation of estrogens by the developing follicle. Follicular development with resultant estradiol production is tightly regulated by pituitary gonadotropins, luteinizing hormone (LH), and follicle-stimulating hormone (FSH). Androgen production by the adrenal gland makes its debut, also known as adrenarche, prior to puberty with development and maturation of the zona reticularis. Adrenarche occurs prior to gonadal sex steroid production, with the adrenal gland secreting dehydroepiandrosterone (DHEA), its sulfate (DHEA-S), and androstenedione. The factor or factors that control onset of adrenarche and increased adrenal androgen production are not completely known. Adrenocorticotropic hormone (ACTH) is necessary, but is not sufficient for initiation of adrenarche and for ongoing adrenal androgen production. There are a number of proposed factors, including cortical androgen-stimulating hormone, which may induce development and maturation of the zona reticularis with subsequent androgen production. Of interest, neither gonadal nor adrenal androgens in the female are under direct feedback control by gonadotropins or ACTH, and it appears that paracrine and autocrine factors may play an important role in their production [2].

Androgens are defined by their ability to bind to the androgen receptor and induce changes in the target cells. The primary circulating androgen is testosterone, which exerts most of its biological activity by being converted to dihydrotestosterone, by 5α-reductase. Dihydrotestosterone binds more avidly and dissociates more slowly from the androgen receptor than testosterone, and hence accounts for a large percentage of androgen activity [3]. Many of the androgens like DHEA, DHEA-S, and androstenedione are really pre-hormones, because they do not primarily bind to the androgen receptor, but must be converted into the more potent androgen, testosterone, prior to binding [4].

In the female, circulating testosterone is derived from the ovary and adrenal gland. Approximately 25% of testosterone is directly secreted by the ovary, 25% by the adrenal gland, and the remaining 50% is derived from peripheral conversion of androstenedione to testosterone. Similar to testosterone, about 25% of the precursor, androstenedione, is made by the ovary and another 25% by the adrenal gland. About 50% of DHEA is primarily secreted by the adrenal gland, and another 30% is produced by conversion of DHEA-S to DHEA. Only 20% of DHEA is made by the ovary. Another important variable is that androgen levels vary throughout the day and throughout the stages of life [5].

Testosterone circulates in the blood in a highly protein-bound state. Only 1–2% of testosterone is free in the circulation, while 60–70% is tightly bound to sex hormone-binding globulin (SHBG) and the rest is loosely bound to serum albumin. SHBG is a high-affinity, low-capacity binding protein which is felt to function as a biological reservoir of testosterone. Albumin is a lower-affinity, high-capacity binding factor in which the testosterone may disassociate easily and enter the cell [6]. When measuring testosterone, one can measure free testosterone, biologically active testosterone (free and albumin bound), and total testosterone (free, albumin bound, and SHBG-bound) levels. SHBG-testosterone has always been considered to be inert and simply a biological reservoir, but recent articles suggest that it may also contribute to androgen activity by binding to SHBG receptors, resulting in cellular internalization of the complex [7].

The study of endocrinology is founded on the principle that hormones secreted by glands circulate in the bloodstream and control the activity of distant targets. Hence, the concentration of the circulating hormones reflects the activity of the specific endocrine gland and its effect on the target organs. Therefore, abnormal levels of a hormone can reflect hypofunction or hyperfunction of the respective endocrine gland. With testosterone, levels are expected to be low in prepubertal children and rise with the onset of puberty, with a much larger increase in boys than girls. Abnormally low levels are found in males and females with hypogonadism, while elevated levels are found in females with clinical evidence of hyperandrogenism, boys with precocious puberty, and females with heterosexual precocious puberty.

Clinical endocrinology and advancements in endocrinology rely on our ability to measure the circulating hormones of interest and our ability to differentiate normal from abnormal levels. In clinical endocrinology, the patient serves as her own bioassay, and during the history and physical examination the clinician searches for signs and symptoms of hypo- or hyperfunction of endocrine glands. An example is the woman with polycystic ovaries who has signs and symptoms of hyperandrogenism. Laboratory assays help to confirm, document the severity, and assist in determining the etiology of the hormonal abnormality.

Assays

Bioassays: in vivo

The first assays for determining testosterone levels were in vivo bioassays, with the most common measurement being growth of the capon comb or male rat accessory sex organs [8]. In 1953, Hershberger et al. described an assay, which is still used to measure androgenic and antiandrogenic activity of compounds, using castrated male rats or mice and measuring the gain in weight of the accessory sexual organs, such as the seminal vesicles, prostate gland, and levator ani muscles [9]. These bioassays are highly specific, but are time intensive and expensive.

Immunoassay

In 1960, a seminal publication by Yalow and Berson in the *Journal of Clinical Investigation* on the immunoassay technique launched a new era in our ability to measure small molecules in the circulation [10]. This innovative technique resulted in Dr. Rosalyn S. Yalow receiving the Nobel Prize in Physiology or Medicine in 1977. Dr. Solomon Berson did not share in this Nobel Prize because he died in 1972 and this prize is not awarded posthumously.

All immunoassays depend upon the biological reaction of an antibody with the antigen of interest. Immunoassays are currently used in a number of clinical tests. These tests include many of our common endocrine assays, tumor and cardiac biomarkers, drugs both therapeutic and those of abuse, and a number of other analytes. There are two major types of immunoassay: competitive and non-competitive, characterized by the immunometric sandwich assay.

The competitive immunoassay is characterized by having a high affinity, but limiting amount of antibody. The antisera are then mixed with a known concentration of labeled antigen and the unknown, unlabeled antigen in the sample. Then there is competition between the labeled and unlabeled analyte proportional to the amount of unlabeled analyte in the sample. The larger the amount of unlabeled analyte in the sample, the less labeled analyte binds to the antibody, and vice versa. The next step classically is separating bound antigen from unbound antigen and then measuring the bound labeled antigen. Hence, there is an inverse relationship between the amount of bound labeled antigen (measured) and the amount of unlabeled antigen. One develops a standardized curve using known amounts of unlabeled antigen and measuring the displacement

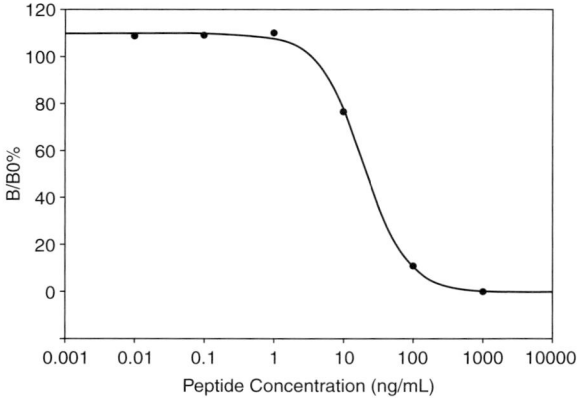

Figure 17.1 Competitive immunoassay. The Y-axis is the % of Bound/Bound (B/B) at 0% concentration of unlabeled antigen, while the X-axis is the concentration of unlabeled antigen. From Abnova (Taiwan) Corporation at www.abnova.com.

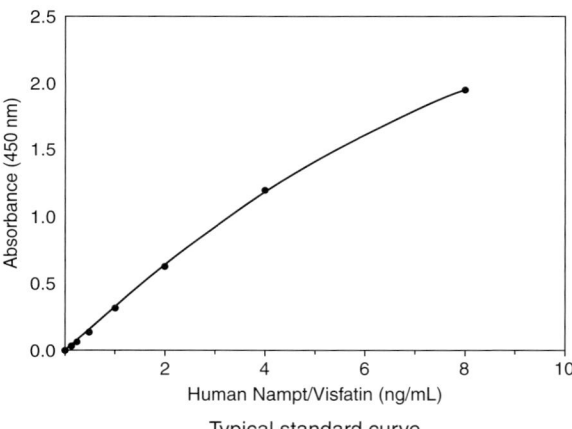

Typical standard curve

Figure 17.2 Standard curve for an immunometric (sandwich) assay. From Creative Diagnostics at www.creative-diagnostics.com.

of labeled analyte (Figure 17.1). One can then use this assay for a sample with an unknown amount of unlabeled antigen to determine the amount of antigen in the sample [11]. The competitive assay was the original type of assay used by Yalow and Berson.

The immunometric assay is characterized by having two or more antibodies, which are directed to different epitopes (antigenic sites) on the target antigen. One of these antibodies is coupled to a solid matrix, while the other antibody is labeled with a signal system, such as an enzyme, fluorescent tag, or chemiluminescent compound. These two antibodies form a "sandwich" with the target antigen caught between the two antibodies. The signal, which is generated by the labeled antibody,

can then be detected by the appropriate detector, whether that is radioactivity or luminescence [12]. In this type of assay, there is a direct relationship between the amount of signal and target antigen in the sample (Figure 17.2).

Mass spectrometry

Mass spectrometry (MS) is a detector system for chromatographic systems which allows one to further evaluate a molecule in terms of its molecular weight and structure, and consequently is able to specifically detect a molecule based on its molecular structure [13]. Gas chromatography–mass spectrometry (GC-MS) is considered the gold standard for evaluating and determining the structure of steroids and their metabolites. However, GC-MS requires prolonged run times and is a complex process, making it difficult to adapt to high throughput systems, such as those needed in clinical laboratories. A large amount of effort has been put into improving MS over the past years to make it more user friendly for the clinical laboratory. Recent refinements in the use of high-pressure liquid chromatography–tandem mass spectrometry (HPLC-MS/MS) have improved the clinical applicability of this technology, by maintaining the sensitivity and specificity of MS with the high throughput needed by the clinical laboratory. HPLC is primarily used to purify and separate the compound of interest from interfering substances. For testosterone, usually a reverse-phase C18 and C8 column are used with water as the polar phase and methanol or acetonitrile as the organic phases. Another method to enhance sensitivity of the assay, especially if there is a small volume, is to use a derivatization procedure to improve ionization and fragmentation. Hydroxylamine is often used as the derivatization agent for testosterone, producing testosterone oxime. Since MS works in vacuum conditions, once the sample is chromatographed the solvent must be removed with ionization of the analyte. Ionization occurs by a process of atmospheric pressure ionization, which generates ions that maintain the structure of the molecule. This ionization process facilitates separation by the MS technique of the steps of ionization from fragmentation of the molecule of interest, which allows for a specific determination of molecular weight and structure of the compound.

Bioassays: in vitro

Bioassays are unique in that they can determine the hormonal activity within a sample, regardless of the structure of the compounds within that sample. Because these assays measure the total activity within a sample, one can also measure anti-hormonal activity within the sample. These assays are indispensable in the search for environmental agents with hormonal activity as well as the biological activity of new "doping agents." These assays may play a future role in the clinical evaluation of male hypogonadism as well as in women and children with signs and symptoms of hyperandrogenism.

These assays use cells or cell lines, often genetically modified, as the primary basis for the assay technique. They are used as a screening tool to evaluate a test sample for biological activity and are improved over in vivo bioassays, with better sensitivity, specificity, higher cost effectiveness, and the potential for a higher volume of samples. However, they do have a number of limitations: they are unable to monitor the in vivo metabolism of a substance because some of the cells used in the assay may metabolize the target substance, altering its biological activity, and the cells used may have different co-regulators than the cells of interest, resulting in variable responses.

The in vitro bioassays can be divided into three major categories: receptor binding assays, cell proliferation assays, and reporter gene assays. Receptor binding assays have been developed to measure compounds that specifically bind to a receptor of interest, which in androgen assays is the androgen receptor [14]. These assays involve combining the sample of interest, labeled receptor ligand, and the receptor. The amount of bound, labeled ligand displaced by the unknown sample corresponds with the receptor binding activity of the unknown.

In the cell proliferation assay, one measures the degree of cellular proliferation that results from addition of an unknown sample to the cell culture. This has been best developed for screening of estrogenic activity in a substance using MCF-7 cells, also known as the E-SCREEN assay. Similar types of assay have been developed for androgens, one a prostate cell line and another MCF-7 cell line transfected with the human androgen receptor. These assays are relatively simple and highly sensitive, but there is a good deal of variation between batches [15].

Reporter gene assays are designed to measure hormonal activity in an unknown sample by measuring the response of a hormone-dependent reporter gene. These types of assay have been developed for measuring serum androgen activity, and are accomplished by transfecting a mammalian cell line or yeast strain with the hormonal receptor of interest as well as a reporter gene. The reporter gene is composed of a promoter, which contains hormone response elements for the hormone receptor of interest coupled to the reporter gene. When an unknown with the hormone of interest is added to this mixture, it binds to the hormone receptor, the ligand–receptor complex then binds to the hormone response elements and activates the reporter gene. With activation, the reporter gene activates transcription and translation of an enzyme such as luciferase or chloramphenicol acetyl transferase. This enzyme then converts the substrate in the mixture into a signal, which can be measured by luminometry, spectrophotometry, or fluorimetry [16]. Reporter gene assays using mammalian cells are becoming more common and are being developed for use with steroids, specifically androgens.

Androgen assays

There are a number of significant challenges in measuring testosterone levels, including the wide range of concentrations between prepubertal children, women, and adult men. Another important aspect is that testosterone secretion follows a circadian rhythm, with highest concentration in the morning and decreasing throughout the day. When measuring testosterone, it is important to consider that only 1–2% of this steroid circulates in a free, biologically active form. There are also other circulating steroids that closely resemble testosterone and may interfere with the assay. Furthermore, there is no universally accepted testosterone standard and validated normal ranges are wanting.

Total testosterone

The gold standard for measuring total testosterone is GC-MS. However, GC-MS is a time-consuming, labor-intensive method limiting its applicability in a high-volume clinical setting. Initial immunoassays involved extraction and chromatography prior to the assay. These initial steps improved accuracy and sensitivity by removing interfering proteins, separating out cross-reacting steroids, and allowing for the processing of a large sample to increase sensitivity. Once again, these initial steps of extraction and chromatography were labor-intensive, cumbersome, and time-consuming, which limits their clinical applicability. To meet the need for high-throughput and speed of turnaround, many direct immunoassays have been developed for measuring total testosterone. While these assays meet the clinical demand for a rapid, high-throughput assay, they have limited accuracy, especially for lower testosterone levels (<300 ng/dL). This is a severe limitation, since the need for a sensitive and accurate assay is to assist in differentiating between pathology and physiology, which for testosterone is differentiating between normal and low levels of the analyte [17,18].

The College of American Pathologists (CAP) has a Quality Control program where they distribute a blinded sample to clinical laboratories, which allows clinical laboratories to compare the accuracy of their assays to other laboratories using the same technology. Previous studies have demonstrated unacceptable variation with large coefficients of variation (c.v.) for lower concentrations, with improvement in variability with increasing testosterone levels. For normal female androgen levels, the c.v. ranged between 13% and 32%, while for the hypogonadal male or androgenized female, there was improvement, but still significant variability in the assays. This large variability significantly limits the use of total testosterone measurement in hypogonadal males and females. These assays had the best reproducibility in measuring ranges of testosterone in eugonadal males. In studies comparing MS to immunoassay of testosterone, most immunoassays overestimated the testosterone level [19,20].

Free testosterone

The gold standard for measuring free testosterone is with equilibrium dialysis. In this method, labeled ^{3}H-testosterone is added to the unknown sample, equilibrium is allowed to occur across a semipermeable membrane, the bound and free ^{3}H-testosterone are separated, and the free portion measured. The total testosterone level is then multiplied by the free fraction of ^{3}H-testosterone to obtain the free testosterone level. Another similar method is ultrafiltration of the sample. These methods are relatively cumbersome, expensive, can be influenced by absorption of analyte to the membrane or by temperature, and are highly dependent on the measurement of total testosterone [21].

Direct immunoassay measurements of free testosterone are attractive in the clinical setting, because of their speed, ease of use, and adaptability to high throughput; however, they have been found to be clinically unreliable with a high variability, poor accuracy, and poor sensitivity. These assays have not been found to be helpful in clinical or research settings [22].

SHBG testosterone can be precipitated by ammonium sulfate, allowing one to measure bioavailable testosterone (free and albumin bound). This assay is performed by adding ^{3}H-testosterone to the unknown sample, precipitating SHGB-testosterone, and measuring the free ^{3}H-testosterone. One can then calculate the bioavailable testosterone by multiplying total testosterone by the non-precipitated ^{3}H-testosterone. This assay is relatively simple for the laboratory, but can be influenced by the degree of precipitation and impurities in the tracer [23].

Other methods for estimating free testosterone include calculations based on the concentration of total testosterone and SHBG, the law of mass action, or empirical equations. The free androgen index (FAI) is calculated by multiplying the quotient testosterone/SHBG by the total testosterone level. This calculation is highly dependent on the accuracy and sensitivity of the assays used to determine total testosterone and SHBG. The FAI has good correlation with equilibrium dialysis measurements of free testosterone in women, but correlates poorly with the same measurements in males. Other calculations for determining free testosterone include algorithms based on the law of mass action and empirical equations using computer modeling [24].

Mass spectrometry

MS after extraction and chromatography allows for the highly accurate measurement of testosterone. The initial extraction and separation phases of GC-MS have increased the time and cost, and made it a labor-intensive assay. However, improvements in HPLC-MS/MS have improved the clinical applicability of this technique, including ease and high throughput. A number of recent studies have focused on using this technology to develop standard ranges for various populations.

Clinical use

The clinical use of testosterone assays depends on a highly accurate, sensitive assay that is reproducible. Before diagnosing a male as hypogonadal it is also important to obtain multiple samples, at least three, between 8 a.m. and 10 a.m. due to the circadian rhythm of testosterone secretion.

In adult males, testosterone measurement is used to differentiate between eugonadism and hypogonadism. Most clinical assays are accurate and sensitive enough to differentiate between these two states in the male. If a level is close to the lower limit of normal, then a calculated free level may be helpful in evaluating the patient.

In adult females, most assays are able to differentiate between the normal range and hyperandrogenism. However, these same assays are not accurate enough to quantify the degree of hyperandrogenism. Free testosterone levels, including the FAI, correlate well with the degree of hyperandrogenism. Current assays usually are not informative in detecting hypoandrogenism in adult females.

When measuring total testosterone in children, both boys and girls, it is vital to have sufficiently sensitive assays with appropriate normal ranges. Currently, these assays are of limited value in children.

Immunoassay interference

There are a number of factors that can interfere with an assay, including exogenous issues, which are not inherent to the specimen, and endogenous factors. Exogenous factors include pre-analytical factors and the assay formulation, while endogenous factors include those inherent to the specimen and can vary between patients and over time (Table 17.1).

Exogenous factors

Pre-analytical factors can have an impact on analyte levels, as well as their interpretation. By strict adherence to patient preparation as well as specimen collection and handling, one can maximize clinical interpretation.

Examples of potential immunoassay interference by exogenous factors include the time of day the specimen was drawn (cortisol), the age of the patient (DHEA-S), the phase of the menstrual cycle (progesterone), and whether the patient is pregnant (human chorionic gonadotropin [hCG]). One example commonly seen in reproductive endocrinology, which illustrates the importance of patient preparation, is measurement of 17-hydroxyprogesterone. 17-Hydroxyprogesterone is often drawn in the patient suspected of having polycystic ovary syndrome to exclude the diagnosis of late-onset congenital adrenal hyperplasia due to 21-hydroxylase deficiency. If the sample is inadvertently obtained in the luteal phase, it will be elevated along with progesterone due to production by the normal corpus luteum. This example highlights the importance of interpreting results based on the physiology of the endocrine gland of interest [25].

The physiological state of the patient at the time the sample was obtained can have a significant impact upon the results. Some conditions which could influence the

Table 17.1 Potential causes of assay interference

Exogenous factors	Endogenous factors
Time	**Protein bound**
Age	Affinity alterations (familial dysalbuminemia)
Circadian rhythm	
Menstrual cycle	Lower concentrations
Pregnancy	
Physiological state	**Matrix of assay**
Infection	pH
Hematologic status	Ionic strength
Immune status	Protein concentration
Fasting	
Exercise, stress, and posture	
Drugs	**Cross-reacting substances**
In vivo	Endogenous
Assay	Exogenous (drugs)
Type of specimen	**Antibodies**
Blood	Heterophile antibodies
Plasma	Human anti-animal antibodies
Serum	Autoantibodies
Saliva	Rheumatoid factor
Urine	
Test tube	**High-dose hook effect**
Glass or plastic	
Serum separator test tube	
Specimen collection	
Hemolysis	
Lipemia	

analyte of interest include an ongoing infection, hematologic and immunologic status, posture, fasting versus fed state, exercise, and stress. By appropriately preparing the patient prior to obtaining the sample, one can minimize this type of interference [26].

Drugs can affect an assay by having in vivo effects on the patient as well as in vitro effects upon the assay itself. Numerous drugs can alter the in vivo environment and affect the results of an assay. A common example of an in vivo effect is that of oral contraceptive pills, which increase the production of a number of circulating binding proteins. This increase in binding proteins can raise the total concentration of an analyte, but not change the free concentration of the hormone (like thyroxine). Drugs can also interact directly with the assay and influence the result; this is especially important where drugs can cross-react in steroid assays [27,28].

The type of specimen collected and prepared for use is also an important pre-analytical aspect of the assay, given that many assays are designed for a specific type of sample, including whole blood, plasma, serum, urine, or saliva. Whole blood is rarely used in immunoassays, owing to the presence of cellular enzymes. These enzymes from whole blood may rapidly destroy the analyte of interest, with one pertinent example being ACTH. For this reason, when planning to measure these unstable analytes, the sample must be collected with special precautions, such as the addition of antiproteolytic agents, rapid centrifugation, and freezing of the sample. Plasma is another type of sample that is often collected and used in the clinical laboratory. When collecting a sample as plasma, it is important to fill the test tube to the manufacturer's predetermined level, otherwise there can be leftover chelating agent, which may interfere. The chelating agent may interfere with the antigen–antibody reaction of the immunoassay, resulting in spurious results. Serum is the matrix of choice for many immunoassays, as it is stable and easily stored. One important exception to the rule for serum is parathyroid hormone, which is rapidly destroyed in serum. Parathyroid hormone needs to be collected in an EDTA test tube and frozen within 30 minutes of collection [29,30].

The type of test tube in which the sample is collected can also influence the results of the assay. Gel serum separator tubes can interfere with numerous peptide and protein assays. For steroid assays, progesterone is reduced by 50% when stored on a gel serum separator test tube for 6 days. In the past, tris(2-butoxyethyl) phosphate was used as a plasticizer but it can inhibit the binding of certain drugs to proteins. Another important example is ACTH, which will bind tightly to glass test tubes resulting in spurious results [31].

While collecting a sample, hemolysis can have a large impact upon certain assays. With hemolysis, there is a release of cellular enzymes that can metabolize a number of peptide hormones and alter their measurement. If one is measuring a substance which has a higher concentration within the cell, then hemolysis will release the analyte of interest and artificially increase levels. An example of hemolysis falsely elevating levels is folate and potassium. A lipemic sample can also interfere with a number of assays, either by directly interfering with a turbidimetric endpoint or displacing a ligand from its binding site [32].

Endogenous factors

There are a number of factors inherent to the specimen which can interfere with the specific assay. Examples of interfering, endogenous factors include heterophile antibodies, autoantibodies, cross-reactive substances, binding proteins, and matrix effects.

Many hormones like testosterone are highly protein bound in the circulation. When one measures the total hormone level, the sample is incubated with a releasing substance, which releases the bound hormone, allowing for measurement of the total concentration of free hormone. In some conditions such as familial dysalbuminemia, the albumin molecule has a very high affinity for the thyroxine molecule, and this disturbed binding can alter free thyroid hormone measurements. Low concentrations of binding proteins can also increase the variability of the assay [33].

The matrix of the assay is the environment in which the assay is run and includes a number of features like pH, ionic strength, and protein concentrations. Changes in the assay matrix can result in spurious results. Usually an assay for an explicit analyte is tailored by the manufacturer for the specific type of sample that one is planning to measure. It is especially important when modifying an assay to use appropriate standards, such as using the manufacturer's diluent for linearity studies or for dilution when a high concentration is suspected. These diluents are usually supplied by the manufacturer or one can develop one's own by using analyte free serum or plasma. Currently, many of the specifics of the matrix are proprietary information guarded by the manufacturer, making it difficult or near impossible to eliminate it as a potentially interfering substance, especially when modifying the assay [34].

Cross-reacting substances occur in an immunoassay because of limited specificity of the antibody used in the assay. This limited specificity results in potential binding of the antibody to other analytes with similar epitopes. This type of interference usually occurs in competitive immunoassays, since in this type of assay only one epitope is required to generate a signal. There are a number of drugs that can interfere with specific immunoassays, including levonorgestrel, which can cross-react in testosterone assays, spironolactone, which can interfere with digoxin assays, and unknown water-soluble substances, which can interfere with steroid hormone assays. These cross-reactants can sometimes be removed from the sample by pre-treatment, such as extracting a sample with organic solvents prior to measurement of a steroid hormone [35].

Antibodies, which are a critical component of the immunoassay, can also interfere with it. In immunoassay interference, heterophile antibodies are defined as anti-animal antibodies when the patient has no known medicinal exposure to the specific antigen. These antibodies may arise from a dietary source and are found with increased frequency in individuals with IgA deficiency. These antibodies are often multi-specific and can interact with the assay antibodies [36]. One case report demonstrated a falsely elevated testosterone level due to the formation of heterophile antibodies from a polyclonal gammopathy in a patient with acute myelogenous leukemia [37]. Human anti-animal antibodies (HAAA) are formed following treatment with a specific animal immunoglobulin or antigen. These antibodies are highly specific, well defined, and may interfere with an assay, if the assay antibodies were formed in the same species. Human anti-mouse antibodies (HAMA) are a specific type of HAAA, and HAMAs are increasing in frequency with the increasing use of murine-derived medicinal products [38].

Autoantibodies can also interfere with immunoassays, with an important example being thyroglobulin antibodies. When measuring thyroglobulin levels, thyroglobulin antibodies can falsely raise or lower levels, depending on the specific assay used. Thyroglobulin is often used to monitor recurrence in patients with thyroid cancer, so a false positive result could have a significant impact on the patient's care. Autoantibodies are estimated to occur in 30-40% of individuals secondary to a history of blood transfusion, infections, immunization, maternal transfer, and intestinal transfer [39].

Autoantibodies can also bind the analyte together into larger complexes, such as with big prolactin, big growth hormone, big-big prolactin, and big-big growth hormone. These large macrocomplexes usually have limited biological activity but can still be detected in varying extents with the immunoassay. These macrocomplexes can usually be detected by precipitation. Polyethylene glycol can precipitate macroprolactin; and if in a sample greater than 50% of prolactin activity is precipitated by polyethylene glycol, it is considered to be consistent with macroprolactinemia [40].

Rheumatoid factor is another important reason for interference in an immunoassay. Rheumatoid factor is found in 1–4% of the general population, 15–20% of the elderly population, and as high as 80% of individuals with rheumatoid arthritis. Rheumatoid factor is directed against the Fc domain of the IgG molecule and, thus, can bind to the immunoglobulins in the immunoassay and either provide a bridge (with a false positive result) or block the assay antibodies (resulting in a false negative result) [41].

The high-dose hook effect occurs in the immunometric, sandwich assay when there is a large excess of antigen in the sample. In this case, there is such a large excess of antigen that a single antigen molecule binds to the capture antibody, while a different antigen molecule binds to the signal antibody. Hence no "sandwich" is formed and no signal is generated. This can usually be detected by washing the assay system prior to adding the second antibody and finding that the assay is saturated. Another method to detect the high-dose hook effect is to dilute the sample, which will result in increasing concentration following this attenuation [42].

Detection of immunoassay interference

The first step in identifying immunoassay interference is having a high index of suspicion when the laboratory data do not support the clinical diagnosis. In cases where there is a large discrepancy between the laboratory data and the diagnosis it is usually obvious, though when there is a minor discrepancy it can be more difficult to identify. Immunoassay interference is estimated to occur in 0.05–2% of the general population and may be as high as 9% in patients with rheumatoid arthritis. It is critical to communicate with the clinical laboratory when one suspects immunoassay interference.

When one suspects immunoassay interference, the first step is to confirm the specimen identity and retest. If this repeat test is also suspicious, then redraw another specimen and test again. In this situation, it is best to send the second specimen to another laboratory, which ideally will perform the assay with an alternative methodology. If there are still concerns about possible interference, then there are a number of methods to investigate and hopefully arrive at an appropriate measurement.

One of the techniques used to detect interference is serial dilution. When performing serial dilutions, it is critical to use diluent that is suitable for the assay, usually provided by the manufacturer. In a sample where there is no interference, one expects the results should dilute proportionally. If interference is giving a spurious result, then serial dilution will usually not give proportional results. An important caveat is that not all analytes dilute linearly, especially some of the more complex tumor markers like CA 19-9 [43].

Another method used to detect immunoassay interference is to spike the questionable sample with a known amount of analyte, and then determine how much is recovered. One expects 90–110% recovery if there are no interfering substances. Percent recovery is determined by the equation: concentration of hormone in the spiked sample–concentration of hormone in the unspiked sample/concentration hormone added [44].

The addition of heterophilic blocking reagents or non-immune animal serum to the sample can remove heterophilic or HAAA. Heterophilic blocking reagents specifically block heterophilic antibodies, while non-immune animal serum will block both heterophilic and the more specific HAAA. The addition of non-immune animal serum allows circulating human antibodies in the sample to bind to the animal serum immunoglobulins, hopefully sparing the assay antibodies [45].

hCG is present in both the circulation and urine. Its presence in urine has allowed for the development of highly sensitive over-the-counter pregnancy tests. If one suspects a falsely positive serum hCG, then a urine sample can be assayed. Because immunoglobulins, including interfering antibodies, do not routinely cross from the blood into the urine, a negative urine hCG with a positive serum hCG confirms the presence of interference, as long as the serum hCG is ≥50 mIU/mL. For testosterone and other steroids, extraction with an organic solvent can remove interfering aqueous substances from the sample [46].

Other methods to detect immunoassay interference include gel filtration chromatography to detect macroprolactin, immunoadsorption chromatography to remove immunoglobulins from the sample, and interference assays. Interference assays use the same antibody as the capture and label, hence only measuring interfering antibodies. Another method, especially for testosterone, is to measure the analyte concentration with MS, which relies on the molecular structure of the analyte.

Studies are currently underway to investigate the prevention of immunoassay interference by preventing it in the person and in the assay. Methods to prevent anti-animal antibody formation in the person receiving treatment with animal antibodies include immunosuppression while the person is given the animal antibody, using antibody fragments instead of the whole antibody in the medicinal, humanizing the antibodies in the medicinal, and polyethylene glycol coating of the therapeutic antibodies. In the assay, methods

being explored to prevent interference by anti-animal antibodies include using modified assay antibodies (remove Fc fragment) or using non-immunoglobulin binding substances (affibodies) [47].

Conclusion

Revolutionary steps in assay technology, including immunoassays, have improved the ability to specifically measure hormones. Improvements in MS and in vitro bioassays offer promise to enhance our ability to measure steroid hormones, specifically androgens. The clinician must maintain a high level of suspicion when the laboratory data do not support the diagnosis, and reconsider the diagnosis or whether interference with the assay may be present.

References

1. McDonough PG. Amount of error in a measurement?…Reliability. *Fertil Steril* 2003;80:1287–1288.

2. Rosenfield R. Ovarian and adrenal function in polycystic ovary syndrome. *Endocrinol Metab Clin North Am* 1999;28:265–293.

3. Deslypere JP, Young M, Wilson JD, et al. Testosterone and 5 alpha-dihydrotestosterone interact differently with the androgen receptor to enhance transcription of the MMTV-CAT report gene. *Mol Cell Endocrinol* 1992;88:15–22.

4. Arlt W, Callies F, van Vliimen JC, et al. Dehydroepiandrosterone replacement in women with adrenal insufficiency. *N Engl J Med* 1999;341:1013–1020.

5. Burger HG. Androgen production in women. *Fertil Steril* 1002;77 (suppl 4):S3–5.

6. Pardridge WM. Plasma protein–mediated transport of steroid and thyroid hormones. *Am J Physiol* 1987;252:E157–164.

7. Rosner W, Hryb DJ, Khan MS, et al. Sex hormone-binding globulin mediates steroid hormone signal transduction at the plasma membrane. *J Steroid Biochem Mol Biol* 1991; 869:481–485.

8. Frank RT, Klempner E, Hollander F, Kriss B. Detailed description of technique for androgen assay by the chick comb method. *Endocrinology* 1942;31:63–70.

9. Hershberger LG, Shipley EG, Meyer RK. Myotrophic activity of 19-nortestosterone and other steroids determined by modified levator ani muscle method. *Exp Biol Med* 1953;83:175–180.

10. Yalow RS, Berson SA. Immunoassay of endogenous plasma insulin. *J Clin Invest* 1960;39:1157–1175.

11. Kasahara Y. Homogeneous enzyme immunoassays. In: Nakamura RM, Kasahara Y, Rechnitz GA (Eds.). *Immunochemical Assays and Biosensor Technology for the 1990s*. Washington, DC: American Society for Microbiology; 1992:169–182.

12. Ekins RP. The estimation of thyroxine in human plasma by an electrophoretic technique. *Clin Chim Acta* 1960;5:453–459.

13. Singh RJ. Validation of a high throughput method for serum/plasma testosterone using liquid chromatography tandem mass spectrometry (LC-MS/MS). *Steroids* 2008;73:1339–1344.

14. Hartig PC, Bobseine KL, Britt BH, et al. Development of two androgen receptor assays using adenoviral transduction of MMTV-luc reporter and/or hAR for endocrine screening. *Toxicol Sci* 2002; 66:82–90.

15. Lambright C, Ostby J, Bobseine K, et al. Cellular and molecular mechanisms of action of linuron: an antiandrogenic herbicide that produces reproductive malformations in male rats. *Toxicol Sci* 2000;56:389–399.

16. Soto AM, Maffini MV, Schaeberle CM, Sonnenschein C. Strengths and weaknesses of in vitro assays for estrogenic and androgenic activity. *Best Pract Res Clin Endocrinol Metab* 2006;20:15–33.

17. Dobs AS. The role of accurate testosterone testing in the treatment and management of male hypogonadism. *Steroids* 2008;73:1305–1310.

18. Bhasin S, Zhang A, Caviello A, et al. The impact of assay quality and reference ranges on clinical decision making in the diagnosis of androgen disorders. *Steroids* 2008;73:1311–1317.

19. Rosner W, Auchus RJ, Azziz R, et al. Position Statement: utility, limitations, and pitfalls in measuring testosterone: an Endocrine Society position statement. *J Clin Endocrinol Metab* 2007;92:405–413.

20. Dorgan JF, Fears TR, McMahon RP, et al. Measurement of steroid sex hormones in serum: a comparison of radioimmunoassay and mass spectrometry. *Steroids* 2002;67:151–158.

21. Vermeulen A, Verdonck L, Kaufman JM. A critical evaluation of simple methods for the estimation of free testosterone in serum. *J Clin Endocrinol Metab* 1999;84:3666–3672.

22. Rosner W. An extraordinary inaccurate assay for free testosterone is still with us. *J Clin Endocrinol Metab* 2001;86:2903.

23. Wheeler MJ. The determination of bioavailable testosterone. *Ann Clin Biochem* 1995;32:345–357.

24. Morley JE, Patrick P, Perry HM 3d. Evaluation of assays available to measure free testosterone. *Metabolism* 2002;51:554–559.

25. Elmlinger MW, Kuhnel W, Ranke MB. Reference ranges for serum concentrations of lutropin (LH), follitropin (FSH), estradiol (E2), prolactin, progesterone, sex hormone-binding globulin (SHBG), dehydroepiandrosterone sulfate (DHEAS), cortisol and ferritin in neonates, children and young adults. *Clin Chem Lab Med* 2005;40:1151–1160.

26. Wilde C. Subject preparation, sample collection and handling. In: Wild D. *The Immunoassay Handbook*. San Diego, CA: Elsevier, Inc; 2005: 443.

27. Davies PH, Franklyn JA. The effects of drugs on tests of thyroid function. *Eur J Clin Pharmacol* 1991;40:439–451.

28. Warner A. Interference of common household chemicals in immunoassay methods for drugs of abuse. *Clin Chem* 1989;35:648–651.

29. Kato K, Umeda Y, Suzuki F, Kosaka A. Improved reaction buffers for solid-phase enzyme immunoassay without interference by serum factors. *Clin Chim Acta* 1980;102:261–265.

30. Shirtcliff EA, Grander DA, Schwartz E, Curran MJ. Use of salivary biomarkers in biobehavioral research: cotton-based sample collection methods can interfere with salivary immunoassay results. *Psychoneuroendocrinology* 2001;26:165–173.

31. Bowen RAR, Chan Y, Ruddel ME, et al. Immunoassay interference by a commonly used blood collection tube additive, the organosilicone surfactant Silwet L-720. *Clin Chem* 2005;51:1874–1882.

32. Wenk RE. Mechanism of interference by hemolysis in immunoassays and requirements for sample quality. *Clin Chem* 1998;44:2554.

33. Sapin R, Gasser F, Schlienger JL. Familial dysalbuminemic hyperthyroxinemia and thyroid hormone autoantibodies: interference in current free thyroid hormone assays. *Horm Res* 1996;45:139–141.

34. Wood WG. "Matrix Effects" in immunoassays. *Scand J Clin Lab Invest* 1991;51:105–112.

35. Miller JJ, Valdes R Jr. Approaches to minimizing interference by cross-reacting molecules in immunoassays. *Clin Chem* 1991;37:144–153.

36. Miller JJ. Towards a better understanding of heterophile (and the like) antibody interference with modern immunoassays. *Clin Chim Acta* 2002;325:1–15.

37. Ramaeker D, Brannian J, Egland K, et al. When is elevated testosterone not testosterone? When it is an immunoassay interfering antibody. *Fertil Steril* 2008;90:886–888.

38. Klee GG. Human anti-mouse antibodies. *Arch Pathol Lab Med* 2000;124:921–923.

39. Spencer CA. Challenges of serum thyroglobulin (Tg) measurement in the presence of Tg autoantibodies. *JCEM* 2004;29:3702–3704.

40. Smith TP, Suliman AM, Fahie-Wilson MN, McKenna TJ. Gross variability in the detection of prolactin in sera containing big prolactin (Maroprolactin) by commercial immunoassays. *JCEM* 2002;87:5410–5415.

41. Despres N, Grant AM. Antibody interference in thyroid assays: a potential for clinical misinformation. *Clin Chem* 1998;44:440–454.

42. Fernando SA, Wilson GS. Studies of the 'hook' effect in the one-step sandwich immunoassay. *J Immunol Methods* 1992;151:47–66.

43. Emerson JF, Ngo G, Emerson SS. Screening for interference in immunoassays. *Clin Chem* 2003;49:1163–1169.

44. Andersen L, Jørgensen PN, Jensen LB, Walsh D. A new insulin immunoassay specific for the rapid-acting insulin analog, insulin aspart, suitable for bioavailability, bioequivalence, and pharmacokinetic studies. *Clin Biochem* 2000;33:627–633.

45. Kricka LJ. Human anti-animal antibody interferences in immunological assays. *Clin Chem* 1999;45:942–956.

46. Olsen TG, Hubert PR, Nycum LR. Falsely elevated human chorionic gonadotropin leading to unnecessary therapy. *Obstet Gynecol* 2001;98:843–845.

47. Bjerner J, Bormer CP, Nustad K. The war on heterophilic antibody interference. *Clin Chem* 2005;51:9–11.

Magnetic resonance imaging of the adrenal gland

Sajal S. Pokharel and Ihab R. Kamel

Introduction

Magnetic resonance imaging (MRI) is an extremely powerful tool for evaluating the adrenal glands. MRI has the unique ability to characterize soft tissues. This ability allows for specific pathologic diagnoses to be made in a large number of adrenal lesions. This is accomplished without the use of ionizing radiation and, oftentimes, without the need for intravenous contrast material. However, certain lesions may remain indeterminate on MRI findings alone. In these cases, correlation with the overall clinical picture will aid in diagnosis. Rarely, biopsy may be required for definitive diagnosis.

MRI techniques

MRI utilizes the property of proton nuclear magnetic resonance coupled with robust spatial localization techniques to generate images. Various pulse sequences can be performed in the course of an MRI study. A pulse sequence is a series of radiofrequency electromagnetic wave excitations, magnetic gradients, and resulting radiofrequency signals (echoes), which can be arranged in a variety of ways. The arrangement determines which tissue property is emphasized. Pulse sequences can be formulated to highlight certain aspects of the tissues being imaged. For example, the sequence can highlight tissue water content (e.g., T2-weighted) or fat content (e.g., Dixon fat-only images).

Lipid detection techniques

In fact, fat or lipid detection turns out to be very important in imaging of the adrenal glands. Several lipid-sensitive sequences exist. The two major categories that are utilized in adrenal imaging are chemical shift selective suppression (CHESS) and chemical shift imaging (CSI). Both of these techniques harness the property that water protons and fat protons resonate (precess, spin) at slightly different frequencies (rates) when placed in a strong external magnetic field. This difference in rate of precession is also called the chemical shift. The way these two sequences utilize this property differs such that the resulting images give complementary information.

Chemical shift imaging

Since there is a precessional (resonance, spin) frequency difference between water and fat protons, only at specific interval periods during nuclear spin precession will a lipid proton and a water proton be in sync, i.e., "in phase." At certain other times, they will be perfectly out of sync, or "out of phase." This phase relationship is cyclical, as these two types of proton continually cycle in and out of phase.

CSI (also known as in- and out-of-phase imaging) takes advantage of this precessional frequency difference to produce two sets of images: "in-phase" and "out-of-phase" images. In-phase images are obtained by timing the MR sequence with an "echo time" (TE), which puts the fat and water protons in sync, e.g., 4.4 ms on a 1.5 Tesla (T) magnet. The out-of-phase images are obtained with a TE which puts the fat and water protons perfectly out of sync, e.g., 2.2 ms on a 1.5 T magnet [1]. On the in-phase images, the signal from water and fat within a single voxel (a term for a three-dimensional pixel) will be additive, and on the out-of-phase images, the water and fat signals will cancel each other. When the two sets of images are compared (Figure 18.1), tissue which loses signal intensity on the out-of-phase images is said to contain significant intracytoplasmic lipid (also referred to as microscopic fat). If the voxel contains no fat or all fat, there will be no drop in signal. CSI is very sensitive to

Androgens in Gynecological Practice, ed. Leo Plouffe and Botros Rizk. Published by Cambridge University Press. © Cambridge University Press 2015.

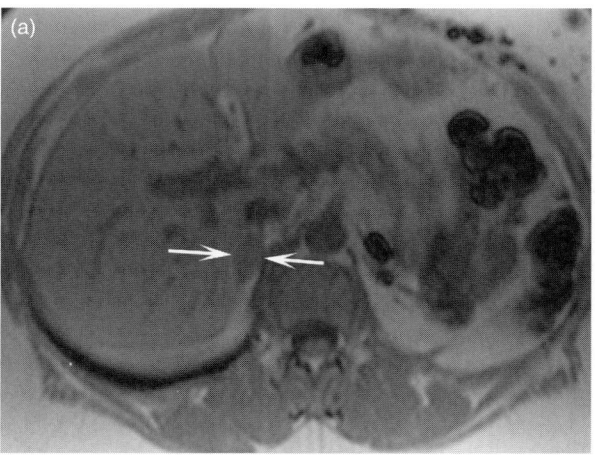 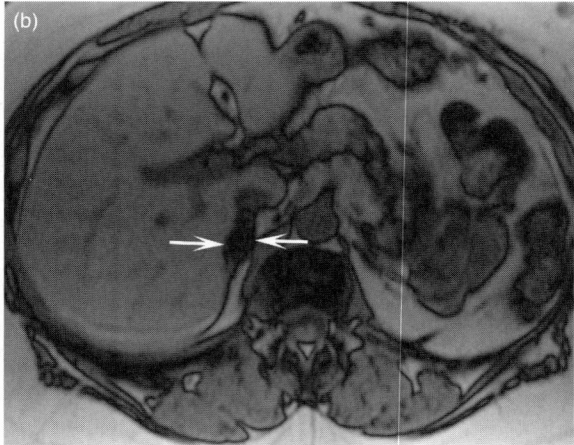

Figure 18.1 Axial T1-weighted in-phase (a) and out-of-phase (b) images demonstrate a 2 cm incidental right adrenal nodule. It is intermediate in signal on the in-phase image (arrows) and becomes dark on the out-of-phase image (arrows). This signal loss is indicative of microscopic fat within the lesion and effectively diagnostic of an adrenal adenoma.

microscopic quantities of fat and has been shown to detect as little as 10% fat content [2].

Out-of-phase images contain an interesting artifact (Figure 18.1b). Dark lines are present at the interface of tissues composed of mainly fat and those composed of mainly water. This phenomenon results from the fact that voxels at these interfaces contain both fat and water, and the signal from these two types of proton cancel each other out, resulting in a low signal voxel. When one looks at opposed phase images the organs appear as if they have been etched with ink and hence, this artifact is referred to as "etching" artifact or "India ink" artifact. Not only is the artifact useful in identifying the out-of-phase set of images, but it also highlights fatty content of a structure that is contained within a mainly water-containing structure.

An extension of CSI is the so called "Dixon method." In this technique, a further step is taken with the in- and out-of-phase image to obtain water-only and fat-only images. If the two sets of images are added, then the fat signal in the in-phase images (water + fat) is cancelled by the subtracted fat signal in the out-of-phase images (water − fat) to give water-only images [(water + fat) − (water − fat) = (water + water)]. If the out-of-phase images are subtracted from the in-phase images then fat-only images are generated [(water + fat) − (water − fat) = (fat + fat)]. This intuitively simple technique turns out to be clinically powerful, especially when modified to overcome some inherent shortcomings of the original method [3,4].

Chemical shift selective suppression of fat

The chemical shift between water and fat is used to perform *macro*scopic fat suppression imaging in CHESS. Macroscopic fat is bulk fat as in adipose cells. This technique is also referred to as fat saturation, or in short "fat sat." In this technique, the macroscopic fat signal is suppressed by first selectively exciting fat with a narrow bandwidth radiofrequency excitation pulse and then nulling these excited fat protons with a "spoiler gradient." Immediately, the imaging sequence is started before the lipid signal can recover. Fat saturation imaging helps to increase image contrast and highlight lesions, such as contrast-enhancing tissue, edema, and blood products by eliminating the bright signal from fat so that the full spectrum of display contrast can be applied to a narrower spectrum of image contrast. Fat saturation is useful for detecting larger quantities of fat (Figure 18.2). The technique, however, is susceptible to magnetic field inhomogeneities such as those caused by foreign bodies, tissue interfaces, etc. This can result in poor fat suppression and may even result in areas of water suppression.

CHESS is commonly utilized in T1-weighted post-contrast images. The frequency selective technique only suppresses fat signals and not the shortened T1 signals due to intravenous contrast, as might be the case with a suppression technique based on tissue T1, such as inversion recovery (IR), which is discussed below. CHESS can be added to virtually any pulse sequence at the expense of slightly increased scan time. Another disadvantage of the fat sat technique is

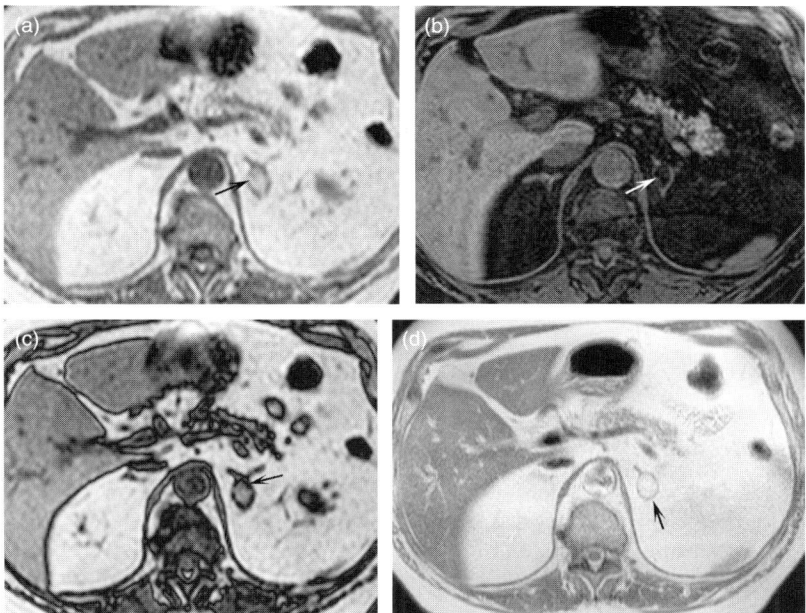

Figure 18.2 Axial T1-weighted in-phase (a), T1-weighted fat saturated (b), out-of-phase (c), and T2-weighted (d) images demonstrate a mass in the left adrenal gland. Note that it is relatively bright on the T1-weighted in-phase image (a, arrow) and becomes dark on the T1-weighted fat saturated image (b, arrow). The signal change follows that of fat in the retroperitoneum and subcutaneous spaces. A macroscopic fat-containing lesion within the adrenal gland is effectively diagnostic of a myelolipoma. Notice on the out-of-phase image (c, arrow) that there is a dark line at the interface of the mass and the normal adrenal gland. Dark lines are also present outlining many of the organs which are in a sea of fat. This is an artifact called "etching" or "India ink" and is found on out-of-phase images. It occurs at interfaces of predominantly fat-containing and predominantly water-containing structures. It can serve as a sign that a lesion contains fat, as in this case. Also note that the T2-weighted non-fat saturated image (d) nicely defines the adrenal lesion (arrow) as well as the other abdominal structures. Non-fat saturated T2-weighted images are optimal for defining the anatomy as they have high signal to noise ratio.

that it is not useful when there is only a small amount of fat within a given voxel. The decrease in signal due to *micro*scopic quantities of fat may be undetectable. Recall, however, that detection of microscopic fat is the strength of in- and out-of-phase imaging. Thus, the two techniques are complementary.

Inversion recovery fat suppression

Another widely utilized method of fat suppression is based on the IR principle. This type of technique nulls the signal from a specific tissue type based on the intrinsic MR decay property of a tissue called T1. The T1 of fat is different (shorter) than that of water. IR sequences, such as short-tau inversion recovery (STIR), take advantage of this difference to suppress the fat signal. This is accomplished by an MRI pulse sequence that first completely inverts the tissue magnetization. As the magnetization is recovering, the imaging excitation pulse is started after a specific "inversion time" (TI) chosen such that the recovering magnetization of fat protons is exactly crossing the null point. Thus,

there is no signal from the lipid protons, but there is signal from other tissue with different T1 values.

The IR technique is useful for obtaining homogeneous tissue suppression, because it is insensitive to magnetic field heterogeneity. However, this method is not tissue specific and has the potential to suppress non-desired tissue types. For example, when suppressing fat, non-fat tissue with similar T1 values, such as blood products or gadolinium-enhanced tissues, may also be suppressed. Therefore, T1-weighted STIR imaging is primarily used to highlight fluid, rather than detect fat.

T2-weighted images

As mentioned, MR images can highlight various tissue properties. T2 is an MR tissue property that relates to the way spinning protons interact with each other (spin-spin relaxation). It turns out that free water has long T2 times, while other soft tissue and fat have relatively shorter T2 times. Since pathology is often associated with inflammation and edema, pathologic tissue tends to have high signal on T2-weighted images (with definite exceptions, including prostate cancer, which has lower T2 signal than surrounding normal prostatic peripheral zone.) MR sequences are termed T2 "weighted" because the images are weighted towards a specific property, such as T2, but retain components of other properties, such as T1 or proton density.

T2-weighted images can be obtained without or with fat suppression. Fat-suppressed T2-weighted images highlight fluid/edema and thus, pathology (Figure 18.3a and b). Non-fat-saturated T2-weighted images are very useful for demonstrating the anatomy, as they tend to have high signal to noise and high contrast to noise (Figure 18.2d).

T1-weighted images

T1-weighted images do not have the tissue contrast of T2-weighted images. However, they are well suited for the detection of fat (both with CSI and CHESS as above). T1-weighted images require relatively little MR scanner time when gradient recalled technique is used. Finally, and most importantly, T1-weighted fat-saturated images are the sequences in which contrast enhancement is assessed.

Post-contrast images

MR intravenous contrast agents are for the most part gadolinium-based. Gadolinium is an element that shortens the T1 time of the water protons that form complexes with it. This property increases the signal from these protons on T1-weighted images. Gadolinium by itself is toxic. Thus, MR contrast agents are actually the gadolinium atom surrounded by certain chelating compounds. These formulations can be given intravenously. The agents then circulate in the blood and reach all tissues in a matter of seconds. Many of the gadolinium-based agents can freely enter the interstitium and equilibrate between the blood pool and the interstitium until the agent is excreted from the body. Timing of the post-contrast imaging, especially during the first minute after contrast injection, will determine where the majority of the contrast is and, thus, what is highlighted. Arterial phase post-contrast images obtained at approximately 20 seconds post-injection highlight the major arteries. Venous phase images obtained approximately 70 seconds after injection demonstrate organ perfusion as well as venous structures. Delayed images obtained, for example, at 3 minutes or later display how the contrast has washed out from certain areas of pathology.

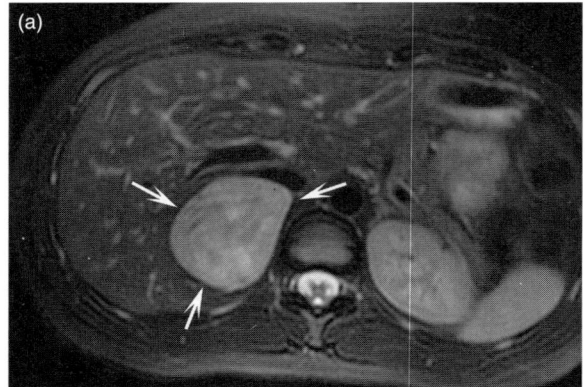

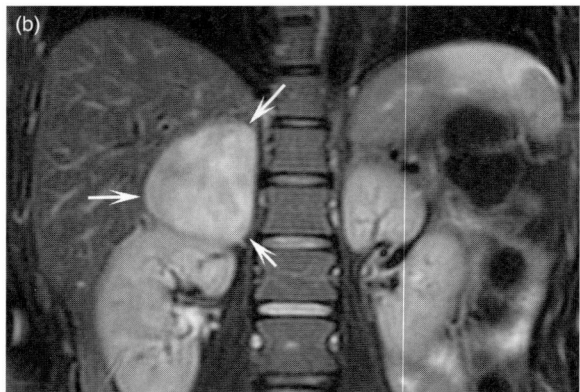

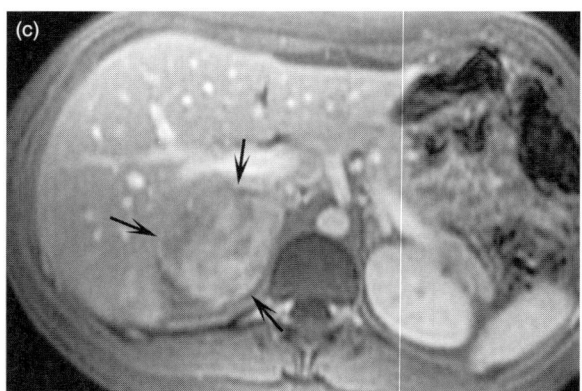

Figure 18.3 A large mass in the right suprarenal space demonstrates bright signal on T2 fat-saturated axial (a) and coronal (b) images. The mass is heterogeneously hyperenhancing on T1-weighted fat-saturated post-contrast image (c). This mass was resected and was demonstrated to be a pheochromocytoma on pathology. Note that the T2 fat-saturated images highlight the pathology and this bright T2 signal is classic for, but not specific or sensitive for, pheochromocytoma.

Multiple phases can be imaged during a study after a single contrast injection to obtain a wealth of information on vascular anatomy and organ and pathology perfusion characteristics.

MRI appearance of normal adrenal glands

The limbs of the normal adrenal glands give them characteristic shapes on cross-sectional imaging. It is this shape and the location in the suprarenal region that identify the adrenal glands on MRI. On both axial and coronal images the adrenal glands take on an inverted "V" or inverted "Y" configuration (Figure 18.4). The adrenal cortex and medulla cannot be differentiated on conventional MRI, and the glands appear homogeneous in signal.

The right adrenal gland is located slightly superior to the left adrenal gland. It is above the right kidney, lateral to the right crus of the diaphragm, medial to the liver, and posterior to the inferior vena cava. The left adrenal gland is above the left kidney, lateral to the left crus of the diaphragm, and posterior to the pancreas.

The normal adrenal glands can sometimes be difficult to identify, especially if there is a paucity of surrounding retroperitoneal fat. One aid is to look for dark structures in the expected locations on out-of-phase images. Because normal adrenal tissue contains a sufficient amount of intracytoplasmic lipids, which are precursors for hormone production [5], there is signal loss on the out-of-phase images and the glands appear dark.

Adrenal adenoma

The most common use of adrenal MRI is in the diagnosis of adrenal adenomas. This may be in the setting of a patient with clinical and/or biochemical evidence of hormonal excess, in the setting of an incidentally detected adrenal nodule on imaging obtained for some other reason, or in the setting of an adrenal nodule detected on a metastatic survey in a patient with known malignancy.

The power of MRI is its ability to diagnose an adenoma without ionizing radiation or intravenous contrast. This ability stems from the fact that a large proportion of adrenal adenomas contain a significant amount of intracytoplasmic lipid, whereas metastatic lesions generally do not [6]. Rare exceptions of microscopic fat-containing metastatic lesions to the adrenal gland from renal cell and hepatocellular carcinoma have been reported [7].

Microscopic intracytoplasmic fat in an adenoma is detected as signal loss on the out-of-phase images relative to the in-phase images. While the signal loss

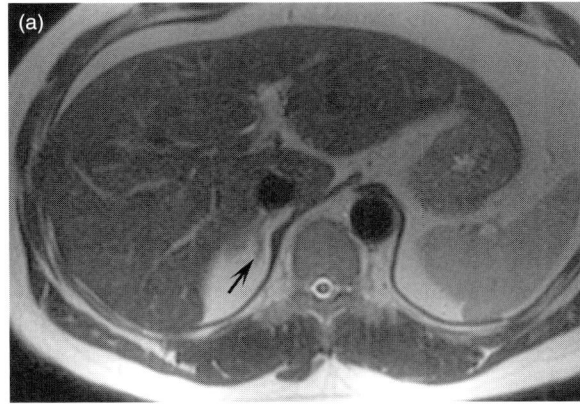
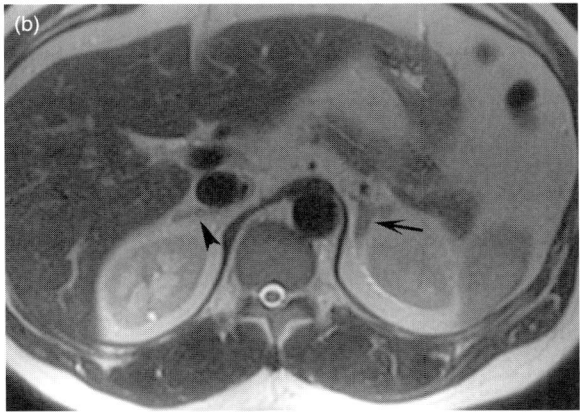
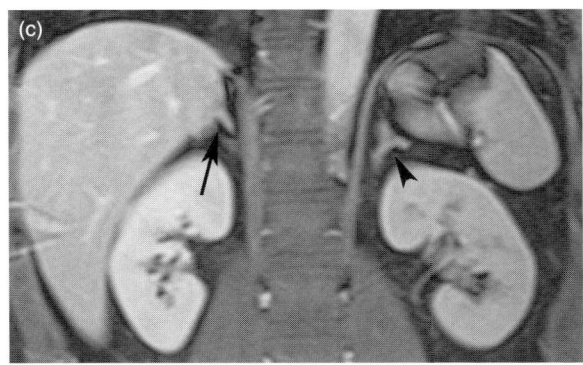

Figure 18.4 Axial T2-weighted (a, b) images and coronal T1-weighted fat saturated post-contrast image (c) demonstrate normal right and left adrenal glands. Note the inverted "Y" configuration on both coronal and axial planes and the expected location in the suprarenal space. The limbs of the glands are thin and elongated. The right adrenal gland is usually located at a level slightly above the left adrenal gland.

is often visually evident (Figure 18.1), it can be quantified by drawing regions of interest (ROI) within the adrenal nodule on both the in-phase and out-of-phase images to calculate the average signal intensity in each sequence. Signal loss of 15–20% on the out-of-phase image is diagnostic of an adenoma [8,9].

The degree of intracytoplasmic fat does not differentiate between hyperfunctioning and non-hyperfunctioning adenomas. However, atrophy of the remaining adrenal gland and the contralateral gland does suggest hyperfunction with feedback ACTH suppression [10].

In a symptomatic patient with hormone excess, a diagnosis of adenoma will provide a target for therapy. In an incidental lesion, the diagnosis is reassuring and may prompt a search for occult hormone excess. In the case of a person with known malignancy, a positive identification of an adenoma may mean the difference between a diagnosis of local organ-confined disease (in the absence of other lesions) versus a diagnosis of distant metastatic disease.

Unfortunately, not all adrenal adenomas contain enough microscopic lipid to cause significant signal loss on the out-of-phase images. In fact, approximately 30% of adenomas are "lipid-poor" [5,11]. Thus, the lack of signal loss on out-of-phase imaging does not completely rule out an adrenal adenoma. In such cases, the next imaging step (if not already obtained) is to perform an adrenal protocol dynamic contrast-enhanced CT. Adenomas tend to enhance early as well as wash out the contrast. This is in contrast to metastatic lesions. Thus, if an adrenal nodule fulfills certain criteria for contrast washout during the dynamic contrast-enhanced CT, a diagnosis of adenoma can be made [12].

Additionally, a non-contrast CT may also detect microscopic lipid in an adenoma. This is done by measuring the attenuation of the lesion. An attenuation of less than 10 Hounsfield Units on the unenhanced CT is generally taken to indicate intracytoplasmic lipid and thus, be diagnostic of an adenoma [12]. This brings up the question of which imaging test to order first when one needs to make the diagnosis of an adrenal adenoma in the three clinical settings outlined above: adrenal protocol MRI or adrenal protocol CT. Given the lack of ionizing radiation and lack of intravenous contrast during the MRI examination, MRI may have the advantage in a setting where it is available and in a patient who can tolerate the MRI, which is a longer exam in a more confined space than CT.

Metastatic disease

If an adrenal nodule is not an adenoma, then the suspicion is for a metastatic deposit, especially in the case of a patient with known malignancy. While an incidental adrenal lesion in a patient without a history of malignancy is unlikely to be metastatic disease, an adrenal mass in a patient with malignancy represents metastatic disease 50–75% of the time [13,14]. The adrenal gland is in general the fourth most common site of metastasis (after lung, liver, and bone). Lung and breast cancer are the two most common primaries to metastasize to the adrenal gland [15].

If an adrenal nodule does not demonstrate signal loss on out-of-phase images, does not demonstrate adequate contrast washout on dynamic contrast-enhanced CT, and is not characteristic of one of the other less common lesions described below, then it is suspicious for a metastatic lesion.

Myelolipoma

Myelolipoma is a primary tumor of the adrenal gland. It is benign and often asymptomatic unless it gets to be large enough to cause mass effect or bleed. It is comprised of a large proportion of fat cells with interspersed myelogenous elements. The fat cells contain macroscopic lipid. Thus, the MRI appearance of the myelolipoma is characteristic. Macroscopic fat is bright on T1-weighted and fast spin echo T2-weighted images. This bright signal becomes dark when CHESS fat saturation is applied. If a nodule in the adrenal gland contains bright signal on the non-fat sat images, which drops out on the fat sat images, this is virtually diagnostic of a myelolipoma (Figure 18.2). The caveat is that very rarely metastatic disease and even adrenal cortical carcinomas can contain small quantities of macroscopic fat [5]. However, the large heterogeneous nature and paucity of the fat in these other lesions should make diagnosis unambiguous.

Note that CSI sequences (in- and out-of-phase) may not make the diagnosis, as the fat cells are nearly all lipid with only small quantities of water. There is not enough water signal to cancel out the fat signal on out-of-phase images. Thus, the myelolipoma remains bright on the out-of-phase images (Figure 18.2c). In the figure, one may notice the thin dark outlining of portions of the myelolipoma. Notice that many of the abdominal structures are outlined similarly on the out-of-phase images. This is the "India ink" or "etching" artifact caused by the canceling of water and fat signal at the interface of two structures, one of which is bulk water and the other bulk fat. In fact, the presence of this artifact at the interface of this adrenal nodule and the background adrenal gland indicates that the nodule contains bulk fat. If this artifact is observed, which is not always the case, indeed a diagnosis of myelolipoma can be made on the out-of-phase images.

Hematoma

What if there is a nodule with bright signal within the adrenal gland on T1-weighted images, and this signal does not suppress on the CHESS fat sat images? This likely represents blood products in an adrenal hematoma. Adrenal hematomas can occur in a variety of settings including trauma, anticoagulation, or stress. They can be large and heterogeneous (Figure 18.5). The heterogeneity may come from varying stages of blood products. In fact, the T1 and T2 signal of the contents of a hematoma can be used to approximately age the hematoma. Acutely (within 1 week) the blood products are T1 iso-hypointense and T2 hypointense. In the subacute stage (1 week to 2 months), blood products are T1 and T2 hyperintense. After that is the chronic stage, when a T1 and T2 dark rim of hemosiderin starts to envelop the lesion.

Hematomas should not enhance. Thus, when a hematoma is suspected, post-contrast imaging with subtraction to evaluate for enhancement (or lack thereof) is helpful. The worry is that the hematoma may be related to an underlying mass in the adrenal gland, such as a metastatic lesion or an adenoma. If any enhancing component is seen, then an underlying mass is likely. Sometimes with large heterogeneous hematomas, it is difficult to exclude a small underlying mass. In these cases, follow-up imaging is helpful as the hematoma should decrease and eventually resolve. Any underlying mass can then be identified.

Pheochromocytoma

Pheochromocytomas are neoplasms derived from chromaffin cells. They can arise from the adrenal medulla as well as from extra-adrenal sympathetic tissue. They tend to secrete catecholamines and can cause paroxysmal refractory hypertension, palpitations, headaches, and diaphoresis. Elevated catecholamine byproducts in the blood and urine often indicate the presence of a pheochromocytoma. In such cases, imaging is performed to locate the lesion for excision.

Pheochromocytomas are said to follow the "10% rule." Ten percent of the time they are bilateral, malignant, or extra-adrenal [16]. One characteristic site

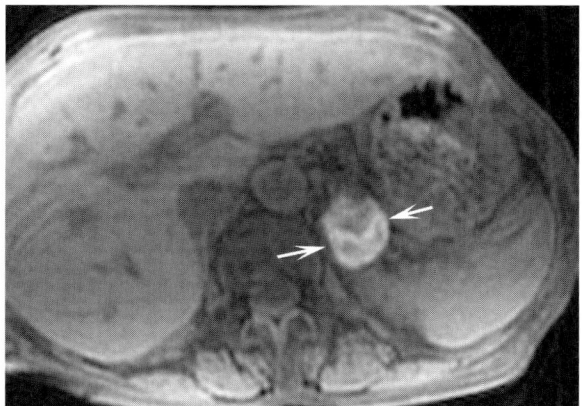

Figure 18.5 Axial T1-weighted fat-saturated non-contrast image demonstrates a large heterogeneous bright lesion involving the left adrenal gland (arrows). In this patient with supratherapeutic international normalized ratio (INR), this is an adrenal hematoma. Given the heterogeneity and size, follow-up imaging should be obtained to ensure that the hematoma resolves without revealing an underlying mass. Post-contrast images with subtraction are also helpful to assess for a small enhancing mass that might have bled.

for extra-adrenal pheochromocytomas is the organ of Zukurkandel, located near the aortic bifurcation. Thus, often when searching for a pheochromocytoma, MRI of the pelvis is added onto the adrenal protocol. Pheochromocytomas can be associated with syndromes such as multiple endocrine neoplasia IIa and IIb, von Hippel–Lindau, tuberous sclerosis, and neurofibromatosis type 1, among others.

On MRI, pheochromocytomas are often T2 hyperintense (Figure 18.3). However, contrary to prior teaching, they do not have to be hyperintense on T2-weighted images. They also enhance avidly and rapidly, and can also fulfill the washout criteria for adrenal adenoma on dynamic contrast-enhanced CT. Fortunately, they are rare tumors and tend to be larger and more heterogeneous than adenomas.

Pheochromocytomas, especially when they become large, can undergo cystic or myxoid degeneration. In the extreme, they can appear as a predominantly cystic lesion that is fluid bright on T2-weighted images. Since true adrenal cysts are very rare, any cystic lesion in the adrenal gland should be regarded with suspicion, especially in the setting of clinical suspicion of pheochromocytoma. Careful search for any thick or nodular enhancement should be performed. If such enhancement is found, then the lesion becomes surgical, as simple cysts and evolved hematomas should not have associated enhancement.

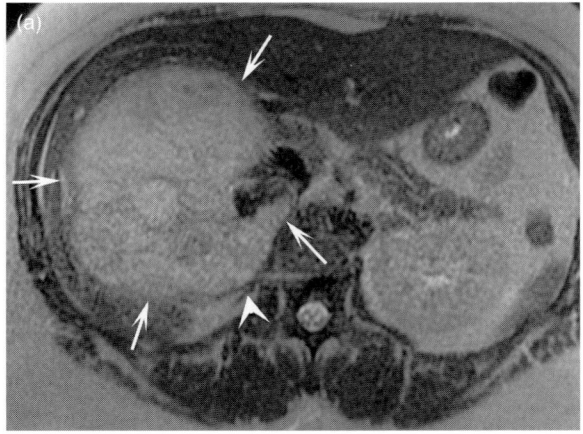

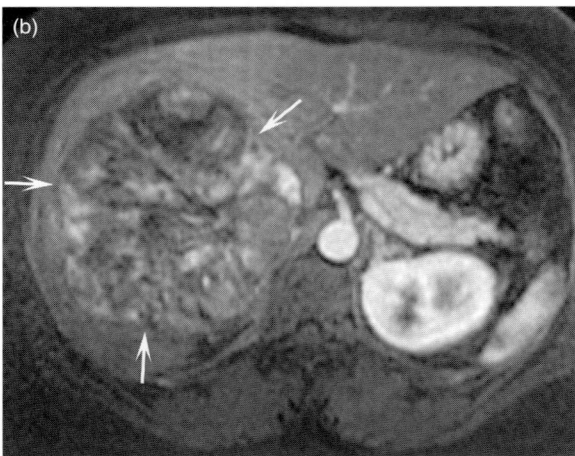

Figure 18.6 Axial T2-weighted (a) and T1-weighted fat-saturated post-contrast (b) images demonstrate a very large mass in the right suprarenal space. A limb of the right adrenal gland can be seen extending to the mass (a, arrowhead), which provides a clue to the adrenal origin of the mass. The mass is heterogeneous in signal on the T2-weighted image and also heterogeneous in enhancement. It invades the liver and extends into the inferior vena cava. This was an adrenal cortical carcinoma. The large size, heterogeneity, and invasion of adjacent structures are typical findings on imaging.

Adrenocortical carcinoma

Another thankfully rare lesion of the adrenal gland is the adrenal cortical carcinoma (ACC). It is an aggressive and malignant tumor arising from the adrenal cortex, as the name implies. A large percentage are hormonally active [17]. ACCs can also be associated with syndromes such as Li–Fraumeni, Beckwith–Wiedemann, and Carney complex.

On diagnosis, these are typically large and have invaded adjacent structures (Figure 18.6). This advanced stage at diagnosis is thought to be related to the rapid rate of growth. The large size, heterogeneity, and enhancement are common imaging findings, which together often suggest the diagnosis, but are not by themselves diagnostic. In fact, ACCs may contain microscopic or macroscopic fat [18]. However, the large size and heterogeneity should suggest that the lesion is not a simple adenoma. If there are ill-defined margins and invasion of adjacent structures, or if there is a tumor thrombus in the adrenal vein, renal vein, or inferior vena cava, then malignancy is likely.

Conclusions

MRI is extremely useful for the detection and characterization of adrenal masses. If an adrenal mass contains microscopic fat as demonstrated on in- and out-of-phase images, then it is an adrenal adenoma. If instead it contains macroscopic fat as demonstrated by fat suppression images, then it is a myelolipoma. In a patient with known malignancy, if an adrenal mass does not fit one of the two categories above and it does not demonstrate the CT contrast washout of an adenoma, then the mass is presumed to be a metastatic lesion. Adrenal hematomas can show heterogeneous hyperintense T1 signal without enhancement. In a patient with proper symptomatology and biochemical evidence of catecholamine excess, an adrenal mass is likely a pheochromocytoma. Finally, if an adrenal mass is very large and heterogeneous and possibly invading adjacent structures, then it is suspicious for an ACC.

References

1. Erturk SM, Alberich-Bayarri A, Herrmann KA, et al. Use of 3.0-T MR imaging for evaluation of the abdomen. *Radiographics* 2009;29(6):1547–1563.

2. Cho CS, Curran S, Schwartz LH, et al. Preoperative radiographic assessment of hepatic steatosis with histologic correlation. *J Am Coll Surg* 2008;206(3):480–488.

3. Dixon WT. Simple proton spectroscopic imaging. *Radiology* 1984;153(1):189–194.

4. Ma J. Dixon techniques for water and fat imaging. *J Magn Reson Imaging* 2008;28(3):543–558.

5. Taffel M, Haji-Momenian S, Nikolaidis P, Miller FH. Adrenal imaging: a comprehensive review. *Radiol Clin North Am* 2012;50(2):219–243, v.

6. Elsayes KM, Mukundan G, Narra VR, et al. Adrenal masses: MR imaging features with pathologic correlation. *Radiographics* 2004;24(Suppl 1):S73–86.

7. Tariq U, Poder L, Carlson D, et al. Multimodality imaging of fat-containing adrenal metastasis from hepatocellular carcinoma. *Clin Nucl Med* 2012;37(6):e157–159.

8. Fujiyoshi F, Nakajo M, Fukukura Y, Tsuchimochi S. Characterization of adrenal tumors by chemical shift fast low-angle shot MR imaging: comparison of four methods of quantitative evaluation. *AJR Am J Roentgenol* 2003;180(6):1649–1657.

9. Halefoglu AM, Yasar A, Bas N, et al. Comparison of computed tomography histogram analysis and chemical-shift magnetic resonance imaging for adrenal mass characterization. *Acta Radiol* 2009;50(9):1071–1079.

10. Choyke PL, Doppman JL. Case 18: adrenocorticotropic hormone-dependent Cushing syndrome. *Radiology* 2000;214(1):195–198.

11. Siegelman ES. Adrenal MRI: techniques and clinical applications. *J Magn Reson Imaging* 2012;36(2):272–285.

12. Berland LL, Silverman SG, Gore RM, et al. Managing incidental findings on abdominal CT: white paper of the ACR incidental findings committee. *J Am Coll Radiol* 2010;7(10):754–773.

13. Lenert JT, Barnett CC, Jr, Kudelka AP, et al. Evaluation and surgical resection of adrenal masses in patients with a history of extra-adrenal malignancy. *Surgery* 2001;130(6):1060–1067.

14. Gillams A, Roberts CM, Shaw P, et al. The value of CT scanning and percutaneous fine needle aspiration of adrenal masses in biopsy-proven lung cancer. *Clin Radiol* 1992;46(1):18–22.

15. Korobkin M. Overview of adrenal imaging/adrenal CT. *Urol Radiol* 1989;11(4):221–226.

16. Boland GW. Adrenal imaging: from Addison to algorithms. *Radiol Clin North Am* 2011;49(3):511–528, vii.

17. Wooten MD, King DK. Adrenal cortical carcinoma. Epidemiology and treatment with mitotane and a review of the literature. *Cancer* 1993;72(11):3145–3155.

18. Ferrozzi F, Bova D. CT and MR demonstration of fat within an adrenal cortical carcinoma. *Abdom Imaging* 1995;20(3):272–274.

Chapter 19

Androgens and DHEA in postmenopausal medicine

Lila E. Nachtigall and Jeffrey A Goldstein

Introduction

The steady decline of serum androgen levels across the lifespan of an adult woman has been well documented. Comparisons between early reproductive age women and women aged 65–75 show mean total testosterone (T) serum levels decreased by 55% and similar reductions are seen for mean values of free T (49%), dehydroepiandrosterone sulfate (DHEA-S) (77%), and androstenedione (A) (77%) [1]. While ovarian aging and the complete cessation of ovarian estradiol (E_2) synthesis produces the dramatic fall in E_2 levels that defines menopause, there is no evidence that the onset of natural menopause influences androgen levels. Rather, the postmenopausal ovarian hilar and stromal cells continue to respond to luteinizing hormone and secrete androgens. A 15-fold difference for T and a fourfold difference for A were noted between ovarian and peripheral veins in postmenopausal women [2], and in four out of five women a gradient for T still appeared more than 10 years after onset of menopause [3]. Additionally, women who underwent surgical menopause with hysterectomy and bilateral oophorectomy had a 40–50% lower serum level of total T and free T than those women with intact ovaries [4]. Other cross-sectional studies have added to the observation that women who had bilateral oophorectomies have significantly lower free T levels [1].

The mechanism of adrenal aging is less well understood, but the age-associated gradual, but marked decline in circulating dehydroepiandrosterone (DHEA) and androgen metabolites has been confirmed by Labrie et al. [5]. DHEA combined with its sulfated conjugate, DHEA-S, is present in the highest amount of all synthesized steroids and is almost totally supplied from the adrenals. While no specific action of its own has been proven, in peripheral tissue DHEA can be converted into active forms of both estrogens and androgens. This conversion plays an important role for postmenopausal women, because it is estimated that 90% of estrogen in these women is derived from peripheral conversion of DHEA. While there is no abrupt cessation of DHEA production mirroring estrogen levels and the onset of menopause, an age-related decline in adrenocorticotropic hormone (ACTH) levels reduces both the amplitude and frequency of DHEA release from the adrenal glands. The result is DHEA levels decline substantially with age as women transition through menopause and beyond. Additionally, the efficiency of peripherally converting exogenously supplied DHEA to androgens and estrogens also decreases after menopause. The approximately 60% or greater decline in serum DHEA and DHEA-S levels associated with menopause may contribute significantly to the development of conditions generally associated with the aging process in women.

Androgen receptors in women have been identified in the ovary, breast, brain, liver, muscle, bone, fat, skin, vulva, and vagina; and for over 60 years the potential benefits of androgen therapy in women's healthcare have been studied [6]. A primary focus has been the treatment of postmenopausal women suffering from low libido and other sexual dysfunctions, but other considerations include the preservation of bone mass, muscle strength, vitality, and the alleviation of vasomotor symptoms and vulvovaginal atrophy. In 2002, a panel comprised of experts from the USA and Australia produced the Princeton consensus statement, which supported the diagnosis of female androgen deficiency syndrome, identified by the concurrent presence of low androgen levels and low libido, and to a lesser extent diminished vitality and general well-being [7]. Controversies arose, as clinical and

Androgens in Gynecological Practice, ed. Leo Plouffe and Botros Rizk. Published by Cambridge University Press. © Cambridge University Press 2015.

scientific challenges to this diagnosis exist based on the uncertainty of what are the normal androgen values by age, the lack of a diagnostic lower limit for any of the androgens that can be defined as androgen deficient, the inconsistency amongst the various assays, and the insensitivity of most assays when detecting low levels of hormones. Additionally, there is great variability in sex hormone-binding globulin (SHBG) levels and their impact on bioavailable or free T. Guidelines published in 2006 by the Endocrine Society recommended against making a diagnosis of androgen deficiency in women for many of these reasons [8].

The anabolic effects of androgens are well established, and gender-based variations in serum levels likely play a significant role in the marked sexual dimorphism that exists in the development and function of muscle, fat, and bone. While the greatest interest for androgen therapy has been to improve or restore libido and reduce symptoms of sexual dysfunction, additional benefits for postmenopausal women may include preservation of bone mass, lean body mass, muscle strength, and general quality of life. Scientific research, including randomized and observational clinical trials and real-world experience with the use of various formulations of testosterone and DHEA for postmenopausal women, has added to our understanding of the therapeutic role for the androgens, and also provided some much needed safety data to address concerns regarding these treatments, including their potential impact on the endometrium and breast.

Bone

Androgens clearly play an important part in bone physiology. Androgen receptors are found in osteoclasts, osteocytes, and most abundantly in osteoblasts [9]. Postmenopausal women with hip fracture were found to have significantly lower free T levels and higher SHBG than age-matched [10] women not in menopause. A large cohort study also showed that low free T increased the risk of hip fractures in women 65 years or older [11]. Barrett-Conner et al. reported that adding T to estrogen increased bone mineral density (BMD) in hip and spine [12]. It is worth noting that the greatest effect of this treatment was in women with surgical menopause.

Most studies of the use of T treatment in postmenopausal women have incorporated this therapy with estrogen support. This addition of T consistently increased bone density. Both Watts et al. [13] in 1995

and Raisz et al. [14] in 1996 showed in double-blinded, placebo-controlled trials that the addition of 2.5 mg per day of methyl T improved bone formation [14] and lumbar spine BMD [13] greater than the treatment of oral conjugated estrogens alone.

Tibolone, a compound simulating estrogen, progesterone, and T with androgenic activity, is not available in the United States but is widely used in Europe. In all studies it prevented bone loss, and even increased bone density in postmenopausal women. In addition, it had excellent suppression of hot flashes. In a study of symptomatic women post-breast cancer, tibolone significantly decreased flushes compared with placebo with a P value of 0.004 at 4 weeks, and a value of 0.0001 at 8 weeks. In a study of osteoporotic women, compared with placebo, tibolone increased both spine and hip bone density. Of great interest is that in the study of the breast cancer survivors, there were more breast cancer recurrences than with placebo. However, in the osteoporotic study, there were fewer new breast cancer cases in the tibolone group than in the placebo group [15].

It had previously been reported in placebo-controlled studies that 50 mg/day of DHEA given to postmenopausal women increased BMD in both the hip and spine [16]. However, the carefully done study by Jancowsky et al. showed these significant increases in hip BMD in older adults undergoing DHEA replacement were mediated primarily by increases in serum E_2, rather than direct effects of DHEA-S [17]. Whatever the mechanism, DHEA does support bone.

Muscle and lean body mass

As important as bone preservation is in postmenopausal woman, there increasing attention is being paid to sarcopenia, the loss of muscle and muscle strength that often begins after menopause and leads to many of the physical disabilities associated with aging. Decreasing levels of T and DHEA add to the absence of estrogen in causing the onset of sarcopenia. DHEA may affect muscle performance as skeletal muscle can convert it to active androgens and estrogens. DHEA also induces the formation of insulin-like growth factor-1, which is important in muscle growth and recovery [18]. There have been several studies of the treatment of postmenopausal women with DHEA to increase their muscle strength. Only when exercise was added to the regimen did muscle strength improve. However, DHEA plus exercise was more successful than exercise alone.

Surprisingly, there have been few studies of T supplementation for muscle development in women; and similar to bone, these studies have been mostly limited to combination estrogen and T treatment. Several randomized controlled studies have confirmed an improvement in lean body mass [19], but contrary to this, in a previously cited study by Davis et al., the addition of T reversed the estrogen effect of decrease in central body fat and resulted in a decrease of total body fat-free mass [20]. The mechanism of this action may be similar to that of progestins, which decrease estrogen's ability to increase insulin sensitivity.

Recent interventions with intermittent T therapy rather than continuous treatment have been shown to be effective in increasing muscle protein synthesis and lean body mass without an increase in side effects in both young athletes and older men [21]. Perhaps this paradigm should be tried in the prevention of sarcopenia in older postmenopausal women.

General quality of life

Androgen receptors have been identified throughout the brain, and the effects of androgens on the brain are mediated through these receptors, but also via the aromatization to E_2 and subsequent E_2 receptor-mediated actions, notably in the hypothalamic and limbic systems. Androgen effects in the brain influence sexual behavior, libido, temperature control, sleep control, assertiveness, cognitive function, and learning capacities, including visual-spatial skills and language fluency. The impact of the postmenopausal hypoestrogenic state on the brain has been more extensively studied, but many brain functions also suffer from the falling androgens levels associated with advancing age. Clinical studies, mostly conducted prior to the initial findings of the Women's Health Initiative (WHI), showed that hormone therapy with estrogen plus androgens provided greater improvement in psychological and sexual symptoms than did estrogen therapy alone for both naturally and surgically menopausal women [22].

The objective of the Study of Women's Health Across the Nation (the SWAN study), a community-based cohort study of 2961 women aged 42–52 years, was to "assess the association between androgens and a variety of end points thought to be affected by androgens." Depressive symptoms were assessed using the 20-item Center for Epidemiologic Studies Depression Scale (CES-D). The Ladder of Life scale was used to assess global quality of life, and physical functioning was measured using the SF-36, a subscale of the Medical Outcome Scale (MOS-SF-36). Circulating testosterone, DHEA-S, and SHBG were measured and a free androgen index (FAI) was calculated. A positive association, at times achieving statistical significance, was noted between levels of T, DHEA-S, and the FAI for higher physical functioning, higher quality of life, and better self-reported health, but a negative association was also noted for CES-D depressive symptomatology [23].

T is almost never used as a primary therapy for symptomatic postmenopausal women, and rarely are general quality of life parameters a primary objective of the studies. In 2011, Glaser and her group reported on the treatment of 300 women, 163 of whom were postmenopausal. They used T pellet implants for one year in doses dependent on weight. The dose range was 75–160 mg with a mean subcutaneous dose of 121 mg. They measured symptom improvement using both the health-related quality of life scale (HRQOL) and the menopause rating scale (MRS). Remarkably, significant improvement was shown for all patients for all 11 categories rated. These included hot flushes, sweating, vaginal dryness, and urogenital complaints, and the higher the dose, the greater the improvement. Although there was no placebo control, the correlation of improvement in symptoms relative to the dose argues against placebo effect. It is also noteworthy that 98% returned for treatment after the study was complete [24].

Many recent studies have focused on the possible beneficial impact of androgen therapy on women with hypoactive sexual desire disorder and related psychological distress. A consistent improvement has been noted for diminished personal distress, but it may be difficult to extrapolate these findings to a more general population and a broader beneficial impact on health-related quality of life.

Endometrium

Androgens have two theoretical pathways to influence the endometrium; directly through androgen receptors and indirectly through peripheral enzymatic conversion by aromatase into E_2. While the ability of the endometrium to respond to either of these pathways has been questioned or minimized, in vitro research supports both as possibly of concern. Immunostaining of androgen receptors in both stromal and endometrial cells has been demonstrated [25].

Tuckerman et al. conducted in vitro studies in fertile women, investigating the effects of A, T, dihydrotestosterone (DHT) and DHEA [26]. A was shown to have a direct inhibitory effect on the endometrium, a finding that was further supported by the ability of the androgen receptor antagonist cyproterone acetate to reverse this inhibition [26]. No significant effect on the endometrium was noted for T, A, and DHEA. The ability of aromatase to convert circulating androgens into E_2 and the stimulatory effect of estrogens on the endometrium are both well recognized. In the endometrium of disease-free patients, gene transcription of aromatase is inhibited and this enzyme is not expressed, eliminating concerns of a pathway toward hyperplasia [27]. However, increased endometrial aromatase activity has been confirmed in gynecologic conditions, including endometriosis, adenomyosis, and fibroids and is thought to play a significant role in the pathophysiology and progression of these diseases. Additionally, the conversion of A to estrone and E_2 was significantly increased in endometrial neoplastic cells [28]. However, the aromatization of androgens appears to have minimal growth-promoting impact on well-differentiated endometrial carcinoma [29].

Several studies have investigated the possible association in postmenopausal women between endogenous serum levels of the various androgens and the risk for endometrial hyperplasia/carcinoma, and have produced conflicting data. Potischman et al. reported significantly higher levels of A and lower levels of SHBG in women with endometrial cancer [30]. In a case-controlled study of 124 postmenopausal women not on hormonal therapy, Lukanova et al. first reported a weak link between androgen levels and endometrial cancer, but after adjustments were made to account for E_2 and estrone levels, the risk was no longer significant [31]. Boman et al. also showed no relationship between serum A levels and proliferation activity in women with endometrial cancer [32].

Various formulations of T have undergone clinical development, either for vasomotor symptom relief or for the treatment of hypoactive sexual desire disorder (HSDD), and their impact on the endometrium has been evaluated. A 6-month, double-blinded randomized trial comparing the effect on endometrial histology of 0.625 mg of oral esterified estrogen, with or without 1.25 mg of methyltestosterone (without progestins), showed estrogen stimulated proliferation in both groups [33]. Zang et al. conducted a randomized, unblinded trial with women receiving either oral T

undecanoate (40 mg every second day), E_2 valerate (2 mg daily), or both for 3 months [34]. Endometrial thickness was measured by ultrasound and endometrial proliferation was assessed by both histology and the expression of Ki-67, a nuclear antigen only expressed by proliferating cells. Endometrial thickness was significantly increased in both the estrogen alone and estrogen plus T arm, but no increase was noted in the T alone arm. Similar findings were noted for endometrial histology, where there was a statistically significant 50% increase in evidence of proliferation in the estrogen alone arm, a non-significant 28% increase for the combination therapy, and no increase for those treated with only T. In the 52-week APHRODITE study, 814 women with HSDD were treated with either 150 µg or 300 µg of patches of T or placebo [35]. While 10.6% of the women on the higher dose of T experienced vaginal bleeding, there were no reported cases of hyperplasia or carcinoma in any of the subjects. The INTIMATE study of the T patch reported no increased risk for vaginal bleeding [36].

Breast

Just as the anti-proliferative effects of progestins on the uterus have been well documented, androgen receptor signaling in the breast exerts a growth-inhibitory, anti-proliferative influence in normal breast tissue. Proof that estrogens stimulate and androgens inhibit breast development is confirmed by studies showing that, independent of genetic sex, the mammary gland fails to develop in females lacking functioning ERα receptors, and develops fully in males lacking functioning androgen receptors [37]. With the onset of menopause, circulating estrogen levels fall dramatically and in the absence of estrogen stimulation, involution of the breast occurs characterized by atrophy of the glandular tissue and fibrosis of the stromal tissue. Since androgen levels more gradually decline in this population, it is theorized that these changes are due to unopposed androgens, a concept supported by evidence that involution is dramatically accelerated by androgen therapy given to women undergoing a female-to-male sex reversal [38].

In postmenopausal woman, T treatment has been shown to inhibit estrogen/progestogen-induced breast cell proliferation [39]. In a 6-month, prospective, double-blinded trial, 99 postmenopausal women were given a combination of oral E_2 2 mg with norethisterone acetate 1 mg, and were randomized to also receive either a T patch releasing 300 µg/day or a placebo patch. The placebo group had more than a five-fold greater increase in breast cell proliferation from baseline, with changes noted in both the epithelial and stromal cells. The addition of T attenuated this proliferation, with no statistical significant change from baseline noted.

Despite the assurances provided by the preclinical science and clinical data described above, breast cancer safety data collected from epidemiologic studies of products containing T have yielded disconcerting results. The Nurses' Health Study, a prospective cohort study with over 100 000 registered nurses between the ages of 30 and 55 years returning an initial questionnaire, was conducted from 1978 to 2002 and represents over 1 million person-years of follow-up. Women who experienced a natural menopause and reported using estrogen plus T had a 1.77 relative risk of breast cancer compared to never-users. This was greater than the 1.15 relative risk for estrogen-only therapy. Of note, the increased risk was only noted for the first 5 years of therapy, and not significantly elevated for longer duration of treatment [40]. As a cohort study, there were numerous limitations to the study, including disparities in prior hormone use and patient demographics that confound interpretation of the results. Most of the observational studies of populations treated with T alone or in combination with estrogens showed no significant increased risk for breast cancer.

The more recent efforts to develop a safe, effective treatment for HSDD, which the recently released DSM-5 has redefined as sexual interest/arousal disorder, and other subsets of female sexual dysfunction have led to renewed interest in exogenous androgen therapy, mostly in the form of transdermal T. The impact on breast tissue and potential risk for breast cancer has been a critical safety signal to be studied. In the APHRODITE study of a transdermal patch containing either 150 or 300 µg of T, efficacy was measured at 24 weeks, but safety was extended to 52 weeks [35]. Two participants in the 300 µg arm were diagnosed with breast cancer, one at 7 months, but history later revealed a bloody nipple discharge before enrollment. One participant in the 150 µg arm was diagnosed with breast cancer 4 months into the trial. Based on accepted doubling times and growth behavior of breast cancer, it is hard to conceive of any of these patients developing de-novo tumors during the study. In the 6-month-long ADORE study of a 300 µg patch, there were no reports of breast cancer and 1.5% of the women complained of

breast tenderness compared with 0 women in the placebo arm [41].

Breast cancer safety data have been obtained for up to 4 years of exposure from the open-label extension of the INTIMATE SM1 and SM2 clinical efficacy trials, each a 6-month efficacy study of a 300 mg/day transdermal testosterone patch (TTP) for the treatment of HSDD in surgically postmenopausal women. Nine hundred and sixty-seven patients received at least one application of the TTP resulting in 1092 patient-years of exposure, and there were three cases of invasive breast cancer, a number consistent with the age-appropriate base-line rate [42]. The BLISS study of a T gel formulation is still ongoing, and upon completion will provide 5 years of safety data with breast cancer a co-primary endpoint.

In 2002, the Endogenous Hormones and Breast Cancer Collaborative Group published results of their epidemiologic study that demonstrated an association between T levels and increased breast cancer risk for postmenopausal women [43]. Other prospective case-controlled studies have also explored the possible association between endogenous androgen levels and breast cancer, often with conflicting results. Interpretation of these data are challenging, as often authors use total T as a measure of androgen exposure and fail to address SHBG levels and free T levels. Formulation and route of therapy may also be an important variable. Oral formulations of both estrogen and T therapy raise SHBG which in turn lowers free T, and thus potentially alters the ratio between E_2 and free T. This elevation of SHBG is not noted with either transdermal estrogen or T therapy, and the clinical significance of this difference and potential impact on breast cancer requires further study.

The varied expression of aromatase activity within both normal and malignant breast tissue further adds to the uncertainty of the impact of androgens. Many studies have shown that postmenopausal E_2 and T levels track together, a logical finding since much of the circulating E_2 is produced by aromatase-based conversion of androgens. Exogenous androgens could serve as precursor for estrogen production in malignant breast tissue. Stromal cells of disease-free breast tissue express low levels of aromatase, but malignant transformation has shown the ability to switch on the promoter that drives the aromatase gene, resulting in increased estrogen production in the breast. This estrogen production may contribute to the tumor growth, and provides part of the rationale for the use of aromatase inhibitors to treat breast cancer patients. However, animal studies of breast cancer have not supported this concern, and in fact androgens have been shown to inhibit breast cancer growth in rhesus monkeys [44].

Perhaps the greatest challenge in assessing the true risk/benefit profile of endogenous androgens levels or the impact of exogenous postmenopausal androgen therapy is the acknowledged complex etiology of breast cancer. Various subtypes of this disease exists, notably differing in receptor status of any and all of the three key sex hormones: estrogen, progesterone, and androgen. This variability has fostered a potential dichotomous, contradictory role for the androgen receptor. In humans, androgen receptors are often expressed in breast cancer, but the full implication of their role in cancer growth is poorly understood. Argument can be made for both the role of androgen excess in the development of some subsets of breast cancer, and the potential for the androgen receptor to serve as a tumor suppressor and growth inhibitor for other subsets.

Conclusion

While ever present, there has been a recent increased interest in preserving women's health, vitality, and sexuality, and slowing down the inevitable aging process. The anabolic effects of androgens on bone, muscle, and body morphology, coupled with their apparent therapeutic benefit on aspects of female sexual dysfunction and libido, make androgen therapy an attractive option for postmenopausal women. Should women receive androgen therapy? The challenge to answering this question lies in better understanding the risks and benefits of exogenous exposure. Based on this evolving profile, we will be better able to determine how liberal or conservative we need to be in identifying the correct patients and defining the diagnostic and clinical parameters of female androgen deficiency syndrome and/or sexual interest/arousal disorder.

References

1. Davison SL, Bell R, Donath S, et al. Androgen levels in adult females: changes with age, menopause, and oophorectomy. *J Clin Endocrinol Metab* 2005;90:3847–53.

2. Judd HL, Judd GE, Lucas WE, Yen SS. Endocrine function of the postmenopausal ovary: concentrations of androgens and estrogens in ovarian and peripheral vein blood. *J Clin Endocrinol Metab* 1974;39:1020–4.

3. Fogle RH, Stanczk FZ, Zhang X, Paulson RJ. Ovarian androgen production in postmenopausal women. *J Clin Endocrinol Metab* 2007;92:3040–3.

4. Laughlin GA, Barrett-Connor E, Kritz-Silverstein D, von Muhlen D. Hysterectomy, oophorectomy and endogenous sex hormone levels in older women: the Rancho Bernardo Study. *J Clin Endocrinol Metab* 2000;85:645–51.

5. Labrie F, Belanger A, Cusan L, et al. Marked decline in serum concentrations of adrenal C19 sex steroid precursors and conjugated androgen metabolites during aging. *J Clin Endocrinol Metab* 1997;82:2396–402.

6. Greenblatt RB, Barfield WE, Garner JF, et al. Evaluation of an estrogen, androgen, estrogen-androgen combination, and a placebo in the treatment of the menopause. *J Clin Endocrinol* 1950;10:1547–58.

7. Bachmann G, Bancroft J, Braunstein G, et al. Female androgen insufficiency: the Princeton consensus statement on definition, classification, and assessment. *Fertil Steril* 2002;77(4):660–5.

8. Weinman ME, Bassoon R, Davis SR, et al. Androgen therapy in women: and Endocrine Society Clinical Practice Guideline. *J Clin Endocrinol Metab* 2006;91(10):3697–710.

9. Abu EO, Horner A, Kusec V, et al. The localization of androgen receptors in human bone. *J Clin Endocrinol Metab* 1997;82(10):3493–7.

10. Dubin NH, Monahan LK, Yu-Yahiro JA, et al. Serum concentrations of steroids, parathyroid hormone, and calcitonin in postmenopausal women during the year following hip fracture: effect of location of fracture and age. *J Gerontol A Biol Sci Med Sci* 1999;54(9):M467–73.

11. Cummings SR, Browner WS, Bauer D, et al. Endogenous hormones and the risk of hip and vertebral fractures among older women. Study of Osteoporotic Fractures Research Group. *N Engl J Med* 1998; 339(11):733–8.

12. Barrett-Connor E, Young R, Notelovitz M, et al. A two-year, double-blind comparison of estrogen-androgen and conjugated estrogens in surgically menopausal women. Effects on bone mineral density, symptoms and lipid profiles. *J Reprod Med* 1999;44(12):1012–20.

13. Watts NB, Notelovitz M, Timmons MC, et al. Comparison of oral estrogens and estrogens plus androgens on bone mineral density. *Obstet Gynecol* 1995;85(4):529–37.

14. Raisz LG, Wiita B, Artis A, et al. Comparison of the effects of estrogen alone and estrogen plus androgen on biochemical markers of bone formation and resorption in postmenopausal women. *J Clin Endocrinol Metab* 1996;81(1):37–43.

15. Kenemans P, Bundred NJ, Foidart JM, et al; LIBERATE Study Group. Safety and efficacy of tibolone in breast-cancer patients with vasomotor symptoms: a double-blind, randomised, non-inferiority trial. *Lancet Oncol* 2009;10(2):135–46.

16. Baulieu EE, Thomas G, Legrain S, et al. Dehydroepiandrosterone (DHEA), DHEA sulfate, and aging: contribution of the DHEA Study to a sociobiomedical issue. *Proc Natl Acad Sci USA* 2000;97(8):4279–84.

17. Jankowski CM, Gozansky WS, Kittelson JM, et al. Increases in bone mineral density in response to oral dehydroepiandrosterone replacement in older adults appear to be mediated by serum estrogens. *J Clin Endocrinol Metab* 2008;93(12):4767–73.

18. Stewart CE, Pell JM. Point:Counterpoint: IGF is/ is not the major physiological regulator of muscle mass. Point: IGF is the major physiological regulator of muscle mass. *J Appl Physiol* 2010;108(6):1820–1;discussion 1823–4; author reply 1832.

19. Floter A, Nathorst-Boos J, Carlstrom K, et al. Effects of combined estrogen/testosterone therapy on bone and body composition in oophorectomized women. *Gynecol Endocrinol* 2005;20(3):155–60.

20. Davis SR, McCloud P, Strauss BJ, Burger H. Testosterone enhances estradiol's effects on postmenopausal bone density and sexuality. *Maturitas* 1995;21(3):227–36.

21. Basaria S, Collins L, Dillon EL, et al. The safety, pharmacokinetics, and effects of LGD-4033, a novel nonsteroidal oral, selective androgen receptor modulator, in healthy young men *J Gerontol A-Biol Sci Med Sci* 2013;68(1):87–95.

22. Sarrel PM. Psychosexual effects of menopause: role of androgens. *Am J Obstet Gynecol* 1999;180:S319–24.

23. Santoro N, Torrens J, Crawford S, et al. Correlates of circulating androgens in mid-life women: the study of women's health across the nation. *J Clin Endocrinol Metab* 2005;90:4836–45.

24. Glaser R, York AE, Dimitrakakis C. Beneficial effects of testosterone therapy in women measured by the validated Menopause Rating Scale (MRS). *Maturitas* 2011;68(4):355–61.

25. Mertens HJ, Heineman MJ, Koudstaal J, et al. Androgen receptor content in human endometrium. *Eur J Obstet Gynecol Reprod Biol* 1996;70:11–13.

26. Tuckerman E, Okon M, Tin-Chiu L, Laird S. Do androgens have a direct effect on endometrial function? An in vitro study. *Fertil Steril* 2000;74:771–9.

27. Maia H, Haddad C, Coelho G, Casoy J. Role of inflammation and aromatase expression in the eutopic endometrium and its relationship with the development of endometriosis. *Women's Health* 2012;8(6):647–58.

28. Yamaki J, Yamamoto T, Okada H. Aromatization of androstenedione by normal and neoplastic endometrium of the uterus. *J Steroid Biochem* 1985;22(1):63–6.

29. Legro RS, Kunselman AR, Miller SA, Satyaswaroop PG. Role of androgens in the growth of endometrial carcinoma: an in vivo animal model. *Am J Obstet Gynecol* 2001;184(3):303–8.

30. Potischman N, Hoover RN, Brinton LA, et al. Case-control study of endogenous steroid hormones and endometrial cancer. *J Natl Cancer Inst* 1996;88:1127–35.

31. Lukanova A, Lundin E, Micheli A, et al. Circulating levels of sex steroid hormones and risk of endometrial cancer in postmenopausal women. *Int J Cancer* 2004;108:425–32.

32. Boman K, Strang P, Backstrom T, Stendahl U. The influence of progesterone and androgens on the growth of endometrial carcinoma. *Cancer* 1993;71:3565–9.

33. Hickok L, Toomey C, Speroff L. A comparison of esterified estrogens with and without methyltestosterone: effects on endometrial histology and serum lipoproteins in postmenopausal women. *Obstet Gynecol* 1993;82:919–24.

34. Zang H, Sahlin L, Masironi B, et al. Effects of testosterone treatment on endometrial proliferation in postmenopausal women. *J Clin Endocrinol Metab* 2007;92(6):2169–75.

35. Davis SR, Moreau M, Kroll R, et al. Testosterone for low libido in postmenopausal women not taking estrogen. *N Engl J Med* 2008;359:2005–17.

36. Shifren J, Davis SR, Moreau M, et al. Testosterone patch for the treatment of hypoactive sexual desire disorder in naturally menopausal women: results from the INTIMATE NM1 Study. *Menopause* 2006;13(5):770–9.

37. Hickey T, Robinson L, Carroll J, Tilley W. Minireview: The androgen receptor in breast tissues: growth inhibitor, tumor suppressor, oncogene? *Mol Endocrinol* 2012; 26:1252–67.

38. Grynberg M, Fanchin R, Dubost G, et al. Histology of genital tract and breast tissue after long term testosterone administration in a female-to-male transsexual population. *Reprod Biomed Online* 2010;20:553–8.

39. Hofling M, Hirschberg A, Skoog L, et al. Testosterone inhibits estrogen/progesterone-induced breast cell proliferation in postmenopausal women. *Menopause* 2007;14(2):183–90.

40. Tamimi R, Hankinson S, Chen W, et al. Combined estrogen and testosterone use and risk of breast

cancer in postmenopausal women. *Arch Intern Med* 2006;166:1483–9.

41. Panay N, Al-Azzawi F, Bouchard C, et al. Testosterone treatment of HSDD in naturally menopausal women: the ADORE study. *Climacteric* 2010;13:121–31.

42. Nachtigall L, Casson P, Lucas J, et al. Safety and tolerability of testosterone patch therapy for up to 4 years in surgically menopausal women receiving oral or transdermal estrogen. *Gynecol Endocrinol* 2011;27(1):39–48.

43. Key T, Appleby P, Barnes I, et al. Endogenous Hormones and Breast Cancer Collaborative Group. Endogenous sex hormones and breast cancer in postmenopausal women: reanalysis of nine prospective studies. *J Natl Cancer Inst* 2002;94(8):606–16.

44. Dimitrakakis C, Zhou J, Wang J, et al. A physiologic role for testosterone in limiting estrogenic stimulation of the breast. *Menopause* 2003;10(4):292–8.

Chapter

20

Lifestyle, diet, and exercise in polycystic ovary syndrome

Elizabeth Burt and Rina Agrawal

Introduction

Subfertility is a diagnosis that affects 80 million people worldwide [1] and approximately 10% of couples in the developed Western world [2]. Polycystic ovary syndrome (PCOS) represents the most common cause of anovulatory infertility. Paradoxically, although recent years have witnessed massive advances in the arena of fertility services, with exciting new interventions and potential treatment strategies for those affected, the incidence of infertility has increased [3]. Whether this is secondary to heightened awareness and public demand or actual changes in disease prevalence is unknown, but one explanation may be the evolution of modern society to a new culture of adverse lifestyle and environmental factors, coupled with commercialism and consumerism. Applicable to all areas of medicine is the ethos "prevention is better than the cure." In this chapter we aim to explore the potentially preventable, modifiable factors that may impact negatively on human fertility in the PCOS population.

With an elusive etiology and pathophysiology, PCOS represents a challenge to the fertility world; however, with over 10–18% of women of reproductive age being affected by PCOS [4], it is imperative that the medical profession is fully conscious of its apparently increasing prevalence and the impact that this may have on both society and the health care delivery systems, such as the UK National Health Service. Focus needs to be placed on conservative methods to maximize the chance of natural conception and alleviating symptomatology, thereby reducing the reliance on pharmacological intervention, circumventing the need for more specialist input and therefore reducing the psychological and physical sequelae that invasive fertility treatment may incur.

Human health and more specifically fertility have been demonstrated to be adversely affected by a number of lifestyle and environmental factors. PCOS is an amalgamation of reproductive dysfunction in addition to unwelcome psychological and metabolic manifestations, and its clinical and biochemical presentation is more pronounced in combination with an unhealthy lifestyle. Often many of these lifestyle factors occur in unison forming an integral part of our "modern society" but unlike genetic factors, these are related to behaviors and circumstances which, once addressed and reformed, may improve reproductive outcomes and lessen symptoms. There has been an increasing awareness of these amongst both patients and service providers, and adjusting these factors by appropriate counselling and intervention may play a crucial and fundamental role in the care pathway of PCOS-affected females.

Obesity

The obesity epidemic is a huge public health concern, the magnitude demonstrated by the statistics of nearly a quarter of adults (22% of men and 24% of women) aged 16 or over being classified as obese in England (body mass index [BMI] >30 kg/m^2) [5]. With associated ramifications in disease prevalence and trajectories, coupled with an increasing burden to healthcare provision and economics and the attributable health expenditure worldwide standing at 2–7% of total health expenditure[6], tackling obesity needs to be at the forefront of global public health initiatives.

Obesity and PCOS prevalence

The relationship between obesity and PCOS is intricate, undeniable, but not straightforward. As obesity

---

Androgens in Gynecological Practice, ed. Leo Plouffe and Botros Rizk. Published by Cambridge University Press. © Cambridge University Press 2015.

levels escalate at an alarming rate, 56% of women in England are classified as overweight or obese [7], and with an apparent increase in the prevalence of PCOS, it could be extrapolated that the two are rising in parallel and a potential causative relationship inferred. Cause and effect is yet to be fully deduced and although pubertal obesity is known to be associated with hyperandrogenism [8], it is not yet clear whether this serves as a precursor to the development of PCOS. Some studies have concurred that the prevalence of PCOS is increased in obese populations [9], whereas others have merely demonstrated a modest incidental increase in risk [10]. This translates into a multifactorial etiology implicating both intrinsic and extrinsic factors, with possibly a genetically susceptible population being influenced by environmental components.

Early intrauterine life

Conception and gestation are vulnerable periods and embryonic cells have a complex system to integrate signals emanating from their environment and can adapt their development accordingly [11]. Certainly, an abnormal metabolic environment may contribute to programming alterations during a crucial window for developmental plasticity and beyond. Ultimately, as these changes in physiology, metabolism, and development are metabolically imprinted they may manifest postnatally and later in adult life. Certainly, it has been appreciated that the presence of maternal obesity in pregnancy renders the female offspring more susceptible to the development of PCOS [12] and likewise, female offspring born to mothers with PCOS have insulin defects [13].

Developmental origins and fetal programming theory of PCOS has been increasingly substantiated from primate and sheep studies demonstrating that, in utero, excess androgen exposure may affect fetal programming leading to the development of PCOS later in life. It has been demonstrated that testosterone-exposed offspring have higher birth weights with associated hyperglycemia and hyperinsulinemia [14–16]. Similar findings are yet to be reproduced in humans but much work is underway [17].

These are important data to consider when counselling women with PCOS of the therapeutic strategies that can be employed, as there is responsibility not only for the women's health but for also the perceivable health of their future generation.

Obesity and the pathophysiology of PCOS

The endocrine and metabolic milieu work in elaborate symbiosis to ensure successful ovulation and reproduction. This balance can be disturbed by many factors and obesity is a dominant perpetrator, as witnessed by the presence of obesity in 35–60% of women with PCOS [18]. Even in the absence of PCOS, obesity upsets the hormonal balance.

Interestingly, more specifically than BMI per se, it appears that the pattern of fat distribution may mediate the clinical consequences to varying extents, with accumulation of fat in the abdomen, indicated by a waist:hip ratio >0.8 or a waist circumference of over 80 cm [19], being associated with worsening reproductive and metabolic sequelae [20]. This has been coined the "abdominal phenotype." Even in the absence of the BMI criteria fulfillment for obesity, women with PCOS will often have an abdominal "roundness" and this pattern of adiposity is associated with more pronounced states of hyperandrogenemia and hyperinsulinemia [21,22].

A postulated mechanism for the development of abdominal obesity in PCOS proposes that with high calorific intake, fat is initially preferentially laid down in the intra-abdominal compartment and with saturation of this area, it is subsequently shunted to the subcutaneous abdominal area. Due to the metabolic activity of the visceral adipocytes there is enlargement of this fat depot, exacerbated by androgen excess and an associated flux of free fatty acids to the liver and muscles. Increasing liver and muscle fat deposition is thought to be responsible for the associated rise in insulin resistance and compensatory insulin release [23,24,]. Testosterone's role and function in the development of insulin resistance is still being fully investigated.

It would appear that insulin is the dominant instigator orchestrating and perpetuating the changes in biochemical parameters, with subsequent rise and fall in testosterone and sex hormone-binding globulin (SHBG) respectively. Hyperinsulinemia impedes the liver's synthesis of SHBG and the negative correlation between abdominal adiposity and SHBG, as a result, leads to elevated levels of free testosterone being presented to the target tissues and higher clearance rates. Women with central, visceral obesity tend to have lower SHBG levels even when compared to their peripherally obese counterparts of the same weight [20] and have a positive net balance of androgens, as production exceeds clearance, leading to a state of relative functional hyperandrogenemia [25].

It should not be overlooked that adipose tissue also has a fundamental storage role affecting steroid hormone bioavailability and is also an extraglandular source of steroidogenesis. Indeed alterations in enzyme systems including 11β-hydroxysteroid dehydrogenase, 5α-reductase, and 5β-reductase have all been investigated [26–28]. Therefore not only is the "reserved" pool of steroid hormones greater in obese women, but also more "machinery is available" with steroidogenetic enzymes present for conversion and metabolism.

It has been speculated that the link between obesity and PCOS may lie in the metabolic role of adipocytes and their role in metabolic regulatory pathways. Adipocytes produce over 50 adipocytokines and many of these have been investigated and are thought to mediate obesity-associated insulin resistance via regulation of lipid and glucose metabolism [29,30].

Leptin, an adipocytokine that plays a fundamental signaling role in metabolism and reproduction, has generated much research attention. At the level of the hypothalamus leptin serves to decrease food intake and in the periphery is antagonistic of insulin in fat and glucose metabolism [30]. Therefore, it appears contradictory that high leptin levels are observed in women with obesity and PCOS. There is, however, a degree of leptin resistance and high circulating leptin levels can modulate the insulin resistance and ovulatory dysfunction shared in obesity and PCOS. In common with insulin, the resistance to leptin does not extend to the ovaries and in the ovary leptin has an inhibitory effect and reduces steroidogenesis, follicular development, and ovulation [19,31,) Furthermore, in the hypothalamus and pituitary, leptin levels can disturb gonadotropin-releasing hormone and gonadotropin secretion [32]. At present, leptin is one of the many adipocytokines incriminated and being explored in the pathogenesis of PCOS and obesity.

Obesity and disease presentation

PCOS is both heterogeneous and unpredictable in its presentation. There is remarkable individual variation and a spectrum of severity. With concurrent obesity there is a detrimental effect on the phenotypic hallmarks of PCOS, as witnessed by the exponential relationship between rising BMI and worsening features of anovulation [33,34]

Women with PCOS and obesity are more prone to menstrual irregularity compared with their low-weight peers and hyperinsulinemia, secondary to a state of insulin resistance, is thought to contribute to this ovulatory dysfunction [35,36]. The reduction in insulin sensitivity is not uniform or applicable to all tissues, and on the contrary, a synergetic action of insulin can be seen with luteinizing hormone acting on thecal cells to augment ovarian androgen production, ultimately culminating with follicular growth arrest, atresia, and anovulation [37]. Furthermore, insulin resistance and obesity have also been associated with detrimental oocyte development [38] and in combination, a degree of endometrial insulin resistance and high circulating insulin levels has been speculated to affect endometrial development and receptivity, with conceivable risks for implantation and fertility[39].

Certainly, for a substantial proportion of women, a PCOS diagnosis will come to fruition in the reproductive years with the inability to conceive naturally, as PCOS is the most common cause of anovulatory infertility. With the UK National Institute for Health and Care Excellence (NICE) guidelines representing evidence-based medicine, it is with reason that they have set their recommendation for women to have BMI of less than 30 kg/m^2 before assisted conception and eligibility for NHS-funded fertility treatment, as the literature delivers a wealth of evidence demonstrating the adverse effects of obesity on reproduction and pregnancy outcome. Independent of any other coexisting pathology, the probability of infertility increases threefold with obesity alone [33]. Certainly, in PCOS with superimposed obesity, there are unfavorable effects on conception rates both spontaneous and after treatment, efficacy of pharmacological agents, and pregnancy outcomes. When undergoing assisted reproduction, obese women have lower pregnancy rates and a greater threat of miscarriage compared with women with a BMI <25 kg/m^2 [38,40]. In addition, other deleterious effects include a relative resistance to ovarian-stimulating medications, fewer oocytes collected, poorer quality and maturation of the oocyte, and ultimately fewer live births. [2,7,41].

Women with PCOS commonly cite hirsutism as being one of the most distressing aspects with both aesthetic and psychological repercussions [42]. In conjunction, obesity and PCOS lend themselves to worsening symptoms associated with androgen excess including hirsutism, alopecia, and acne [43].

A relatively high proportion of women with PCOS (20–49%) will present with glucose intolerance, reiterating the importance of glucose tolerance tests in clinical practice for the workup of PCOS. Obese women with

PCOS tend to have a more severe insulin-resistant state. In comparison with normal weight women with PCOS, obese PCOS women have higher concentrations of both fasting and glucose-stimulated insulin[43]. The stigma of insulin resistance acanthosis nigricans is more clinically apparent when obesity accompanies PCOS [32].

PCOS represents a chronic disease in its own right but, importantly, it is also interrelated to other diseases, affecting both the morbidity and mortality of this cohort of women in the long term. PCOS, secondary to its metabolic disturbances, is associated with the metabolic syndrome with a predisposition to cardiovascular disease and diabetes risk [44]. Obesity is also a well-recognized independent risk factor for these comorbidities and therefore, if occurring in combination, the risk is summative with a cluster of risk factors including dyslipidemia, altered blood vessel function, hypertension, and glucose intolerance. In the absence of obesity, PCOS rarely heralds glucose intolerance and likewise the lipogenic effects in PCOS are exaggerated with obesity, with a pattern vulnerable to atherosclerosis of high total cholesterol levels and triglycerides and lower levels of high-density lipoproteins [43].

PCOS and psychological stress

Common to PCOS, obesity and other chronic illness, quality of life can be severely compromised and be affected by depression, anxiety, and other psychiatric ailments [45,46]. Stress has an important role in PCOS, both as cause and consequence, and can often fuel a vicious cycle.

Many mechanisms have been proposed associating stress with the pathophysiology of PCOS. Stress is associated with food cravings and "stress eating," often with poor nutritional choice and high-calorific foods [47], can compound obesity and this, in addition to psychological stress itself, can aggravate the endocrine disequilibrium of PCOS to a greater extent. Obesity alone is linked with a reduction in the frequency of sexual activity but, in the presence of overeating which is itself associated with a decreased sexual drive, can contribute to reduced fecundity with psychosocial and relationship struggles.

Both acute and chronic stress can provoke and alter aspects of the hypothalamic–pituitary–adrenal axis and this seems to be heightened in PCOS. In response to stress there appears to be excessive and prolonged cortisol release, which in turn propagates worsening obesity and enhances insulin resistance [48].

Conversely, although stress can heighten PCOS, the emotional distress that can ensue from the inability to conceive with an infertility diagnosis secondary to PCOS can be vast [49]. Moreover, the cosmetic and body image consequences of obesity, hyperandrogenemia with hirsutism, alopecia, and acne can be the overriding complaint especially in younger females, which may precipitate social, psychological, and sexual impairment [42].

Despite optimism being given in the form of medical intervention, the pressures and expectations associated with treatment and required lifestyle changes can lead to immense strain, relationship breakdown, and treatment dropout. Indeed 15–20% of couples undertaking reproductive treatment require counseling, with psychological interventions demonstrating successful outcomes in pregnancy rates [50].

Not to be overlooked, stress constitutes a risk factor for cardiovascular disease and diabetes independent of PCOS and obesity and therefore recognition and alleviation of stress prior to, during treatment, and long term is paramount and should be strongly advocated.

Lifestyle interventions

With concomitant obesity having such profound effects on all aspects of PCOS it is with justification that lifestyle interventions comprise the recommended first line of its treatment [51]. Indeed modest amounts of weight loss (>5% of body weight) have been shown to have significant beneficial effects clinically, and in terms of cost effectiveness [24,51,52]. For women with PCOS, weight loss can be even more difficult to achieve, the effects of hyperinsulinemia and psychological barriers being to blame, and therefore specialized help must be at hand.

Obesity is a challenge as its pathophysiology is complex, involving both physical and psychological dimensions. Interventions, therefore, require a multidisciplinary approach utilizing the expertise of psychologists, dieticians, and exercise specialists, providing competent support in all aspects of the complex issues that constitute weight management. The objectives should include the avoidance of further weight gain and the attainment of a degree of weight loss, all whilst endorsing habitual alterations in nutrition and physical activity. This should be in concordance with the approach advocated by NICE who, in their guidance on tackling obesity, emphasize the need for weight management programs.

Emphasis needs be placed on a comprehensive, systematic approach to weight control and symptom modification. To aid compliance and motivation, it should be reiterated that small adaptations can be enough to lead to beneficial effects with restoration of ovulation and increased pregnancy rates [53] A complete upheaval of lifestyle may seem daunting to both patient and physician but, by adopting a holistic multifaceted approach, potentially through group or supervised programs, benefits have been observed with dietary modification and this, in combination with exercise, can lead to cumulative improvement and long-term weight maintenance [54]. Behavior modification interventions, in addition to addressing and alleviating psychological problems, have proven beneficial in complementing dietary and exercise advice to help achieve sustained weight loss and enduring change [55].

Dietary advice: With regard to specific dietetic advice and particular eating patterns, there is a paucity of data and although many diets have been tested, superiority of a specific diet plan has not been fully determined. With no official guidelines for dietary advice in PCOS, dietician referral should be considered and initial counseling should encourage a high glycemic index diet and reduced calorie intake incorporating a healthy balanced diet [56]. Equally, rapid weight loss and extreme dieting should not be promoted as this too can be related to poor reproductive outcome.

The intake of fats, especially saturated fats, is of particular concern in PCOS as they can further upset insulin signaling causing worsening of insulin resistance. It therefore seems prudent that low-saturated fat diets should be encouraged in PCOS given their more explicit effect on hyperinsulinemia [57]. Focus on the role of particular nutrients in the pathogenesis of PCOS has led to the analysis of advanced glycosylated end products (AGEs). The higher levels of AGEs found in PCOS stimulated interest in this subject. AGEs, which are found exogenously from cooking at high temperatures, can promote insulin resistance and are positively related to androgen levels [58]. Reducing AGEs by dietary modification may serve as an adjunct in lifestyle interventions for the control of PCOS.

Exercise: The metabolic cascade, implicated in the pathogenesis of PCOS, is amplified and exacerbated by increased adiposity. It therefore would seem intuitive that by engaging in lifestyle modification with the aim of weight loss and avoidance of weight gain, a degree of reversal can be achieved, with alleviation in insulin resistance, reduced androgen levels, and resumption in ovulation. With calorific restriction and exercise it has been theorized, but not concluded, that there is initial loss of intra-abdominal visceral fat due to its increased metabolic composition and this, in contrast to its superficial counterpart, contributes most to symptom improvement due to its modulation in insulin resistance [23,24]. A 5–10% reduction in weight is associated with both lower fasting and glucose-induced insulin levels, and notably, the effects of weight loss appear to be superior to those of insulin-sensitizing medications [59]. The beneficial effect is perhaps derived from altered secretion in adipocytokines corroborated by the positive effects of weight loss on leptin levels [29,31].

It could be logical to assume that the beneficial effects of lifestyle changes are secondary to weight reduction induced by dietary control with or without exercise, but interestingly evidence also suggests that exercise can improve some aspects of PCOS independent of weight loss. Improvements in insulin sensitivity and body glucose control can be demonstrated after exercise alone [22]. In addition, exercise can modulate sympathetic nerve activity and reduce androgen production. This demonstrates that the scope of lifestyle intervention goes beyond that of only weight loss, which is of clinical relevance as, although not all women with PCOS are overweight and some may find it difficult to lose weight, they may still benefit from lifestyle modification with introduction of exercise.

Lifestyle interventions can impact positively on many components that make up PCOS and overall can lead to improvement in psychological wellbeing and quality of life [60]. Many studies have demonstrated a normalization of menstrual irregularity and improvement of hyperandrogenemia symptoms, with decreased testosterone levels and higher SHBG levels secondary to lifestyle interventions in both the adolescent and adult populations [4,61]. However, the favorable effects demonstrated on features of the metabolic syndrome, with associated projected long-term health benefits, seen secondary to interventions in obese adolescents are not consistently replicated in the comparable adult population [4,61]. This stresses the importance of intervention early in life.

Delineating the precise mechanisms and causal relationships between obesity and PCOS may be arduous, but clinically is of major importance considering the alarming increase in both adult and childhood obesity over the past three decades [55]. PCOS typically presents during the early adolescence period, and importantly early atherosclerosis harboring dangerous repercussions for later life can be seen even at this seemingly early age. Beneficial effects on testosterone levels are also seen with peripubertal weight loss [61]. Therefore efforts need to be targeted towards this vulnerable population and maybe even earlier for preventative and therapeutic lifestyle interventions. It has been reported that 70% of children who are obese during adolescence become obese adults [62] and therefore we need to preempt and impede this transition to adult obesity to prevent the considerable ramifications that this may have on PCOS presentation, infertility prevalence, and long-term health comorbidities in the future.

Conclusion

Currently with obesity, in combination with unhealthy living, taking a provocative stance in the world of media, and with projected figures for obesity set to rise even further, it is imperative that we, as the medical profession, take advantage and seize these opportunities with new guidelines being produced and resources channeled into prevention and lifestyle intervention targeting particularly the younger generation. At all ages, many engage in behaviors without regard for and ignorant to the health implications, and with education and long-term goal setting we should aim to give insight and enable personal responsibility for health. It should also be reiterated that the clinical presentation of PCOS evolves and changes throughout the course of a woman's life and is not a condition exclusive to adolescence and reproductive years but also has long-term effects. By manipulating the natural course of PCOS at an early stage, via influencing factors that are known to have disruptive effects, we may be able to avoid, or at least anticipate, the disease trajectory, and therefore a holistic lifestyle approach should be favored in clinical practice.

References

1. World Health Organization (WHO). (2002) Current practices and controversies in assisted reproduction: a report of a WHO meeting on medical, ethical and social aspects of assisted reproduction.

2. Grainger DA, Frazier LM, Rowland CA. (2006) Preconception care and treatment with assisted reproductive technologies. *Matern Child Health J* 10:S161–164.

3. Barrett JR. (2006) Fertile grounds of inquiry: environmental effects on human reproduction. *Environ Health Perspect* 114:A644–649.

4. Moran L, Hutchinson S, Norma R, Teede H. (2011) Lifestyle changes in women with polycystic ovary syndrome, *Cochrane Database Syst Rev* 2:CD007506.

5. NHS. Statistics on obesity, physical activity and diet: England, 2011.

6. World Health Organization (WHO). Consultation on Obesity (2000) Obesity: preventing and managing the global epidemic. WHO technical report series 894. Geneva: World Health Organization.

7. Pandey S, Pandey S, Maheshwari A, Bhattacharya S. (2010) The impact of female obesity on the outcome of fertility treatment. *J Hum Reprod Sci* 3(2):62–67.

8. McCartney C, Blank S, Prendergast K. (2007) Obesity and sex steroid changes across puberty: evidence for marked hyperandrogenaemia in pre- and early pubertal obese girls. *J Clin Endocrinol Metab* 92:5588–5595.

9. Alvarez-Blasco F, Botella-Carretero J, San Millan J, Escobar-Morreale H (2006) Prevalence and characteristics of the polycystic ovary syndrome in over-weight and obese women. *Arch Intern Med* 166(19):2081–2086.

10. Yildiz B, Knochenhauer E, Azziz R. (2008) Impact of obesity on the risk for polycystic ovary syndrome. *J Clin Endocrinol Metab* 93(1):162–168.

11. Maloney C.A, Rees W.D. (2005) Gene–nutrient interactions during fetal development. *Reproduction* 130(4):401–410.

12. Cresswell J, Barker D, Osmomnd C, Egger P, Phillips D, Fraser R. (1997) Fetal growth, length of gestation, and polycystic ovaries in later life. *Lancet* 350:1131–1135.

13. Kent S, Gnatuk C, Kunselman A. (2008) Hyperandrogenism and hyperinsulinaemia in children of women with PCOS: a controlled study. *J Clin Endocrinol Metab* 93:1662–1669.

14. Robinson J, Forsdike R, Taylor J. (1999) In utero exposure of female lambs to testosterone reduces the sensitivity of the gonadotropin-releasing hormone neuronal network to inhibition to progesterone. *Endocrinology* 140:5797–5805.

15. Abbott D, Tarantal A, Dumesic D. (2009) Fetal, infant, adolescent and adult phenotypes of polycystic ovary syndrome in prenatally androgenized female rhesus monkeys. *Am J Primatol.*71:776–784.

16. Pasquali R, Stener-Victorin E, Yildiz B, Duleba A, Hoeger K, Mason H, Homberg R, Hickey T, franks S, Tapamaien J, Balan A, Abbott D, Diamanti-Kandarakis E, Legro R. (2011) PCOS Forum: research in polycystic ovary syndrome today and tomorrow. *Clin Endocrinol.* 74:424–433.

17. Anderson H, Fogel N, Grebe S (2010) Infants of women with polycystic ovary syndrome have lower cord blood androstenedione and estradiol levels. *J Clin Endocrinol Metab* 61:946–951.

18. Badawy A, Elnashar A. (2011) Treatment options for polycystic ovary syndrome. *Int J Women's Health* 3:25–35.

19. Tamer Erel C, Senturk L (2009). The impact of body mass on assisted reproduction. *Curr Opin Obstet Gynaecol* 21:228–235.

20. Pasquali R, Gambineri A, Pagotto U. (2006) The impact of obesity on reproduction with polycystic ovary syndrome. *BJOG* 113:1148–1159.

21. Lord J, Thomas R, Fox B, Acharya U, Wilkin T. (2006) The central issue? Visceral fat mass is a good marker of insulin resistance in polycystic ovary syndrome. *BJOG* 113:1203–1209.

22. Hutchison S, Stepto N, Harrison C, Moran L, Strauss B, Teede H. (2011) Effect of exercise on insulin resistance and body composition in overweight and obese women with and without polycystic ovary syndrome. *J Clin Endocrinol Metab* 96(1):E48–E56.

23. Park H, Lee K (2005) Greater beneficial effects of visceral fat reduction compared with subcutaneous fat reduction on parameters of the metabolic syndrome: a study of weight reduction programmes in subjects with visceral and subcutaneous obesity. *Diabet Med* 2005 22:266–272.

24. Kuchenbecker W, Groen H, Van Asselt S, Bolster J, Zwerver J, Slart R, VdJagt E, Land J, Hoek A. (2011) In women with polycystic ovary syndrome and obesity, loss of intra abdominal fat is associated with resumption of ovulation. *Hum Reprod* 26(9):2505–2512.

25. Kirschner M, Samojlik E, Drejka M, Szmal E, Schneider G, Ertel N (1990) Androgen-estrogen metabolism in women with upper body versus lower body obesity. *J Clin Endocrinol Metab* 70:473–479.

26. Rodin A, Thakkar H, Taylor N. (1994) Hyperandrogenism on polycystic ovary syndrome. Evidence of dysregulation of 11 beta-hydroxysteroid dehydrogenase. *N Engl J Med* 330:460–465.

27. Stewart P, Shackleton C, Beastall G. (1990) 5 alpha – reductase activity in polycystic ovary syndrome. *Lancet* 335:431–433.

28. Gambineri A, Forlani G, Murnarini A. (2009) Increased clearance of cortisol by 5 beta–reductase in a subset of women with adrenal hyperandrogenism in polycystic ovary syndrome. *J Endocrinol Invest* 32:210–218.

29. Carmina E, Bucchieri S, Mansueto P, Rini G, Ferin M, Lobo R. (2009) Circulating levels of adipose products and differences in fat distribution in the ovulatory and anovulatory phenotypes of polycystic ovary syndrome. *Fertil Steril* 91:1332–1335.

30. Hahn S, Haselhorst U, Quasbeck D, Tan S, Kimming R, Mann K, Janssen O. (2006) Decreased soluble leptin receptor levels in women with polycystic ovary syndrome. *Eur J Endocrinol* 154:287–294.

31. Gosman G, Kathcher H, Legro RS. (2006) Obesity and the role of gut adipose hormones in female reproduction. *Hum Reprod Update* 12:585–601.

32. Pasquali R, Gambineri A. (2006) Metabolic effects of obesity on reproduction. *Reprod Biomed Online* 12:542–511.

33. Rich-Edwards J, Goldman M, Willett W, Hunter D, Stampfer M, Colditz G. (1994) Adolescent body mass index and infertility caused by ovulatory disorder. *Am J Obstet Gynecol* 171:171–7.

34. Grodstein F, Goldman M, Cramer D. (1994) Body mass index and ovulatory infertility. *Epidemiology* 5:247–250.

35. Kiddy D, Hamilton-Fairley D, Bush A, Short F, Anyaoku V, Reed M, Franks S. (1992) Improvement in endocrine and ovarian function during dietary treatment of obese women with polycystic ovary syndrome. *Clin Endocrinol* 36:105–111.

36. Balen A, Homberg R, Franks S (2009) Defining polycystic ovary syndrome. *BMJ* 338:a2968.

37. Willis D, Mason H, Gilling-Smith C, Frans S. (1996) Modulation by insulin of follicle stimulating hormone and luteinizing hormone actions in human granulosa cells of normal and polycystic ovaries. *J Clin Endocrinol Metab* 81:302–309.

38. Brewer C, Balan A. (2010) The adverse effects of obesity on conception and implantation. *Reproduction* 140:347–364.

39. Levens E, Skarulis M. (2008) Assessing the role of endometrial alteration among obese women undergoing assisted reproduction. *Fertil Steril* 89:1606–1608.

40. Department of Health. (2009) Regulated fertility services: a commissioning aid.

41. Jungheim E, Lanzendorf S, Odem R, Moley K, Chang A, Ratts V (2009) Morbid obesity is associated with lower clinical pregnancy rates after in vitro fertilization in women with polycystic ovary syndrome. *Fertil Steril* 92:256–261.

42. Sonino N, Fava G, Mani E. (1993) Quality of life in hirsute women. *Postgrad Med J* 69:615–661.

43. Gambineri A, Pelusi C, Vicennati V, Pagotto U, Pasquali R. (2002) Obesity and the polycystic ovary syndrome. *Int J Obes Rel Metab Disord* 26:883–896.

44. Moran L, Misso ML, Wild RA, Norman RJ (2010) Impaired glucose tolerance, type 2 diabetes and metabolic syndrome in polycystic ovary syndrome: a systematic review and meta-analysis. *Hum Reprod Update* 16(4):347–363.

45. Janssen O, Hahn S, Tan S, Benson S, Elsenbruch S. (2008) Mood and sexual function in polycystic ovary syndrome. *Semin Reprod Med* 26:45–52.

46. Coffey S, Mason H. (2003) The effect of polycystic ovary syndrome on health-related quality of life. *Gynecol Endocrinol* 17(5):379–386.

47. Lim S, Norman R, Clifton P, Noakes M. (2009) Hyperandrogenaemia, psychological stress and food craving in young women. *Physiol Behav* 98(3):276–280.

48. Benson S, Arck P, Tan S, Hahn S, Mann K, Rifaie N, Janssen O, Schedlowski M, Elenbruch S. (2009) Disturbed stress responses in women with polycystic ovary syndrome. *Psychoneuroendocrinology* 34(5):727–735.

49. Deeks A, Gibson-Helm M, Teede H.(2010) Anxiety and depression in polycystic ovary syndrome: a comprehensive investigation. *Fertil Steril* 93(7):2421–2423.

50. Cousineau T, Domar A (2007). Psychological impact of infertility. *Best Pract Res Clin Obstet Gynaecol* 21(2):293–308.

51. Moran L, Pasquali R, Teede HJ, Hoeger KM, Norman RJ. (2009) Treatment of obesity in polycystic ovary syndrome: a position statement of the Androgen Excess and Polycystic Ovary Syndrome Society. *Fertil Steril* 92(6):1966–1982.

52. Nicholson F, Rolland C, Broom J, Love J. (2010) Effectiveness of long term (twelve months) nonsurgical weight loss interventions for obese women with polycystic ovary syndrome: a systematic review. *Int J Women's Health* 2:393–399.

53. Crosignani P, Colombo M, Vegetti W, Somigliana E, Gessati A, Ragni G. (2003) Overweight and obese anovulatory patients with polycystic ovaries: parallel improvements in anthropometric indices, ovarian physiology and fertility rate induced by diet. *Hum Reprod* 18:1928–1932.

54. Nybacka A, Carlstrom K, Stahle A, Nyren S, Hellstrom P, Hirschberg A. (2011) Randomised comparison of the influence of dietary management and/or physical exercise on ovarian function and metabolic parameters in overweight women with polycystic ovary syndrome. *Fertil Steril* 96(6):1508-1513.

55. Bennett B, Sothern MS. Diet, exercise, behavior: the promise and limits of lifestyle change. *Semin Pediatr Surg* 2009;18:152–158.

56. Barr S, Hart K, Reeves S, Sharp K, Jeanes Y. (2011) Habitual dietary intake, eating pattern and physical activity of women with polycystic ovary syndrome. *Eur J Clin Nutr* 65(10):1126–1132.

57. Reaven G. (2005) The insulin resistance syndrome: definition and dietary approach to treatment. *Annu Rev Nutr* 25:391–406.

58. Diamanti-Kandarakis E, Katsikis I, Piperi C. (2008) Increased serum advanced glycation end-products is a distinct finding in lean women with polycystic ovary syndrome (PCOS). *Clin Endocrinol* 69:634–641.

59. Gambineri A, Pelusi C, Genghini S, Morselli-Labate A, Cacciari M, Pagotto E. (2004) Effect of flutamide and metformin administered alone or in combination in dieting obese women with polycystic ovary syndrome. *Clin Endocrinol (Oxf)* 60:241–249.

60. Thomson R, Buckley J, Brinkworth G. (2011) Exercise for the treatment and management of overweight women with polycystic ovary syndrome: a review of the literature. *Obes Rev* 12:e202–210.

61. Lass N, Kleber M, Winkel K, Wunsch R, Reinehr T. (2011) Effect of lifestyle intervention of features of polycystic ovarian syndrome, metabolic syndrome and intima-media thickness in obese adolescent girls. *J Clin Endocrinol Metab* 96(11):3533-3540.

62. Jan S, Bellman C, Barone J, Jessen L, Arnold M. (2009) Shape it up: a school-based education programme to promote healthy eating and exercise development by a healthy plan in collaboration with a college of pharmacy. *J Manag Care Pharma* 15 (5):403–13.

Polycystic ovary syndrome ovulation induction

Carolyn J. Alexander

Introduction

Polycystic ovary syndrome (PCOS) is one of the most common causes of ovulatory dysfunction and identifiable causes of infertility. Clomiphene citrate, aromatase inhibitors such as letrozole, metformin, the combination of metformin and clomiphene citrate, and gonadotropins are the most commonly used medications for ovulation induction in patients with PCOS. The following chapter discusses the background and indications for use of each of these medications.

Clomiphene citrate

Clomiphene citrate, a selective estrogen receptor modulator, has been used as a first-line medical ovulation induction agent since 1967. Clomiphene citrate is administered for 5 days beginning on any spontaneous or progestin-induced menstrual cycle day from 2 to 5, starting with 50 mg/day and increasing to 150 mg/day if anovulatory. If ovulation cannot be achieved at doses of 150 mg/day, the patient is deemed to have clomiphene citrate resistance. The dose of clomiphene citrate can be increased up to 250 mg. If pregnancy cannot be achieved after six ovulatory cycles, then the patient is deemed a clomiphene citrate failure. Most conceptions occur within the first six ovulatory cycles and at doses of less than 150 mg a day, but the fecundity rate decreases dramatically with age (<4% at >41 years of age) [1–3].

Studies with clomiphene citrate have shown an ovulation rate of 60–85% and a pregnancy rate of 30–50% after six ovulatory cycles, with an increased risk of multiple pregnancy of 5–7% [4]. A Cochrane systematic review and meta-analysis of three randomized controlled trials comparing clomiphene citrate with placebo demonstrated that clomiphene citrate improves

both ovulation and pregnancy rate [5] The addition of an ovulatory trigger dose of human chorionic gonadotropin (hCG) to clomiphene citrate ovulation induction therapy does not improve ovulation, pregnancy, or miscarriage rates [6].

Aromatase inhibitors

Aromatase inhibitors were first proposed as ovulation-inducing drugs in anovulatory women in 2001 [7]. There are concerns regarding the potential teratogenic effect of letrozole for infertility treatment [8], but two subsequent publications did not find an increased risk of fetal anomaly [6,9,10]. Letrozole is typically prescribed at a starting dose of 2.5–5 mg and can be increased by increments of 2.5 mg, but the optimum dose range has not been established. Letrozole may have a better side effect profile than other aromatase inhibitors and result in fewer multiple pregnancies [1,11].

There are no published randomized controlled trials (RCTs) comparing aromatase inhibitors with placebo or no treatment in therapy-naive PCOS women. A single RCT comparing letrozole with placebo in clomiphene citrate-resistant (CCR) PCOS women found no difference in pregnancy or live birth rate per patient [12]. Two recent systematic reviews and meta-analyses of six RCTs comparing letrozole with clomiphene citrate in PCOS women who were therapy naive, CCR, or where the type of PCOS women was not reported, demonstrated a higher ovulation rate per patient with letrozole, but no difference in ovulation rate per cycle or pregnancy rate per patient [6,13]. Individually, the RCTs in either therapy-naive or CCR PCOS women demonstrated no difference in

Androgens in Gynecological Practice, ed. Leo Plouffe and Botros Rizk. Published by Cambridge University Press. © Cambridge University Press 2015.

pregnancy rates between letrozole and clomiphene citrate [6].

A study by Nahid and Sirous [14] showed that the effects of letrozole and clomiphene citrate on ovulation were almost the same, though clomiphene citrate caused endometrial thinning more often than letrozole. Also the side effects reported by patients in the group receiving clomiphene citrate were higher, while in the group receiving letrozole no complication was reported. Based on these findings, letrozole can be considered an appropriate alternative for clomiphene citrate, especially with a decreased side effect profile.

One disadvantage of clomiphene citrate is depletion of the estrogen receptors throughout the body and its long half-life. In contrast, an aromatase inhibitor blocks the conversion of androgens to estrogens in the ovarian follicles, peripheral tissues, and in the brain. This results in a decrease in circulating and local estrogens and a rise in intra-ovarian androgens. The fall in estrogen levels releases the hypothalamo-pituitary axis from the negative feedback of estrogens. Thus, there is a surge in follicle-stimulating hormone (FSH) release, which results in follicular growth. Since the feedback mechanism is intact, normal follicular growth, selection of dominant follicle, and atresia of smaller growing follicle occurs, thereby facilitating monofollicular growth and ovulation [15–21].

Another likely mechanism of action of the aromatase inhibitors is by the increasing intra-ovarian androgens. This likely increases the follicular sensitivity to FSH. Recent data shows the role of androgens in early follicular development by augmenting FSH receptors and stimulating insulin-like growth factor (IGF)-1; FSH and IGF-1 act synergistically to promote follicular growth [15,22–24].

Metformin

Treatment with insulin-sensitizing agents, such as metformin, thiazolidinedione, and D-chiro-inositol, has been suggested to be useful for ovulation induction in some women with PCOS [25–31]. The proposed mechanisms of action are similar to those observed with weight loss, such as a decrease in insulin levels, which may indirectly decrease ovarian androgen production, and a potential direct effect on the hypothalamus [32].

Pharmacology and dosing

Metformin is a biguanide antihyperglycemic agent approved by the United States Food and Drug Administration (FDA) for the treatment of type 2 diabetes mellitus. It reduces blood glucose levels by decreasing hepatic gluconeogenesis and intestinal absorption of glucose, as well as enhancing peripheral glucose uptake and use by increasing insulin sensitivity [33]. It is rapidly absorbed from the small intestine, with peak plasma levels occurring 2 hours after ingestion. Food decreases the rate of drug absorption as well as peak drug concentration. Metformin is not metabolized and is largely excreted in the urine; its plasma elimination half-life is approximately 6 hours.

When used in anovulatory women with PCOS, it acts to decrease insulin levels and luteinizing hormone (LH). This decrease in insulin and LH levels subsequently leads to an increase in the levels of sex hormone-binding globulin (SHBG), resulting in a decreased level of androgens. The decrease in androgens is in part because of the increased level of SHBG, but also because of the decrease in LH levels. Women with PCOS also benefit from metformin, because it typically causes a slight reduction in weight [34–36], which results in decreased serum testosterone concentration, resumption of ovulation, and pregnancy [37–40]. Studies have found that approximately 50% of women with PCOS or anovulatory cycles resume regular menses after taking metformin for 6 months.

The optimum dose of metformin to restore ovulatory menses has not been determined. The target dose, in anovulatory patients, is typically 1500–2500 mg in divided daily doses. Its most common adverse effects are gastrointestinal in nature, including diarrhea, nausea, emesis, flatulence, indigestion, and abdominal discomfort. The reported discontinuation rate is 5%, secondary to side effects [33]. These side effects can be sometimes diminished or avoided by a gradual dose escalation, beginning treatment with 500 mg taken with a meal, and if tolerated, increased to 500 mg twice daily with meals and then three times daily with meals. One to two weeks should elapse between increases in dose [41]. Another alternative is to switch to metformin XL or the liquid form, which typically have fewer side effects, before abandoning metformin therapy [42].

After initiating metformin therapy, it may take up to 6 months for ovulatory cycles to ensue. During the initial phase of oligomenorrhea, it may be difficult to determine whether ovulation has taken place. Although cumbersome, measuring serum progesterone every 10 days may be one approach to identify whether ovulation has occurred following initiation of metformin therapy [41].

Precautions and monitoring

Metformin therapy may have hepatic toxicity or be complicated by lactic acidosis; as such, liver and renal functions must be evaluated before treatment and monitored periodically thereafter. Metformin should not be prescribed in patients with conditions that may increase the risk of lactic acidosis, such as renal insufficiency, congestive heart failure, or sepsis.

Caution must be taken when patients taking metformin are undergoing intravascular contrast studies with iodinated materials, as this may lead to changes in renal function and has been associated with lactic acidosis. Per FDA current recommendations, metformin should be discontinued at the time of or prior to any contrast procedure and withheld for 48 hours after the procedure, and resumed only after renal function has been shown to be normal [41].

Metformin should also be temporarily suspended for all surgical procedures that involve restriction of fluid intake and should not be restarted until normal fluid intake has resumed and renal function has been shown to be normal.

Indications

As previously mentioned, the only FDA-approved indication for metformin is for type 2 diabetes mellitus; however, it has been used "off-label" to treat or help prevent several clinical problems associated with PCOS.

For women with oligomenorrhea due to PCOS, endometrial protection from hyperplasia is needed, and estrogen-progestin contraceptives are considered first line, followed by continuous or cyclic progestin-only when combined contraceptives are contraindicated. Metformin can also be used as a second-line therapy, restoring ovulatory menses in approximately 50% of women with PCOS [43,44]. However, when metformin is used, cyclic progestin therapy is added for the first 6 months of metformin treatment until regular cycles are established. Metformin has not been proven to be endometrial protective.

In one trial, 23 women with PCOS were randomly assigned to receive metformin, 500 mg three times per day, or placebo for 6 months [43]. Approximately 50% of the women achieved normalization of menstrual function, which was confirmed by intermenstrual interval and luteal phase serum progesterone monitoring. In a meta-analysis of 13 trials, women treated with metformin had a fourfold higher chance of ovulating when compared to placebo [45]. A similar increase was seen for the combination of metformin plus clomiphene versus clomiphene alone. Although metformin may restore ovulation in some women, it appears to be less effective than clomiphene for pregnancy and live birth rates.

In women with hirsutism secondary to PCOS, an estrogen-progestin contraceptive, as recommended by the 2008 Endocrine Society Guidelines, is the first line of treatment. If after 6 months the response is suboptimal, an antiandrogen is then added. Several clinical trials have studied the effects of metformin and other insulin-lowering agents on circulating androgen concentrations. Even though metformin has been shown to cause improvement in plasma insulin and insulin sensitivity, with a reduction in serum free testosterone and an increase in serum HDL cholesterol, there has not been a significant reduction in hirsutism, despite the lower free serum testosterone concentration. The best available evidence comes from a meta-analysis of nine placebo-controlled trials of insulin-lowering drugs [46]. When the eight metformin trials were analyzed separately, no significant benefit was seen in hirsutism when compared to placebo. Based upon these findings, the Endocrine Society Clinical Practice Guidelines suggest against the routine use of metformin for the treatment of hirsutism [47].

Anovulatory infertility is one of the main concerns for patients with PCOS. A structured approach is warranted for its treatment, starting with low-cost interventions and advancing to high-resource interventions. As such, the initial management steps are weight loss through caloric restriction and increased exercise for women with a body mass index (BMI) >27 kg/m^2, followed by medical therapy [41].

In 2003, a meta-analysis of studies involving treatment with metformin in women with PCOS concluded that its efficacy was similar for treating anovulatory infertility when compared to clomiphene [48]. However, subsequent randomized multicenter trials comparing both metformin and clomiphene, alone and in combination, have found clomiphene clearly superior to metformin and observed that combined treatment offers no significant additional benefit [49–51].

The Pregnancy in Polycystic Ovary Syndrome I (PPCOS I) trial, a large multi-center RCT that evaluated the use of metformin versus clomiphene, versus both combined, shows a significantly lower live birth rate with metformin compared with clomiphene, only offering an advantage when added to clomiphene in women with BMI over 35 kg/m^2 [50].

In the largest trial, clomiphene yielded a significantly higher live birth rate than metformin (22.5% vs. 7.2%). Results for combined therapy, with metformin and clomiphene, were not significantly better (26.8%), when compared to clomiphene alone (22.5%) [50]. A few small studies, however, involving clomiphene-resistant anovulatory patients with PCOS, have shown that combined treatment has increased ovulation and pregnancy rates over those achieved with clomiphene alone [52–55].

Patients that fail to ovulate in response to clomiphene, clomiphene failure or clomiphene resistant, may respond to supplemental or combination treatment regimens. Some of the options include adjuvant treatment with glucocorticoids, exogenous hCG, aromatase inhibitors, and metformin. These alternatives are helpful as many couples may be unable to pursue the obvious alternative of gonadotropin treatment due to associated costs, and others may be reluctant to do so once fully advised of the risks [56].

Combined metformin and clomiphene citrate

In women with PCOS and anovulatory infertility, clomiphene citrate is recommended as the first-line pharmacological therapy as previously described. If ovulation cannot be achieved with clomiphene citrate administration, and resistance is reached, other agents are recommended such as metformin, aromatase inhibitors, or gonadotropins [56].

The combination of metformin and clomiphene citrate seems to be an effective supplementation in clomiphene-resistant women. In Nestler and colleagues' study, sixty PCOS women with a mean BMI of 32.3 were randomized to metformin (500 mg 3 times daily) versus placebo [31]. Women who failed to ovulate on metformin or placebo were given clomiphene citrate (50 mg on days 5 to 9). After addition of clomiphene citrate to metformin, 90% of the women ovulated versus 8% of the women who received clomiphene citrate plus placebo. Two high-quality reviews found that a combination of metformin and clomiphene citrate was better than clomiphene citrate alone for ovulation and for pregnancy rate in women with clomiphene citrate resistance [57].

In a few small studies involving clomiphene-resistant anovulatory women with PCOS, combined treatment with metformin and clomiphene increased ovulation and pregnancy rates over those achieved with clomiphene alone [52,53,55]. A 2008 meta-analysis, including 17 randomized trials, concluded that combined treatment achieves higher ovulation and pregnancy rates than treatment with clomiphene alone [27]. Although there is no convincing evidence that combined treatment with metformin and clomiphene can increase live birth rates over those achieved with clomiphene alone [57], it seems like a reasonable attempt for women who have few alternatives besides ovarian drilling or treatment with exogenous gonadotropins.

Limited evidence indicates that combined treatment with metformin or clomiphene and rosiglitazone is no more effective than metformin alone [29,58]. Also, the safety alert issued by the FDA concerning a possible increased risk of ischemic cardiovascular events in patients receiving treatment with thiazolidinediones argues against their adjuvant use for ovulation induction [59].

In a recent Cochrane review from 2012, analyzing insulin-sensitizing drugs in women with PCOS, when comparing metformin combined with clomiphene citrate, there was a significant effect when combined versus clomiphene alone (18 RCTs, number of cycles = 3265; OR 1.74, 95% CI 1.50–2.00), with a moderate degree of heterogeneity ($I^2 = 62\%$). The analysis was stratified based on sensitivity to clomiphene and BMI. A significant heterogeneity was observed in a group of trials when the status of clomiphene sensitivity was not defined ($I^2 = 67\%$); in contrast, as might be expected, the clomiphene-resistant group appeared to be homogeneous ($I^2 = 0\%$) [57].

We recommend that metformin is used in combination with clomiphene citrate, primarily in women with PCOS who are also clomiphene resistant, to aid in ovulation, pregnancy, and live birth [60]. This combination deserves consideration in women who prove clomiphene resistant, before proceeding to ovarian drilling or treatment with gonadotropins [56].

Gonadotropins

Gonadal function regulation is mediated by gonadotropin-releasing hormone (GnRH) and by the two gonadotropins, LH and FSH, which are biosynthesized in and secreted by gonadotropes in the pituitary. Placentally derived hCG also plays a role in regulation. These three gonadotropins regulate reproductive endocrinology, including steroid production, follicle and sperm maturation, ovulation, as well as maintenance of early pregnancy.

Structurally, these three gonadotropins exist as heterodimers; they all share a common alpha subunit and have a homologous hormone-specific beta subunit. They are secreted and circulate as such, having different bioactivity and pharmacokinetics based on varying degrees of glycosylation and sulfonation [68]. These circulating isoforms are present in different proportions in the different available commercial preparations [62,63].

Since their introduction into clinical practice in 1961 [49,50], exogenous gonadotropins have played an important role in ovulation induction. They have been used particularly in women who fail to respond to other forms of treatment as well as women who are gonadotropin deficient. Gonadotropins are highly effective, but are also associated with potentially serious complications including ovarian hyperstimulation syndrome and multiple pregnancies. They have a high cost, require close monitoring with ultrasound and determination of estradiol levels, and lack an oral formulation, reasons for which gonadotropins are considered a second-line treatment for PCOS patients with infertility.

Gonadotropin preparations

Gonadotropin preparations have significantly evolved since their initial discovery and first use. The initial crude preparations, obtained from postmenopausal female urine, have been gradually refined to highly purified urinary extracts; and finally, since 1996, the recombinant preparations we have available today [64,65].

These recombinant preparations are easier to administer, subcutaneous versus intramuscular, and have a higher grade of purity and consistency of preparation when compared to earlier forms. They have also made it possible to increase our understanding of their specific actions in follicular development and oocyte maturation, as well as allowing us to tailor stimulation regimens to the needs and requirements of individual patients in an attempt to optimize cycle outcomes [66].

Indications for gonadotropin treatment

The choice of gonadotropin preparation and treatment regimen will vary according to the type of ovulatory disturbance the patient has. The World Health Organization (WHO) has adopted a classification system for the causes of anovulation that provides a practical guide to appropriate therapeutic intervention.

The WHO classification system comprises group I: hypogonadotropic, hypogonadal anovulation; group II: normogonadotropic, normoestrogenic anovulation; and group III: hypergonadotropic, hypoestrogenic anovulation, and hyperprolactinemic anovulation. Gonadotropin treatment is indicated for WHO groups I and group II, for patients who have failed to ovulate or conceive with clomiphene citrate. For the purpose of this review we will focus on the latter as it pertains to the PCOS patient population.

When clomiphene citrate treatment, alone or in combination (glucocorticoids, exogenous hCG, aromatase inhibitors, or metformin), fails to achieve ovulation, gonadotropins become a suitable option. Ovarian stimulation with clomiphene citrate followed by gonadotropins in patients with clomiphene resistance in PCOS patients showed a cumulative live birth rate of 60% at 1 year and 78% at 2 years [67].

However, failure with other treatments is not a prerequisite for use of gonadotropins, rather it is a logical stepwise approach, as use of gonadotropins implies an increase in costs, risks, and logistical demand. A recent RCT evaluating clinical outcome of gonadotropin versus clomiphene citrate as first-line therapy for ovarian stimulation for ovulation induction showed a higher pregnancy and live birth rate with low-dose FSH [68].

Clomiphene-resistant anovulatory women with PCOS usually have a normal range of gonadotropin concentration, and many patients will have elevated LH levels. These patients benefit from exogenous gonadotropins. Theoretically, purified FSH preparations do not affect or impact endogenous LH production, thus avoiding further worsening of the ongoing LH hypersecretion. However, there is no evidence to support this theory and both human menopausal gonadotropin (hMG) and FSH may be used.

A meta-analysis including 14 RCTs comparing purified urinary FSH versus hMG for ovulation induction in clomiphene-resistant PCOS patients found that the only advantage of FSH over hMG is a reduced risk of ovarian hyperstimulation syndrome (OR = 0.20, 95% CI = 0.08–0.46) [69]. When comparing FSH forms, recombinant versus purified urinary, or different treatment regimens, there was no difference in the rates of ovulation, pregnancy, miscarriage, multiple pregnancy, or in the incidence of ovarian hyperstimulation syndrome [70,71].

The doses needed to achieve the desired response are relatively low; in fact, some patients might be extremely sensitive to gonadotropins and even subtherapeutic doses can cause ovarian hyperstimulation syndrome. This requires frequent adjustments in dosage and careful monitoring. Even though unifollicular ovulation is the goal, this might not be the case in many patients, and the risk of multiple pregnancy is high. These patients will seldom require luteal phase support, as women with PCOS usually have elevated endogenous LH, which provides adequate support. For those patients that exhibit poor luteal function and require support, progesterone therapy is preferable over hCG as the latter increases the risk for ovarian hyperstimulation syndrome [67].

Monitoring and administration

Monitoring during gonadotropin stimulation to predict adequate, but not excessive response is essential. Gonadotropin treatment regimens in patients with PCOS should start conservatively, since the risk of developing ovarian hyperstimulation syndrome, or having a multiple pregnancy, is high in these patients. Proper monitoring for these patients requires integration of hormonal and physical data. Follicle size during a treatment cycle needs to be followed with serial ultrasounds and response needs to be assessed with serum estradiol measurements.

Circulating estradiol levels peak approximately 12 hours after the preceding dose of gonadotropins, therefore times for drug administration and serum evaluation should be standardized, usually in the evening. Estradiol level is typically obtained in the morning, so the results can be interpreted by the afternoon and any dose changes can be communicated to the patient before her administration of medication in the evening. The estradiol levels should sustain a constant rise during stimulation, doubling every 2–3 days.

Ultrasound monitoring for follicular size and endometrial thickness should occur concomitantly with the serum estradiol measurements. Uniform follicular growth should be seen with constant growth. Follicles have a greater chance of ovulating as they increase in size; virtually all large follicles (>20 mm) will ovulate. Follicles 14 mm or smaller still have approximately a 40% chance of ovulating.

Typical monitoring for WHO group II patients includes an estradiol level and ultrasound on day 5 of stimulation. Monitoring these patients on the third day of stimulation is also an option for earlier adjustments if subsequent dosing is expected. Repeat assessment is dictated by the rate of rise in estradiol and growth of the developing follicles from ultrasound examination. The therapeutic window for patients with PCOS is notoriously narrow. The high cohort of available follicles in these patients can lead to a greater follicular responsiveness [72]. Development of ovarian overstimulation syndrome can occur with gonadotropin doses that start too high or are increased too frequently. Low initial doses of gonadotropin with small increases is a safe strategy for eliciting follicular response in these patients [73–77].

Low-dose step-up protocols typically start with a dose of 75 IU or less and dosing increments are increased as needed in 37.5 IU intervals [78–80]. hCG is injected when the leading follicle is >18 mm in diameter [81]. Ovarian hyperstimulation syndrome (1.4%) and multiple pregnancy rates (5.7%) have been reported to be very low with this regimen [89]. Another treatment regimen for PCOS patients is the step-down protocol. The FSH starting dose is 150 IU/day until a follicle of >10 mm is seen. The dose is decreased by 37.5 IU/day and further to 75 IU/day 3 days later until hCG is administered [83].

Ovarian hyperstimulation syndrome is one of the major risks of ovulation induction. It refers to a combination of ovarian enlargement due to multiple ovarian cysts and an acute fluid shift out of the intravascular space.

Exogenous gonadotropins administered for ovarian stimulation will override normal feedback mechanisms, thereby resulting in recruitment of multiple antral follicles, as opposed to monofollicular development that occurs in spontaneous cycles. These large numbers of follicles release vasoactive substances that are secreted during maturation and luteinization, causing increased capillary permeability. This will subsequently lead to fluid shifts out of the intravascular space, hemoconcentration, and third-space accumulation of fluid. Clinical symptoms usually appear 5–10 days following the first dose of the ovulatory trigger. Some serious complications include renal failure, hypovolemic shock, thromboembolic episodes, acute respiratory distress syndrome, and death [83,84].

Some patients can be identified as being at higher risk for ovarian hyperstimulation syndrome and in this way changes in stimulation regimens can be made and preventative measures can be implemented. Risk factors include young age, previous ovarian hyperstimulation syndrome, sensitivity to gonadotropins, PCOS, high basal anti-müllerian hormone, and high antral follicle count (>14) [85,86]. Even though complete prevention is still not possible, identifying risk factors can significantly reduce its incidence.

Prevention strategies are divided into two types: primary prevention for those patients identified to be at higher risk based on the presence of previously mentioned risk factors on initial screening, and secondary prevention for patients in whom risk factors arise from excessive response to stimulation protocols. Primary prevention strategies include reducing exposure to gonadotropins by reducing the dose as well as duration of exposure, addition of GnRH antagonists, avoidance of hCG for luteal phase support and use of progesterone instead, in vitro maturation, and use of insulin-sensitizing agents [87–90]. Secondary prevention strategies include coasting, cryopreservation of embryos for reimplantation when the patient's serum hormone levels are not elevated, alternative trigger agents, and lastly, cycle cancellation.

Conclusion

In conclusion, this chapter highlights the medications used for ovulation induction in patients with PCOS. Clomiphene citrate is a selective estrogen receptor modulator. Aromatase inhibitors block the conversion of androgens to estrogens, ultimately resulting in an increased FSH surge. Metformin likely induces ovulation with a similar mechanism to weight loss, such as a decrease in insulin levels, which may indirectly decrease ovarian androgen production, and it has a potentially direct effect on the hypothalamus. Lastly, gonadotropins are commonly used for patients with PCOS who fail to respond to other forms of treatment, as well as women who are gonadotropin deficient. We have discussed the side effect profiles, indications, risks, monitoring, administration, and mechanisms of action of each of these treatment modalities.

References

1. Bates GW, Propst AM. Polycystic ovarian syndrome management options. *Obstet Gynecol Clin North Am* 2012;39(4):495–506. Use of clomiphene citrate in women. *Fertil Steril* 2006;86(5):187–93.

2. Dovey S, Sneeringer R, Penzias A. Clomiphene citrate and intrauterine insemination: analysis of more than 4100 cycles. *Fertil Steril* 2008;90(6):2281–6.

3. Imani B. A nomogram to predict the probability of live birth after clomiphene citrate induction of ovulation in normogonadotropic oligoamenorrheic infertility. *Fertil Steril* 2002;71(1):91–7.

4. Kafy S, Tulandi T. New advances in ovulation induction. *Curr Opin Obstet Gynecol* 2007;19(3):248–52.

5. Brown J, Farquhar C, Beck J, et al. Clomiphene and anti-oestrogens for ovulation induction in PCOS. *Cochrane Database Syst Rev* 2009;4(4). Art. No.: CD002249.

6. Costello MF, Misso ML, Wong J, et al. The treatment of infertility in PCOS: a brief update. *Aust N Z J Obstet Gynaecol* 2012;52(4):400–3.

7. Forman R, Gill S, Moretti M, et al. Fetal safety of letrozole and clomiphene citrate for ovulation induction. *J Obstet Gynaecol Can* 2007;29(8):668–71.

8. Tulandi T, Martin J, Al-Fadhli R, et al. Congenital malformations among 911 newborns conceived after infertility treatment with letrozole or clomiphene citrate. *Fertil Steril* 2006;85(6):1761–5.

9. Brown MM, Hemmings R, Brassard N. The outcome of 150 babies following the treatment with letrozole or letrozole and gonadotropins. *Fertil Steril* 2005;84 (Suppl 1); O–231, Abstract 1033.

10. Mitwally MF, Casper RF. Use of an aromatase inhibitor for induction of ovulation in patients with an inadequate response to clomiphene citrate. *Fertil Steril* 2001;75(2):305–9.

11. Pritts EA, Yuen AK, Sharma S, et al. The use of high dose letrozole in ovulation induction and controlled ovarian hyperstimulation. *ISRN Obstet Gynecol* 2011;2011:242864.

12. Kamath MS, Aleyamma TK, Chandy A, George K. Aromatase inhibitors in women with clomiphene citrate resistance: a randomized, double-blind, placebo-controlled trial. *Fertil Steril* 2010;94(7):2857–9.

13. Misso ML, Wong J, Teede HJ, et al. Aromatase inhibitors for PCOS: a systematic review and meta-analysis. *Hum Reprod Update* 2012;18(3):301–12.

14. Nahid L, Sirous K. Comparison of the effects of letrozole and clomiphene citrate for ovulation induction in infertile women with polycystic ovary syndrome. *Minerva Ginecol* 2012; 64(3):253–8.

15. Kar S. Current evidence supporting "letrozole" for ovulation induction. *J Hum Reprod Sci.* 2013;6(2):93–8.

16. Kamat A, Hinshelwood MM, Murry BA, Mendelson CR. Mechanisms in tissue-specific regulation of estrogen biosynthesis in humans. *Trends Endocrinol Metab* 2002;13(3):122–8.

17. Mason AJ, Berkemeier LM, Schmelzer CH, Schwall RH. Activin B: precursor sequences, genomic structure and in vitro activities. *Mol Endocrinol* 1989;3(9):1352–8.

18. Naftolin F, MacLusky NJ. Aromatization hypothesis revisisted. In: Serio M (Ed.). *Differentiation Basic and Clinical Aspects*. New York: Raven Press; 1984. pp. 79–91.

19. Naftolin F, MacLusky NJ, Leranth CZ, et al. The cellular effects of estrogens on neuroendocrine tissue. *J Steroid Biochem* 1988;30(1–6):195–207.

20. Naftolin F. Brain aromatization of androgens. *J Reprod Med* 1994;39(4):257–61.

21. Roberts V, Meunier H, Vaughan J, et al. Production and regulation of inhibin subunits in pituitary gonadotropes. *Endocrinology* 1989;124(1):552–4.

22. Vendola KA, Zhou J, Adesanya OO, Weil SJ, Bondy CA. Androgens stimulate early stages of follicular growth in the primate ovary. *J Clin Invest* 1998;101(12):2622–9.

23. Weil SJ, Vendola K, Zhou J, et al. Androgen receptor gene expression in the primate ovary: cellular localization, regulation, and functional correlations. *J Clin Endocrinol Metab* 1998;83(7):2479–85.

24. Weil S, Vendola K, Zhou J, Bondy CA. Androgen and follicle-stimulating hormone interactions in primate ovarian follicle development. *J Clin Endocrinol Metab* 1999;84(8):2951–6.

25. Nestler JE, Jakubowicz DJ, Reamer P, et al. Ovulatory and metabolic effects of D-chiro-inositol in the polycystic ovary syndrome. *N Engl J Med* 1999;340(17):1314–20.

26. Azziz R, Ehrmann D, Legro RS, et al. Troglitazone improves ovulation and hirsutism in the polycystic ovary syndrome: a multicenter, double blind, placebo-controlled trial. *J Clin Endocrinol Metab* 2001;86(4):1626–32.

27. Creanga AA, Bradley HM, McCormick C, Witkop CT. Use of metformin in polycystic ovary syndrome: a meta-analysis. *Obstet Gynecol* 2008;111(4):959–68.

28. Ghazeeri G, Kutteh WH, Bryer-Ash M, et al. Effect of rosiglitazone on spontaneous and clomiphene citrate-induced ovulation in women with polycystic ovary syndrome. *Fertil Steril* 2003;79(3):562–6.

29. Baillargeon JP, Jakubowicz DJ, Iuorno MJ, et al. Effects of metformin and rosiglitazone, alone and in combination, in nonobese women with polycystic ovary syndrome and normal indices of insulin sensitivity. *Fertil Steril* 2004;82(4):893–902.

30. Carmina E, Lobo RA. Does metformin induce ovulation in normoandrogenic anovulatory women? *Am J Obstet Gynecol* 2004;191(5):1580–4.

31. Nestler JE, Jakubowicz DJ, Evans WS, Pasquali R. Effects of metformin on spontaneous and clomiphene-induced ovulation in the polycystic ovary syndrome. *N Engl J Med* 1998;338(26):1876–80.

32. Perales-Puchalt A, Legro RS. Ovulation induction in women with polycystic ovary syndrome. *Steroids* 2003;78(8): 767–72.

33. Barbieri RL. Metformin for the treatment of polycystic ovary syndrome. *Obstet Gynecol* 2003;101(4):785–93.

34. Haas DA, Carr BR, Attia GR. Effects of metformin on body mass index, menstrual cyclicity, and ovulation induction in women with polycystic ovary syndrome. *Fertil Steril* 2003;79(3):469–81.

35. Maciel GA, Soares Junior JM, Alves da Motta EL, et al. Nonobese women with polycystic ovary syndrome respond better than obese women to treatment with metformin. *Fertil Steril* 2004;81(2):355–60.

36. Nestler JE, Jakubowicz DJ. Decreases in ovarian cytochrome P450c17 alpha activity and serum free testosterone after reduction of insulin secretion in polycystic ovary syndrome. *N Engl J Med* 1996;335(9):617–23.

37. Niskanen L, Uusitupa M, Sarlund H, et al. The effects of weight loss on insulin sensitivity, skeletal muscle composition and capillary density in obese nondiabetic subjects. *Int J Obes Relat Metab Disord* 1996;20(2):154–60.

38. Pasquali R, Antenucci D, Casimirri F, et al. Clinical and hormonal characteristics of obese amenorrheic hyperandrogenic women before and after weight loss. *J Clin Endocrinol Metab* 1989;68(1):173–9.

39. Bates GW, Whitworth NS. Effect of body weight reduction on plasma androgens in obese, infertile women. *Fertil Steril* 1982;38(4):406–9.

40. Guzick DS, Wing R, Smith D, et al. Endocrine consequences of weight loss in obese, hyperandrogenic, anovulatory women. *Fertil Steril* 1994;61(4):598–604.

41. Barbieri R, Ehrmann D, Snyder P, et al. Metformin for treatment of the polycystic ovary syndrome. www.uptodate-com.libproxy.usc.edu/contents/metformin-for-treatment-of-the-polycystic-ovarysyndrome?detectedLanguage=en&source=search_result&search=metformin+pcos&selectedTitle=1~150&provider=noProvider#H12.

42. Blonde L, Dailey GE, Jabbour SA, et al. Gastrointestinal tolerability of extended-release metformin tablets compared to immediate-release metformin tablets: results of a retrospective cohort study. *Curr Med Res Opin* 2004;20(4): 565–72.

43. Ikeda T, Iwata K, Murakami H. Inhibitory effect of metformin on intestinal glucose absorption in the perfused rat intestine. *Biochem Pharmacol* 20001;59(7):887–90.

44. DeFronzo RA. Pharmacologic therapy for type 2 diabetes mellitus. *Ann Intern Med* 1999;131(4):281–303.

45. Lord JM, Flight IH, Norman RJ. Insulin-sensitising drugs (metformin, troglitazone, rosiglitazone, pioglitazone, D-chiro-inositol) for polycystic ovary syndrome. *Cochrane Database Syst Rev* 2003;(3):CD003053.

46. Cosma M, Swiglo BA, Flynn DN, et al. Clinical review: insulin sensitizers for the treatment of hirsutism: a systematic review and metaanalyses of randomized controlled trials. *J Clin Endocrinol Metab* 2008;93(4):1135–42.

47. Martin KA, Chang RJ, Ehrmann DA, et al. Evaluation and treatment of hirsutism in premenopausal women: an Endocrine Society Clinical practice guideline. *J Clin Endocrinol Metab* 2008;93(4):1105–20.

48. Lord JM, Flight IH, Norman RJ. Metformin in polycystic ovary syndrome: systematic review and meta-analysis. *BMJ*. 2003;327(7421):951–3.

49. Moll E, Bossuyt PM, Korevaar JC, et al. Effect of clomifene citrate plus metformin and clomifene citrate plus placebo on induction of ovulation in women with newly diagnosed polycystic ovary syndrome: randomised double blind clinical trial. *BMJ.* 2006;332(7556):1485.

50. Legro RS, Barnhart HX, Schlaff WD, et al. Clomiphene, metformin, or both for infertility in the polycystic ovary syndrome. *N Engl J Med* 2007;356(6):551–66.

51. Zain MM, Jamaluddin R, Ibrahim A, Norman RJ. Comparison of clomiphene citrate, metformin, or the combination of both for firstline ovulation induction, achievement of pregnancy, and live birth in Asian women with polycystic ovary syndrome: a randomized controlled trial, *Fertil Steril* 2009;91(2):514–21.

52. Vandermolen DT, Ratts VS, Evans WS, et al. Metformin increases the ovulatory rate and pregnancy rate from clomiphene citrate in patients with polycystic ovary syndrome who are resistant to clomiphene citrate alone. *Fertil Steril* 2001;75(2):310–15.

53. Kocak M, Caliskan E, Simsir C, Haberal A. Metformin therapy improves ovulatory rates, cervical scores, and pregnancy rates in clomiphene citrate-resistant women with polycystic ovary syndrome, *Fertil Steril* 2002;77(1):101–6.

54. Malkawi HY, Qublan HS. The effect of metformin plus clomiphene citrate on ovulation and pregnancy rates in clomiphene-resistant women with polycystic ovary syndrome, *Saudi Med J* 2002;23(6):663–6.

55. Hwu YM, Lin SY, Huang WY, Lin MH, Lee RK. Ultra-short metformin pretreatment for clomiphene citrate-resistant polycystic ovary syndrome, *Int J Gynaecol Obstet* 2005;90(1):39–43.

56. Fritz MA., Speroff L. Induction of ovulation. In: *Clinical Gynecologic Endocrinology and Infertility*, 8th edn. Baltimore, MD: Lippincott Williams & Wilkins, 2011.

57. Tang T, Lord JM, Norman RJ, et al. Insulin-sensitising drugs (metformin, rosiglitazone, pioglitazone, D-chiro-inositol) for women with polycystic ovary syndrome, oligo amenorrhoea and subfertility. *Cochrane Database Syst Rev*. 2012;5:CD003053.

58. Rouzi AA, Ardawi MS. A randomized controlled trial of the efficacy of rosiglitazone and clomiphene citrate versus metformin and clomiphene citrate in women with clomiphene citrate-resistant polycystic ovary syndrome. *Fertil Steril* 2006;85(2):428–35.

59. Kaul S, Bolger AF, Herrington D, Giugliano RP, Eckel RH. Thiazolidinedione drugs and cardiovascular risks: a science advisory from the American Heart Association and American College Of Cardiology Foundation, *J Am Coll Cardiol* 2010;55(17):1885–94.

60. Misso ML, Teede HJ, Hart R, et al. Status of clomiphene citrate and metformin for infertility in PCOS. *Trends Endocrinol Metab* 2012;23(10):533–43.

61. Wide L, Naessen T, Sundstrom-Poromaa I, Eriksson K. Sulfonation and sialylation of gonadotropins in women during the menstrual cycle, after menopause, and with polycystic ovarian syndrome and in men. *J Clin Endocrinol Metab* 2007;92(11):4410–17.

62. Barrios-De-Tomasi J, Timossi C, Merchant H, et al. Assessment of the in vitro and in vivo biological activities of the human follicle-stimulating isohormones. *Mol Cell Endocrinol* 2002;186(2):189–98.

63. Andersen CY, Leonardsen L, Ulloa-Aguirre A, et al. Effect of different FSH isoforms on cyclic-AMP production by mouse cumulus-oocyte-complexes: a time course study. *Mol Hum Reprod* 2001;7(2):129–35.

64. Lunenfeld B. Historical perspectives in gonadotropin therapy. *Hum Reprod Update* 2004;10(6):453–67.

65. Casper RF. Are recombinant gonadotrophins safer, purer and more effective than urinary gonadotrophins?, *Reprod Biomed Online* 2005;11(5):539–40.

66. Filicori M. The role of luteinizing hormone in folliculogenesis and ovulation induction. *Fertil Steril* 1999;71(3):405–14.

67. Whelan JG, 3rd, Vlahos NF. The ovarian hyperstimulation syndrome, *Fertil Steril* 2000;73(5):883–96.

68. Schmidt GE, Kim MH, Mansour R, et al. The effects of enclomiphene and zuclomiphene citrates on mouse embryos fertilized in vitro and in vivo, *Am J Obstet Gynecol* 1986;154(4):727–36.

69. Nugent D, Vandekerckhove P, Hughes E, et al. Gonadotrophin therapy for ovulation induction in subfertility associated with polycystic ovary syndrome. *Cochrane Database Syst Rev* 2000;(4):CD000410.

70. Bayram N, van Wely M, van Der Veen F. Recombinant FSH versus urinary gonadotrophins or recombinant FSH for ovulation induction in subfertility associated with polycystic ovary syndrome, *Cochrane Database Syst Rev* 2001;(2):CD002121.

71. Van Wely M, Bayram N, van der Veen F. Recombinant FSH in alternative doses or versus urinary gonadotrophins for ovulation induction in subfertility associated with polycystic ovary syndrome: a systematic review based on a Cochrane review. *Hum Reprod* 2003;18(6):1143–9.

72. Van Der Meer M, Hompes PG, De Boer JA, et al. Cohort size rather than follicle-stimulating hormone threshold level determines ovarian sensitivity in polycystic ovary syndrome. *J Clin Endocrinol Metab* 1998;83(2):423–6.

73. Shoham Z, Patel A, Jacobs HS. Polycystic ovarian syndrome: safety and effectiveness of stepwise and low-dose administration of purified follicle-stimulating hormone. *Fertil Steril* 1991;55(6):1051–6.

74. Hamilton-Fairley D, Kiddy D, Watson H, et al. Low-dose gonadotrophin therapy for induction of ovulation in 100 women with polycystic ovary syndrome. *Hum Reprod* 1991;6(8):1095–9.

75. Grigoriou O, Antoniou G, Antonaki V, et al. Low-dose follicle-stimulating hormone treatment for polycystic ovarian disease. *Int J Gynaecol Obstet* 1996;52(1):55–9.

76. Ergur AR, Yergok YZ, Ertekin A, et al. Clomiphene citrate-resistant polycystic ovary syndrome. Preventing multifollicular development. *J Reprod Med* 1998;43(3):185–90.

77. Homburg R, Levy T, Ben-Rafael Z. A comparative prospective study of conventional regimen with chronic low-dose administration of follicle-stimulating hormone for anovulation associated with polycystic ovary syndrome. *Fertil Steril* 1995; 63(4):729–33.

78. White DM, Polson DW, Kiddy D. Induction of ovulation with low-dose gonadotropins in polycystic ovary syndrome. *J Clin Endocrinol Metab* 1996;81(11):3821–4.

79. Kamrava MM, Seibel MM, Berger MJ, et al. Reversal of persistent anovulation in polycystic ovarian disease by administration of chronic low-dose follicle-stimulating hormone. *Fertil Steril* 1982;37(4):520–3.

80. Polson DW, Mason HD, Saldahna MB, Franks S. Ovulation of a single dominant follicle during treatment with low-dose pulsatile follicle stimulating hormone in women with polycystic ovary syndrome. *Clin Endocrinol (Oxf)* 1987;26(2):205–12.

81. Messinis IE, Milingos SD. Current and future status of ovulation induction in polycystic ovary syndrome. *Hum Reprod Update* 1997;3(3):235–53.

82. Homburg R, Howles CM. Low-dose FSH therapy for anovulatory infertility associated with polycystic ovary syndrome: rationale, results, reflections and refinements. *Hum Reprod Update* 1999;5(5):493–9.

83. Macklon NS, Fauser BC. The step-down protocol. In Tarlatzis B (Ed.) *Ovulation Induction*. Paris, Elsevier, pp. 108–18.

84. Elchalal U, Schenker JG. The pathophysiology of ovarian hyperstimulation syndrome–views and ideas. *Hum Reprod* 1997;12(6):1129–37.

85. Kaiser UB. The pathogenesis of the ovarian hyperstimulation syndrome. *N Engl J Med* 2003;349(8):729–32.

86. Lee TH, Liu CH, Huang CC, et al. Serum anti-mullerian hormone and estradiol levels as predictors of ovarian hyperstimulation syndrome in assisted reproduction technology cycles. *Hum Reprod* 2008;23(1):160–7.

87. Esinler I, Bayar U, Bozdag G, Yarali H. Outcome of intracytoplasmic sperm injection in patients with polycystic ovary syndrome or isolated polycystic ovaries. *Fertil Steril* 2005;84(4):932–7.

88. North American Ganirelix Study Group. Efficacy and safety of ganirelix acetate versus leuprolide acetate in women undergoing controlled ovarian hyperstimulation. *Fertil Steril* 2001;75(1):38–45.

89. Rabinovici J, Kushnir O, Shalev J, et al. Rescue of menotrophin cycles prone to develop ovarian hyperstimulation. *Br J Obstet Gynaecol* 1987;94(11):1098–102.

90. Wada I, Matson PL, Troup SA, et al. Outcome of treatment subsequent to the elective cryopreservation of all embryos from women at risk of the ovarian hyperstimulation syndrome. *Hum Reprod* 1992;7(7):962–6.

Ovulation induction in women with polycystic ovary syndrome

Shawky Z. A. Badawy and Botros R. M. B. Rizk

Introduction

Our knowledge of polycystic ovary syndrome (PCOS) dates back to the clinical work and publications by two gynecologists: Irving Stein and Michael Leventhal from Chicago. Their publications in 1935 were about this syndrome and its treatment. The syndrome was named after them and it was common to write about "Stein–Leventhal syndrome," where the ovaries are morphologically enlarged with multiple cysts. These women presented with irregular cycles in the form of oligomenorrhea, hirsutism, infertility, and obesity [1]. At that time, there were no hormonal studies, but the clinical picture was evident. Stein and Leventhal also treated their patients with wedge resection of the ovaries [2]. Their published report of several pregnancies that occurred following this procedure introduced this procedure as basic treatment for PCOS, which was carried out for several decades until the endocrinology of this syndrome was evaluated when the hormonal assays became available about 40 years ago.

The endocrinology studies that were done certainly explain many of the abnormalities that affect the hypothalamic–pituitary–ovarian axis. These include high luteinizing hormone (LH) levels, low follicle-stimulating (FSH) levels, and an increase in androgen production by the ovaries and/or the adrenal glands. These increased androgens will result in increased pulsatility of gonadotropin-releasing hormone (GnRH), leading to more LH secretion. The increase in estrogens as a result of conversion of androgens leads to suppression of FSH. As a result of this disproportionate LH and FSH pattern, these women become anovulatory, leading to the main symptoms of this syndrome of oligomenorrhea and amenorrhea

[3,4]. Hormonal studies facilitated the recognition of the mechanisms by which wedge resection of the ovary helps those patients to achieve ovulation and pregnancy. Essentially, removal of a segment of the ovary in the form of a wedge, taking out a good part of the stroma, results in changes in the androgen levels and consequently, reversal of the effect on the hypothalamic–pituitary axis, leading to good levels of FSH to stimulate folliculogenesis and then ovulation accordingly. This is evident from the work done by Howard Judd and Sam Yen, in which they evaluated hormonal levels before wedge resection and immediately after wedge resection for a period of time. They also discovered that the effect of wedge resection is not a permanent one, because several months after surgery, the levels of androgens will start to rise again and these patients will become anovulatory [5].

PCOS might mimic other clinical conditions that lead to anovulation. It therefore became important to define diagnostic criteria for this syndrome. This was evident from the 1990 NIH-NICHD conference on PCOS at which diagnostic criteria were agreed which included (1) chronic anovulation after exclusion of other factors, including thyroid disease, adrenal disease, and hyperprolactinemia; and (2) hyperandrogenism that could be biochemical or clinical [6].

In 2003, an international conference was held in Rotterdam and the committee agreed on the inclusion of chronic anovulation, hyperandrogenemia, and polycystic appearance on ultrasound evaluation [7,8]. It was stated that two out of these three criteria will make the diagnosis. The inclusion of polycystic appearance criteria by ultrasound suggests that some of these patients may be ovulatory. Therefore, the criteria of polycystic appearance need to be seen either

Androgens in Gynecological Practice, ed. Leo Plouffe and Botros Rizk. Published by Cambridge University Press. © Cambridge University Press 2015.

as part of the diagnosis or as a separate phenotype of PCOS, which is somewhat confusing. In 2006, the Androgen Excess Society met and confirmed the NIH criteria, and added ultrasound polycystic appearance as a separate phenotype [9].

Furthermore, the wedge resection that was initially started by Stein and Leventhal has been deleted from our management of patients with PCOS, since we in the field of human reproduction have many other medical advances with good treatment of these patients, helping them to achieve ovulation and pregnancy without surgical intervention.

Management of PCOS depends on the patient's needs. If pregnancy is not the goal, management should be directed towards suppression of androgens and cycling of the endometrium to prevent hyperplasia and cancer, which is common if these patients are not treated properly. Therefore, in this category, cyclic progestins or steroidal oral contraceptives, if there are no contraindications, will be the line of treatment. On the other hand, for those patients who desire pregnancy, induction of ovulation should be the principal mode of management.

The incidence of PCOS has been about 6–12% of reproductive age women. However, with the use of the modern criteria, this incidence could be much higher [10,11]. This constitutes a large part of the female population who will need evaluation and management for their infertility at some point. The management course includes the use of many of the available fertility medications. In this chapter we will discuss the merits, indications, and contraindications for the use of every medication.

Pharmacologic agents for induction of ovulation in PCOS patients

Clomiphene citrate

Clomiphene citrate is the first estrogen receptor modulator to be used for fertility purposes. The pharmacology indicates that it has two isomers: the Z isomer and the E isomer. The Z is zuclomiphene and the E is enclomiphene. The Z clomiphene has a longer half-life than the E clomiphene. Studies have shown that the Z clomiphene remains in the body for some time after ovulation.

It is metabolized in the gastrointestinal tract and excreted through the intestines, and traces of clomiphene have been detected for 6 weeks following the administration of this medication [12]. The starting dose for induction of ovulation is 50 mg from days 3–7 of the cycle for a total of 5 days. If there is no response, then the dose is increased in increments of 50 mg up to 150–200 mg daily for 5 days of each cycle. If ovulation is not achieved by this dose, then the medication is not effective and the treating physician should move to a different type of fertility medication. The success rate of induction of ovulation with clomiphene citrate reaches up to 85%. However, only half of these patients will achieve pregnancy [13,14]. If other factors are normal, i.e., patency of the fallopian tubes and normal semen analysis, then the failure to achieve pregnancy, despite ovulation, might be due to the antiestrogenic effect of clomiphene citrate. This antiestrogenic effect is mainly on the endometrium and also on the cervical mucus [15]. The antiestrogenic effect on the endometrium will interfere with the production of the various adhesive molecules, pinopodes, resulting in poor decidualization, which will not be conducive to implantation. If pregnancy happens, the miscarriage rate will be high.

Clomiphene citrate acts by attaching itself to the estrogen receptors in the hypothalamus, thus interfering with the estrogen mechanism at the hypothalamic level. The end result will be an increase in the secretion of the GnRH pulses, both in frequency and amplitude. This will lead to secretion of both FSH and LH. FSH will be secreted at a higher level than LH, thus stimulating the process of folliculogenesis and eventually ovulation [16].

Treatment with clomiphene citrate increases the incidence of multiple births, mainly twins by 7–9% and triplets by 0.3%. Other side effects include antiestrogen effects, leading to vasomotor symptoms and visual symptoms in the form of blurring of vision in 1–2% of the patients, which might be due to the effect of clomiphene citrate on the visual cortex. It should be noted that these symptoms will disappear after the course of treatment is completed. The incidence of congenital malformations in pregnancies following clomiphene citrate is the same as in the general population [17].

Some patients with PCOS may have an elevation of dehydroepiandrosterone sulfate (DHEA-S) from the adrenal gland. This might affect the response of these patients to clomiphene citrate. In these patients, dexamethasone is given in combination with clomiphene citrate during induction of ovulation and for several days after that. This has been shown to increase the response to clomiphene citrate, and consequently the pregnancy rate [18].

Similarly, some patients with PCOS have elevated prolactin levels without any evidence of a space-occupying lesion in the pituitary gland. These patients may benefit from receiving a dopamine agonist in addition to clomiphene citrate to potentiate its effect [19].

Letrozole for induction of ovulation

Letrozole is an aromatase inhibitor that was approved by the FDA in 1997 for chemotherapeutic treatment of postmenopausal patients with a history of breast cancer [20]. The trials for the use of letrozole for induction of ovulation started in 2000, with the first report presented at the American Society of Reproductive Medicine Meeting in 2001 [21]. The mode of action in induction of ovulation is to inhibit aromatase and therefore, the conversion of testosterone to estrogen is decreased. The negative feedback mechanism of estrogen on the hypothalamic–pituitary axis will be diminished and this will result in an increase in secretion of gonadotropins, which will stimulate the process of folliculogenesis and ovulation. Several doses were used, starting with 2.5 mg and increments of 2.5 mg up to 12.5 mg dosage starting on day 3 of the cycle till day 7.

Several comparative studies were done comparing letrozole to clomiphene citrate, and it was found that the ovulation rate was nearly identical. However, the pregnancy rate is higher with letrozole than with clomiphene citrate. It should also be noted that many of the side effects of clomiphene citrate do not occur with the use of letrozole. Specifically, endometrial thickness is much better with the use of letrozole than with clomiphene citrate. The investigators also found that the incidence of congenital malformations in resulting pregnancies was less in patients who received letrozole, as compared hose administered clomiphene citrate for induction of ovulation.

It appears that the use of letrozole is effective in induction of ovulation in PCOS, as well as in other patients who have failed clomiphene citrate treatment. It might also be used as the first line of treatment for induction of ovulation, rather than using clomiphene citrate, in PCOS patients. Other studies also demonstrated the effectiveness of letrozole when used in combination protocols with gonadotropins to treat the poor responders in cycles for assisted reproduction. This is really a great advance and a new addition to our medications that are used for induction of ovulation in anovulatory infertility patients [22,23].

Ovarian drilling in the management of PCOS

In 1935, Stein and Leventhal introduced this syndrome of anovulation, hirsutism, and enlarged ovaries. The management they started and continued for several decades was wedge resection of the ovaries. Some of these patients achieved pregnancy, and therefore, surgical management became a standard type of management at that time and for some time after that. It was then realized that with the surgical management of these ovaries, the patients are subjected to development of adhesions, peri-ovarian and peritubal, which complicates the matter and interferes with the management of infertility. With the advances in reproductive medicine and the introduction of fertility medications, the medical management of PCOS gained the upper hand and a good percentage of these patients respond to the induction of ovulation using medications like clomiphene citrate and human menopausal gonadotropins.

Today, we have more medications on trial such as the use of aromatase inhibitors. In addition, human menopausal gonadotropins have been purified and widely used. Furthermore, recombinant gonadotropins have been introduced for induction of ovulation and are widely used now. Certainly, there is a very good response by PCOS patients to these medications, with very good rates of ovulation and pregnancy. However, some of these patients fail to respond to the fertility medications and this might be due to the fact that their androgen levels are too high or there is an associated hyperprolactinemia or thyroid disease. Therefore, it is imperative to do a thorough endocrine evaluation of these patients and to treat endocrinopathies to facilitate the response of the patients' ovaries to fertility medications.

In the past three decades, the surgical management of PCOS to facilitate induction of ovulation has been reintroduced. This is in the form of a laparoscopic type of surgical treatment in which the ovaries are subjected to several unipolar electrode punctures. Some studies evaluated the response of these ovaries after surgery to ovulation-inducing medications and found them to be reasonably successful. They attributed this success to decrease in the size of the ovaries and decrease in the level of testosterone and LH [24]. This was associated with improvement in FSH levels, and ovulation and pregnancy after that. It is evident from some studies that the pregnancy rate after ovarian drilling is equal to the pregnancy rate with the use of fertility medications to induce ovulation in these patients [25].

The main drawback of ovarian drilling is the development of adhesions, and sometimes excessive thermal effect, which can lead to the development of amenorrhea in such patients [26].

Another study compared the effect of ovarian drilling on pregnancy rate and occurrence of diabetes during pregnancy in women with and without metabolic syndrome. The study found no difference in pregnancy rate or the occurrence of diabetes in pregnancy in patients with/without metabolic syndrome [27].

The use of insulin-sensitizing agents for induction of ovulation in PCOS patients

Patients with PCOS have also been described as highly susceptible to the development of metabolic syndrome [28]. In this syndrome, the tissues are insulin resistant resulting in an increase in the level of blood sugar and frequently hemoglobin A_{1c}, suggesting a prediabetic or diabetic status. High insulin levels in these patients also act on the insulin-like growth receptors in the ovary; and together with high LH levels, these will lead to high levels of testosterone. These elevated levels of testosterone will result in anovulatory cycles by their effect on the hypothalamic–pituitary axis. Many of these patients are overweight or obese. Some may not be in this category, but they still might have the metabolic syndrome. In some cases, these patients might have hyperlipidemia and may develop raised blood pressure and increased incidence of heart disease [29].

Metformin is essentially a dimethyl biguanide, and is used in non-insulin-dependent diabetes mellitus. It inhibits glucose production by the liver, its half-life is 4–8 hours, and it is excreted through the kidneys. Historically, metformin was discovered in 1920 to be useful in reduction of blood sugar, and was introduced in the United States in 1995 [30,31].

Metformin is widely used for induction of ovulation in PCOS. It can result in ovulatory cycles when it is given alone or in addition to clomiphene citrate. There have been various trials, all which suggest that the combination of clomiphene citrate and metformin might lead to a slight increase in the pregnancy rate as compared with clomiphene citrate alone. However, in some of these studies, the difference may not be statistically significant. Nevertheless, the use of metformin

for induction of ovulation in these patients will reduce the number resistant to medical treatment for this purpose [32,33].

Use of gonadotropin preparations for ovulation induction in PCOS patients

The use of gonadotropins for induction of ovulation has been practiced for over six decades, since 1960. The first preparation used was a mixture of FSH and LH extracted from the urine of postmenopausal women [34,35] The original preparations of urinary gonadotropins contained 75 units of FSH and less than 1% of LH activity [36]. In the early 1990s, the urinary gonadotropins were purified and the contents of the ampule contained 75 units of FSH and less than 0.1% of LH. In the late 1990s, recombinant FSH became available containing FSH alone with no LH contamination [37]. There have been discussions about the efficacy of highly purified urinary gonadotropins and recombinant gonadotropins in induction of ovulation. According to the Cochrane database reviews [38], there are no significant differences in ovulation rate, pregnancy rate, miscarriage rate, multiple pregnancy rate, and ovarian hyperstimulation syndrome. This was also confirmed by previous committees of the American Society of Reproductive Medicine [39].

The use of gonadotropins for induction of ovulation in anovulatory women, and particularly for PCOS, needs experience for follow-up of these patients in order to avoid hyperstimulation syndrome. Furthermore, the use of gonadotropins needs proper monitoring in the form of estradiol levels as well as sonograms to check follicular maturity during the process of controlled ovarian stimulation. Gonadotropin treatment is usually reserved for those cases of PCOS that fail to respond to other ovulation induction agents, including clomiphene citrate, aromatase inhibitors, and metformin.

There has been discussion about the protocol of ovarian stimulation using gonadotropins. This has to be taken into consideration because of the risk to these patients of developing hyperstimulation syndrome. The most acceptable protocol is to start on a small dose of gonadotropins (75 units) and then to do a step-up, with small amounts of gonadotropins added on a weekly basis until mature follicles are detected by ultrasound. Then, the triggering hormone is given in the form of human chorionic gonadotropin. This step-up protocol starting with a small dose of FSH has been shown to be effective and with a pregnancy rate comparable to the classic protocol, in which a higher dose is used from the start. In some cases, the initial low-dose protocol may need to go a step down to be safe and to produce the necessary follicle growth. It was found that this protocol has a pregnancy rate comparable to the step-up protocol [40,41].

There has been discussion about the use of a GnRH agonist concomitant to gonadotropin stimulation for the induction of ovulation in PCOS patients. The theory is to determine whether GnRH agonist use can reduce the incidence of hyperstimulation syndrome. The results of randomized controlled studies did not support that concept and, therefore, GnRH agonist use is beneficial for neither increased pregnancy rate nor decrease in ovarian hyperstimulation syndrome [42].

The use of insulin-sensitizing agents like metformin in combination with gonadotropins may reduce the incidence of ovarian hyperstimulation and multiple pregnancy rate. However, these are preliminary studies and more studies are needed in order to evaluate this modality of treatment [43].

It is suggested that patients with PCOS who may be going for in vitro fertilization cycle are given the long protocol with downregulation of the hypothalamic–pituitary–ovarian axis using GnRH agonist. In addition, it is suggested that the step-up protocol of ovarian stimulation with gonadotropins could be used for these IVF patients [44].

Conclusion

PCOS is a major cause of anovulation and infertility. Induction of ovulation in such cases has shifted towards use of fertility medications. These agents are successful with resultant good pregnancy rates. There is no increase in congenital malformations as compared to pregnancies in the general population. It is essential to treat any associated endocrine factors, including adrenal hyperandrogenism, thyroid dysfunction, and hyperprolactinemia, to increase the success rate. The use of insulin-sensitizing agents must be considered in patients with the metabolic syndrome [28].

Furthermore, patients with PCOS who are obese must be counseled for weight loss through dieting and exercise. Weight loss in these cases has shown good results, leading to spontaneous ovulation in some cases, and facilitates the response to fertility medications in other cases [45].

References

1. Stein IF, Leventhal ML. Amenorrhea associated with bilateral polycystic ovaries. *Am J Obstet Gynecol* 1935;29:181–91.

2. Stein IF. Duration of infertility following ovarian wedge resection. *West J Surg* 1964;72:237.

3. Goldzieher JW, Axelrod LR. Clinical and biochemical features of polycystic ovarian disease. *Fertil Steril* 1963;14:631–53.

4. Rebar R, Judd HL, Yen SSC, et al. Characterization of the inappropriate gonadotropin secretion in polycystic ovary syndrome. *J Clin Invest* 1976;57:1320–9.

5. Judd H, Anderson D, Yen S. The effect of ovarian wedge resection on circulating gonadotropin and ovarian steroids levels in patients with polycystic ovary syndrome. *J Clin Endocrinol Metab* 1976;42:347–55.

6. Zawadski JK, Dunaif A. Diagnostic criteria for polycystic ovary syndrome towards a rational approach. In: Dunaif A, Givens JR, Haseltine FP, Merriam GR (Eds.) *Polycystic Ovary Syndrome.* Boston, MA: Blackwell Scientific Publications, 1992:377–84.

7. The Rotterdam ESHRE/ASRM-sponsored PCOS Consensus Workshop Group. Revised 2003 Consensus on Diagnostic Criteria and Long-Term Health Risks Related to Polycystic Ovarian Syndrome. *Human Reprod* 2004;19:41–7.

8. The Rotterdam ESHRE/ASRM-sponsored PCOS Consensus Workshop Group. Revised 2003 Consensus on Diagnostic Criteria and Long-Term Health Risks Related to Polycystic Ovarian Syndrome. *Fertil Steril* 2004; 81(1):19–25.

9. Azziz R, Carmina E, Dewailly D, et al. The Androgen Excess and PCOS Society Criteria for the Polycystic Ovary Syndrome: The Complete Task Force Report. *Fertil Steril* 2009;91:456–88.

10. Farah L, Lozenby AJ, Boots LR, Azziz R. Prevalence of polycystic ovary syndrome in women seeking treatment for community electrologists. Alabama Professional Electrology Association Society Group. *J Reprod Med* 1999;44:870–4.

11. Azziz R, Woods KS, Reyna R, et al. The prevalence and features of the polycystic ovary syndrome in an unselected population. *J Clin Endocrinol Metab* 2004;89:2745–9.

12. Szulc M, Morgan DJ, McLeish M, et al. Pharmacokinetics of intravenous clomiphene isomers. *Br J Clin Pharmacol* 1989; 27(5):639–40.

13. Veuse TD, Chung AP. Ovulation induction in polycystic ovary syndrome. *J Obstet Gynecol Can* 2010;32(5),495–502.

14. Elkind-Hirsch K, Darensbourg C, Creasy G, Gipe D. Conception rates in clomiphene citrate cycles with and without hormone supplementation: a pilot study. *Curr Med Res Opin* 2005;21(7):1035–40.

15. Gysler M, March CM, Mishell DR Jr, Bailey E. A decade's experience with an individualized clomiphene treatment regimen including its effect on the post coital test. *Fertil Steril* 1982;37:161–7.

16. Roy S, Greenblatt RB, Mahesh VB, Jungck EC. Clomiphene citrate: further observations on its use in induction of ovulation in the human and on its mode of action. *Fertil Steril* 1963;14:575–95.

17. Dicky RP, Hotkamp DE. Development, pharmacology and clinical experience with clomiphene citrate. *Hum Reprod Update* 1996;2:483–506.

18. Elnashar A, Abdelmageed E, Fayed M, Sharaf M. Conception rates in clomiphene citrate cycles with and without hormone supplementation: a pilot study. *Curr Med Res Opin* 2005;21(7):1035–40.

19. Persanezhad ME, Alborzi SA, Jahroni BN. A prospective double blind randomized placebo controlled clinical trial of bromocriptine in clomiphene resistant patients with polycystic ovarian syndrome and normal prolactin level. *Arch Gynecol Obstet* 2004;269:125–9.

20. Bayar U, Basaran M, Kiran S, et al. Use of an aromatase inhibitor in patients with polycystic ovarian syndrome: a perspective randomized trial. *Fertil Steril* 2006;86:1447–51.

21. Mitwally MFM, Casper RF. Use of aromatase inhibitor for induction of ovulation in patients with an inadequate response to Clomiphene citrate. *Fertil Steril* 2001;75:305–9.

22. Al-Omari W, Al-Hadithi N, Izat B, Sulaiman W. The effect of an aromatase inhibitor on ovulation induction and endometrial receptivity in clomiphene resistant women with polycystic ovary syndrome. *Mid East Fertil Soc J* 2001;6:52–5.

23. Metawie MH. Comparative study of aromatase inhibitor, letrozole with clomiphene citrate for induction of ovulation. *Mid East Fertil Soc J* 2001;6:57–9.

24. Amer SA, Li TC, Ledger WL. Ovulation induction using laparoscopic ovarian drilling in women with polycystic ovarian syndrome: predictors of success. *Hum Reprod* 2004;19:1719–24.

25. Bayram N, VanWely M, Kaaijk EM, et al. Using an electrocautery strategy or recombinant follicle stimulating hormone to induce ovulation in polycystic ovary syndrome: randomized controlled trial. *BMJ* 2004;328(7433):192.

26. Mercorio F, Mercorio A, DiSpiezio Sardo A, et al. Evaluation of ovarian adhesion formation after laparoscopic ovarian drilling by second look mini-laparoscopy. *Fertil Steril* 2008;89:1229–33.

27. Kong GWS, Cheung LP, Lok IH. Effects of laparoscopic ovarian drilling in treating infertile anovulatory polycystic ovarian syndrome patients with and without metabolic syndrome. *Hong Kong Med J* 2011;17:5–10.

28. Dunaif A. Insulin resistance and the polycystic ovary syndrome: mechanism and implications for pathogenesis. *Endocr Rev* 1997;18(6):774–800.

29. Essah PA, Nestler JE. Metabolic syndrome in women with polycystic ovary syndrome. *Fertil Steril* 2006;86:S18–19.

30. Scheen AJ. Clinical pharmacokinetics of metformin. *Clin Pharmacokinet* 1996;30(5):359–71.

31. Glucophage label and approval history. U.S. Food and Drug Administration. Retrieved January 8, 2007. Data available for download on FDA Website.

32. Lord JM, Flight IHK, Norman RJ. Metformin in polycystic ovary syndrome: systematic review and meta-analysis. *BMJ* 2003; 327(7421):951–3.

33. Legro RS, Barnhart HX, Schlaff WD, et al. Clomiphene, metformin or both for infertility in the polycystic ovary syndrome. *N Engl J Med* 2007;356:551–66.

34. Lunenfeld B. Treatment of anovulation by human gonadotropins. *J Int Fed Gynaecol Obstet* 1963;1:153.

35. Rosenberg E, Coleman J, Damani M, Garcia CR. Clinical effect of postmenopausal gonadotropins. *J Clin Endocrinol Metab* 1962;23:181–9.

36. Giudice E, Crisci C, Eshkol A, Papoian R. Composition of commercial gonadotropin preparation extracted from human post menopausal urine: characterization of nongonadotropin proteins. *Hum Reprod* 1994;9:2291–9.

37. Howles CM. Genetic engineering of human FSH (Gonal F). *Hum Reprod Update* 1996;2:172–91.

38. Bayram N, Van Wely M, van Der Veen F. Recombinant FSH versus urinary gonadotropins or recombinant FSH for ovulation induction in subfertility associated with polycystic ovary syndrome. *Cochrane Database Syst Rev* 2001;(2):CD002121.

39. The Practice Committee of the American Society for Reproductive Medicine. Gonadotropin preparations: past, present, future prospective. *Fertil Steril* 2008;90:S13–20.

40. Balash J. Inducing follicular development in anovulatory patients and normally ovulating women: current concepts and the role of recombinant gonadotropins. In: Gardner DK, Weissman A, Howles CM, Shoham Z (Eds.) *Textbook of Assisted Reproductive Techniques: Laboratory and Clinical Perspective*. London, UK: Martin Dunitz, Ltd, 2001:698.

41. Sagle MA, Kiddy DS, Franks S. A comparative randomized study of low dose human menopausal gonadotropin and follicle stimulating hormone in women with polycystic ovarian syndrome. *Fertil Steril* 1991;55(1):56–60.

42. Nugent D, Vandekerckhove P, Hughes E, et al. Gonadotropin therapy for ovulation induction in subfertility associated with polycystic ovary syndrome. *Cochrane Database Syst Rev* 2000;(4):CD000410.

43. DeLeo V, la Marca A, Ditto A, et al. Effects of metformin on gonadotropin induced ovulation in women with polycystic ovary syndrome. *Fertil Steril* 1999;72:282–5.

44. Jacobs HS, MacDongall BA. Polycystic ovaries and ART. In: Gardner DK, Weissman A, Howles CM, Shoham Z (Eds.) *Textbook of Assisted Reproductive Techniques: Laboratory and Clinical Perspective.* London, UK: Martin Dunitz, Ltd, 2001:1040.

45. Clark AM, Thomley B, Tomlinson L, et al. Weight loss in obese infertile women results in improvement in reproductive outcome for all forms of fertility treatment. *Hum Reprod* 1998;13:1502–5.

Index

ABCC11 gene, 82
abdominal obesity, 6
abdominal phenotype, 209
acanthosis nigricans, 6, 19, 28, 122
    in PCOS, 164–165
acetazolamide, 172
acne, 82–84
    diagnosis, 82
    in PCOS, 163
    treatment, 33–34, 83
acrochordon, *see* skin tags
acromegaly, 29, 30
    differential diagnosis, 172
ACTH, 111, 118, 180
    Cushing's syndrome, 127–129
    ectopic ACTH syndrome, 119
    excess, 118
ACTH stimulation test, 6, 114
adipocytes, metabolic role, 210
adipocytokines, 210, 212
adipokines, 91
adiponectin, 91
adrenal aging, 200
adrenal disorders, 119–120
    hematoma, 197
    PPNAD, 120
    tumors, *see* adrenal tumors
adrenal MRI, 191–199
    inversion recovery fat suppression, 193
    lipid detection techniques, 191–193
        chemical shift imaging, 191–192
        CHESS, 192–193
    normal adrenal glands, 195
    post-contrast images, 194
    T1-weighted images, 194
    T2-weighted images, 193–194
    see also adrenal disorders, adrenal tumors
adrenal tumors, 29
    adenoma, 195–196
    adrenocortical carcinoma, 198
    and hirsutism, 4, 18
    metastatic disease, 196
    myelolipoma, 196–197
    pheochromocytoma, 197
adrenalectomy, 116, 130, 131–132
adrenocorticotropic hormone, *see* ACTH
adult granulosa cell tumor, 139–141

advanced glycation end products, 92, 212
Aging Male Symptoms (AMS) score, 155
alopecia, 11–24
    androgenetic, 12, 84–85
    female pattern hair loss, *see* female
        pattern hair loss
    in PCOS, 164
alopecia areata, 14
amenorrhea, 6, 104, 165
    differential diagnosis, 172
    see also PCOS
American Society for Reproductive
    Medicine (ASRM), 74, 89
aminoglutethimide, 130, 131
anabolic steroids, 157
anastrozole, 32
androblastomas, *see* Sertoli-stromal
    cell tumors
AndroFeme, *see* transdermal
    testosterone patch
AndroGel system, 43
androgens, 180
    anabolic effects, 201
    and hair growth, 11–12
    deficiency/excess, *see* androgen
        deficiency, androgen excess
    extraovarian, 65
    genital tract effects, 109
        development and
            differentiation, 97–99
        structure and function, 99–103
    metabolism, 1
    ovarian production, 67–69
    in postmenopausal women, 64–74, 105,
        106, 200–207
        bone, 201
        breast, 203–205
        endometrium, 202–203
        muscle and lean body mass, 201–202
        quality of life, 202
    skin effects, 79–82
        hair follicles, 81
        sebaceous glands, 81
        sweat glands, 81–82
    skin metabolism, 80–81
    skin-related, 79
    see also individual hormones

androgen antagonists, *see* antiandrogens
androgen assays, 180–190
    clinical use, 184–185
    in vitro, 182–183
    in vivo, 181
        immunoassay, 181–182, 185–188
        mass spectrometry, 182
    testosterone
        free, 184
        total, 183–184
        see also individual assays
androgen deficiency, 38, 39, 41, 104–106
    men, 152–153
        patient-reported outcome tools,
            155–156
    premenopausal women, 104
        idiopathic, 105
        oral contraceptives, 104
    treatment, *see* testosterone replacement
        therapy
Androgen Deficiency of Aging Men
    (ADAM) questionnaire, 155
androgen destabilizers, 61
androgen excess, 1, 14, 19, 26, 29, 34, 83, 85,
    89, 103–104, 136, 209
    diagnosis, 75
    drugs causing, 172
    etiology, 5
    family history, 6
    gestational, 6
    idiopathic, 4
    and PCOS, 209
    see also congenital adrenal hyperplasia,
        PCOS
Androgen Excess and Polycystic Ovary
    Syndrome Society, 1
Androgen Excess Society (AES), 74, 89
androgen exposure
    levels of, 71
    markers of, 69–71
androgen insensitivity syndrome, *see*
    complete androgen insensitivity
    syndrome (CAIS)
androgen metabolites, 69–71
androgen receptors, 55–63, 79–80, 200
    agonists, 56, 57, 61
    antagonists, *see* antiandrogens

233

androgen receptors (*cont.*)
  bone, 201
  co-activators and co-repressors, 58–59
  historical context, 55
  inactive state, 57
  ligand binding and activation, 57–58
  molecular disorders linked to, 59–60
    androgen insensitivity
      syndrome, 59–60
    breast cancer, 60
    Kennedy disease, 60
  partial agonists, 61
  structure, 55–57
    ligand-binding domain, 56
  therapeutic interventions, 60–61
androgen replacement therapy, *see*
      testosterone replacement therapy
androgen response elements, 58
androgen-producing ovarian tumors,
      136–145
  classification, 139
  diagnosis, 137–139
    interventional testing, 138
    laboratory tests, 137
    radiology, 137–138
  embryology, 139
  history and examination, 136–137
  idiopathic, 143
  pregnancy-associated, 143
  treatment, 143–144
androgen-related skin disorders, 82–87
  acne, 82–84, 163–164
  androgenetic alopecia, 12, 84–85
  hidradenitis suppurativa, 86
  hirsutism, *see* hirsutism
androgen-secreting tumors, 6
androgenetic alopecia, 12, 84–85
  diagnosis, 84
  treatment, 85
androgenic compounds, 57
androstenedione, 11, 29, 42, 79, 118, 180
  normal range, 105
anorgasmia, 40
antagonists, 57
antiandrogens, 8, 56, 61
  CAH, 115
  FPHL, 15–16
  hidradenitis suppurativa, 87
  PCOS, 33
  see also individual drugs
antibodies, effects on immunoassay, 186
anti-müllerian hormone, 30, 75, 98
anxiety disorders
  in PCOS, 169
APHRODITE study, 204
apolipoprotein CIII, 49
apoptosis, 11
arachidonic acid, 122
arginase, 101
aromatase, 98, 204

fetal, 98
aromatase inhibitors, 32, 216, 228
  PCOS, 216–217
  see also individual drugs
asymmetric dimethylarginine, 92
autoantibodies, 187
aversion disorder, 38

basic fibroblast growth factor (bFGF), 118
Beckwith–Wiedemann syndrome, 198
beta-endorphin, 118
beta-lipoprotein, 118
bicalutamide, 61
bilateral inferior petrosal sinus
      sampling, 129
bleaching of hair, 7
BLISS study, 204
body mass index (BMI), 19, 26, 209
bone
  androgen effects, 201
  androgen receptors in, 201
  mass, 154
  mineral density (BMD), 154
Boston Area Community Health (BACH)
      Survey, 153
breast
  androgen effects, 203–205
  Tanner staging, 28
breast cancer
  androgen receptor role in, 60, 205
  and androgen replacement therapy, 45,
      204
  PCOS, 171
"buffalo hump" of Cushing's syndrome, 121

CA 19-9, 187
CAH, *see* congenital adrenal hyperplasia
camouflage for hair loss, 17
cardiovascular risks
  androgen replacement therapy, 45–49
  PCOS, 89–96, 168
Carney complex, 123, 198
carotid intima–media thickness, 92
cell proliferation assay, 183
Center for Epidemiologic Studies
      Depression Scale (CES-D), 202
central centrifugal cicatricial alopecia, 14
chaperone proteins, 57
chemerin, 92
chemical shift imaging (CSI), 191–192
chemical shift selective suppression, *see*
      CHESS
CHESS, 191, 192–193
chlormadinone acetate, 8, 86
clitoris, 98
  androgen effects, 99–100
clitoromegaly, 6, 19, 28
clomiphene citrate, 31, 220
  mechanism of action, 227
  with metformin, 219

PCOS, 216
  ovulation induction, 227–228
  resistance, 32
  side effects, 217
co-chaperone proteins, 57
cognitive-behavioral impact of androgen
      disorders, 146–150
complete androgen insensitivity syndrome
      (CAIS), 59–60, 98, 146
  cognitive-behavioral impact, 147
  gender identity/role, 147
  mental health status, 148
computed tomography (CT) of
      androgen-producing ovarian
      tumors, 137
congenital adrenal hyperplasia, 6, 13, 85,
      103, 111–117, 146
  clinical presentation, 113
  cognitive-behavioral impact, 147
  diagnosis, 114–115
    hormonal, 114
    molecular genetics, 114–115
    newborn screening, 115
  gender identity/role, 147
  hirsutism, 18
  21-hydroxylase deficiency, *see*
      21-hydroxylase deficiency
  late onset, 29
  mental health status, 148
  pathophysiology, 111–113
  pregnancy, 116
  treatment, 115–116
    medical, 115
    prenatal, 116
    surgical, 115–116
  visuo-spatial processing, 147
Congenital Adrenal Hyperplasia Adult
      Study Executive (CaHASE), 115
coronary heart disease, 92
corticotropin stimulation test, 114
corticotropin-releasing hormone (CRH),
      112, 118
  ectopic CRH syndrome, 119
corticotropin-releasing hormone
      stimulation test, 129
cortisol, 112
  anabolic effects, 121
  excess, 118, 122, 123, 136
    diagnosis, 123–127
      see also Cushing's syndrome
    late-night salivary, 126
    midnight serum, 127
    urinary free, 124–126
cosyntropin, 172
cranial prosthesis, 17
C-reactive protein, 92
CRH, *see* corticotropin-releasing hormone
CT, *see* computed tomography
Cushing, Harvey, 119
Cushingoid habitus, 121

Cushing's disease, 119
Cushing's syndrome, 19, 26, 29, 85, 89, 118–135
  biochemical testing, 123–129
    ACTH dependence, 127
    ACTH source, 127–129
    cortisol excess, 123–127
  clinical presentation, 120–122
    hirsutism, 18
  diagnosis, 123–124
  differential diagnosis, 122, 123, 172
  epidemiology, 120
  etiology, 118–119
    adrenal disorders, 119–120
    ectopic ACTH syndrome, 119
    ectopic CRH syndrome, 119
    glucocorticoid therapy, 115
    iatrogenic, 119, 122, 129
    McCune–Albright syndrome, 120
    PPNAD, 120
  female predominance, 120
  morbidity and mortality, 122
  pathophysiology, 118
  screening for, 30
  treatment, 129–132
    ACTH-dependent Cushing's syndrome, 130–132
    ACTH-independent Cushing's syndrome, 130
    iatrogenic Cushing's syndrome, 129
cyclosporin, 19, 172
cyclosporin A-binding protein (Cyp40), 57
CYP11A, 69
CYP17, 69
CYP19, 69
CYP19A1, 98
CYP21A2, 111
cyproterone acetate, 8, 20, 61
  acne vulgaris, 84
  CAH, 115
  FPHL, 16
  hidradenitis suppurativa, 87
  hirsutism, 21, 86
  PCOS, 34
cystitis, 40

danazol, 18, 19, 172
dehydroepiandrosterone, see DHEA
dehydroepiandrosterone sulfate, see DHEA-S
11-deoxycorticosterone, 118
depilatory agents, 7, 21
desogestrel, 8
dexamethasone, 115
dexamethasone suppression test, 30, 126–127
  1 mg overnight, 126
  2 mg/day 48-hour low-dose, 126
  2 mg/day 48-hour low-dose with CRH, 127

high-dose, 128–129
DHEA, 11, 42, 64, 79, 98, 180
  circulating levels, 67–69, 70
  postmenopause, 66–68, 106
  post-ovariectomy effects, 99
  vaginal epithelial growth, 101
DHEA-S, 11, 14, 29, 42, 79, 80, 98, 118, 180
  adrenal tumors, 29
  androgen-producing ovarian tumors, 137
  circulating levels, 68
  measurement, 20, 104
  normal range, 105
  in PCOS, 227
  postmenopausal women, 200–207
diabetes mellitus
  gestational, 170
  type 2, 89
    and PCOS, 167
    and testosterone deficiency, 154
diazoxide, 172
dietary interventions in PCOS, 212
dihydrotestosterone, 2, 11, 69, 79, 101
  female pattern hair loss, 12
  and hair growth, 12
disorders of sex development
  cognitive-behavioral impact, 146–150
  see also individual disorders
Dixon method, 192
doping agents, 183
drospirenone, 8, 20, 61
dutasteride, 16
dysgerminomas, 143
dyslipidemia, 89, 91
dyspareunia, 38, 40–41
  vaginal atrophy, 41

ectopic ACTH syndrome, 119
ectopic CRH syndrome, 119
eflornithine, 6
  hirsutism, 21, 34, 86
electroepilation, 22
electrolysis, galvanic, 7
Endogenous Hormones and Breast Cancer Collaborative Group, 204
endometrial cancer, 33
  and androgen replacement therapy, 45
  and PCOS, 31
endometrium, androgen effects, 202–203
endothelin-1, 92
enzyme regulation, androgen-mediated, 100–101
epicardial fat, 92
epidermal growth factor receptor inhibitors, 82
E-SCREEN assay, 183
estradiol, circulating levels, 221
estrogens, 61
estrogen receptor, 55, 56
estrogen replacement therapy, see hormone replacement therapy

ethinylestradiol, 8
  hidradenitis suppurativa, 87
etomidate, 130, 131
European Male Aging Study (EMAS), 153
European Society of Human Reproduction and Embryology (ESHRE), 74, 89
exercise, 212–213
external genitalia, 98–99

familial dysalbuminemia, 186
Favre–Racouchot syndrome, 83
FDG-PET/CT, 138
female androgen deficiency syndrome, 42, 200
female orgasmic disorder, 38, 40
female pattern hair loss, 12
  clinical features and classification, 13
  evaluation, 14
  pathogenesis, 12–13
  treatment, 15–17
    androgen-dependent therapies, 15–16
    androgen-independent therapies, 15
    see also specific drugs
female sexual differentiation, see sexual differentiation
female sexual dysfunction, see sexual dysfunction
Ferriman–Gallwey score, modified, 2, 4, 17, 19, 26, 163
fibrinogen, 92
fibroblast growth factor receptor 2 (FGFR2), 82
fibrosing alopecia, 14
finasteride, 8
  androgenetic alopecia, 85
  CAH, 115
  FPHL, 16
  hirsutism, 21, 86
  PCOS, 34
FK506-binding proteins, 57, 58
9α-fludrocortisone acetate, 115
fluorine-18-deoxyglucose positron emission tomography/computed tomography, see FDG-PET/CT
flutamide, 8, 61
  acne vulgaris, 83
  CAH, 115
  FPHL, 16
  hirsutism, 21, 33, 86
foldosome, 57, 58
follicle-stimulating hormone, see FSH
follicular miniaturization, 84
follicular unit transplantation, 17
FPHL, see female pattern hair loss
frailty, 154
free androgen index (FAI), 44, 184, 202
free testosterone, 184
FSH, 30, 180, 219, 226

gel filtration chromatography, 188
gender identity, 146

gender role, 146
genital tract, androgen effects, 97–109
    development and differentiation, 97–99
        structure and function, 99–103
            clitoris and vagina, 99–100
            sexual arousal response, 101–103
            vasomotor and trophic enzyme
                regulation, 100–101
germ cell tumors, 139, 143
gestational diabetes mellitus, 170
gestodene, 8
glucocorticoids, 18, 19, 115
    hypertrichosis, 172
glucose tolerance test, 6
GnRH agonists, 230
gonadotropins, 219–222
    indications, 220–221
    monitoring and administration, 221–222
    ovarian hyperstimulation syndrome, *see*
        ovarian hyperstimulation syndrome
    ovulation induction, 229–230
    preparations, 220
    see also individual hormones
gonadotropin-releasing hormone, *see* GnRH
        agonists
granulosa cell tumors (GCT), 140
    adult, 139–141
    juvenile, 141
guanylyl cyclase, 101
gynecomastia, 60

hair follicles, androgen effects, 81
hair growth, 1, 2
    anagen phase, 11
    catagen phase, 11
    physiology, 11
    sex steroid effects, 11–12
    telogen phase, 11
hair transplantation, 16
Hamilton–Norwood classification of FPHL, 14
hCG, 188, 219
HDL cholesterol, 91
health-related quality of life scale
        (HRQOL), 202
heat shock proteins, 57, 79
    see also specific proteins
hemolysis, effect on androgen assay, 186
heterodimers, 58
hidradenitis suppurativa, 86–87
    diagnosis, 86
    treatment, 87
high-dose dexamethasone suppression test,
        128–129
HIP, 57
hirsutism, 1–10, 17, 148, 174
    diagnosis, 4–6, 85
    epidemiology, 2–4
    etiology, 17–19
        adrenal tumors, 18
        androgen-related, 85–86

congenital adrenal hyperplasia, 18, 85
Cushing's syndrome, 18
hyperprolactinemia, 19
hyperthecosis, 18
iatrogenic, 18
idiopathic, 1, 4, 18–19
ovarian tumors, 18
PCOS, 18, 162–163, 210
evaluation, 19
laboratory investigations, 19–20
management, 6–8
modified Ferriman–Gallwey score, 2, 4,
    17, 19, 26, 163
pathophysiology, 1–2
prevalence, 2–4
treatment, 20–22, 33–34, 86
    cosmetic, 21–22
    medical, 20–21
    see also individual drugs
hMG, 220
    PCOS treatment, 33
homocysteine, 92
homodimers, 58
HOP, 57
hormone action, mechanism of, 56
hormone replacement therapy, 151–152
    estrogen-only trial, 152
    estrogen plus progestin trial, 152
    risk–benefit ratio, 152
    see also testosterone replacement therapy
hormone response element, 56
HSDD, *see* hypoactive sexual desire disorder
HSP40, 57
HSP56, 79
HSP70, 57, 79
HSP90, 57, 58, 79
human anti-animal antibodies (HAAA), 187
human chorionic gonadotropin, *see* hCG
human menopausal gonadotropin,
        see HMG
hydrocortisone, 115
16-hydroxy DHEA-S, 98
17α-hydroxylase, 111
21-hydroxylase deficiency, 29, 111, 113–114
    treatment, 115
17-hydroxyprogesterone, 6, 30, 113, 114
3α-hydroxysteroid dehydrogenase, 80
3β-hydroxysteroid dehydrogenase, 80, 111
17β-hydroxysteroid dehydrogenase, 80
hyperandrogenism, *see* androgen excess
hypercortisolemia, *see* cortisol, excess
hyperinsulinemia, 209, 210, 212
hyperlipoproteinemia, 60
hyperprolactinemia, 19, 89
    differential diagnosis, 172
hypertension
    in PCOS, 91, 167
    pregnancy-induced, 171
hyperthecosis, 18, 34
hypertrichosis, 19

drugs causing, 172
    see also hirsutism
hypertriglyceridemia, 33
hypoactive sexual desire disorder, 38, 39–40
    treatment, *see* testosterone replacement
        therapy
hypoandrogenism, *see* androgen deficiency
hypogonadism, 122
    men, 153–155
    symptoms and signs, 153–155
Hypogonadism in Males (HIM) study, 153
hypokalemia, 128
hypophysectomy, 130–131
hypopituitarism, 131
hypothalamic–pituitary–adrenal axis, 118

immunoadsorption chromatography, 188
immunoassay, 181–182
    interference, 185–188
        detection of, 187–188
        endogenous factors, 186–187
        exogenous factors, 185–186
    serial dilutions, 187
impaired glucose tolerance, 89
    prevalence, 89
in vitro fertilization (IVF), 33
in- and out-of-phase imaging, *see* chemical
        shift imaging (CSI)
incidentaloma, 127
inferior petrosal sinus sampling,
        bilateral, 129
infertility, 6, 208
    PCOS, 25, 170, 218
inocoterone, 84
insulin resistance, 28, 89, 90, 122, 163, 164,
        209, 210
    in PCOS, 167
insulin sensitizers, 8
    hirsutism, 21
    PCOS, 229
insulin-like growth factor-1 (IGF-1), 82,
        118
insulin-like growth factor-2 (IGF-2), 118
interference assays, 188
interferon-alpha, 172
International Agency for Research on
        Cancer (IARC), 139
INTIMATE trials, 204
intracrinology, 65, 67
Intrinsa, *see* transdermal testosterone patch
inversion recovery fat suppression, 193
isosexual pseudo-precocity, 141
isotretinoin, 84

juvenile granulosa cell tumor, 141

Kennedy disease, 60
keratosis pilaris, 83
ketoconazole, 130, 131
Krukenberg tumors, 6

Ladder of Life scale, 202
lanugo hair, 11
late-night salivary cortisol, 126
LDL cholesterol, 91
lean body mass, androgen effects, 201
leptin, 91, 210, 212
letrozole, 32, 216
    ovulation induction, 228
levonorgestrel, 8, 172
Leydig cells, 139
    tumors of, 140, 142
LH, 30, 180, 219, 226
lichen sclerosis et atrophicus, 40
lifestyle interventions in PCOS, 211–213
    diet, 212
    exercise, 212–213
Li–Fraumeni syndrome, 198
ligand-binding domain, 56
lipid panel screen, 30
lipodystrophy, 28
lipoprotein(a), 92
Livensa, *see* transdermal testosterone patch
Ludwig's classification of FPHL, 13
luteinizing hormone, *see* LH
luteomas of pregnancy, 143
17,20-lyase, 111

macroprolactin, 187, 188
macroprolactinemia, 29
magnetic resonance imaging of adrenal
        gland, *see* adrenal MRI
mass spectrometry, 64, 182
    testosterone, 64, 184
McCune–Albright syndrome, 120
Medical Outcome Scale (MOS-SF-36), 202
melanocyte-stimulating hormone
        (MSH), 118
men
    androgen deficiency, 152–153
        patient-reported outcome tools,
            155–156
    androgen replacement therapy, 151–160
    hypogonadism, 153–155
menopause, 151–152
    see also postmenopausal women
menopause rating scale (MRS), 202
mental health status
    PCOS, 169–170
    sex differences, 148
metabolic syndrome, 25, 26, 31, 69, 89, 90,
        121
    definitions, 90
    diagnosis, 166
    and PCOS, 90, 165–166
    prevalence, 90
    and testosterone deficiency, 154
    treatment, 34–35
metformin, 21, 33, 217–219
    with clomiphene citrate, 219
    indications, 218–219

ovulation induction, 229
    pharmacology and dosing, 217
    precautions and monitoring, 218
methyltestosterone, 43, 61
metyrapone, 130, 131, 172
midnight serum cortisol, 127
Million Women Study, 152
minocycline, 172
minoxidil, 15, 19, 85, 172
miscarriage in PCOS, 170
mitotane (o,p'-DDD), 130, 131
"moon face" of Cushing's syndrome, 121
MRI, *see* magnetic resonance imaging
MRL-41, *see* clomiphene citrate
multiple endocrine neoplasia type 1
        (MEN1), 119
muscle, androgen effects, 201
myelolipoma, 196–197

Nelson's syndrome, 132
neonate
    complications of PCOS, 171
    screening for CAH, 115
neoplasms in PCOS, 171
    breast cancer, 171
    endometrial cancer, 31, 33
nitric oxide, 101
nitric oxide synthase, 101
nonalcoholic fatty liver disease, 91, 168
nonalcoholic steatohepatitis, 168
norethindrone, 172
norgestreol, 172
nuclear receptors, 56
Nurses' Health Study, 204

obesity, 91, 92, 208
    abdominal, 209
    central, 121
    in PCOS, 166–167, 208–209
        disease presentation, 210–211
        pathophysiology, 209–210
    treatment
        diet, 212
        exercise, 212–213
obstructive sleep apnea, 91
    in PCOS, 168
oligomenorrhea, 6, 165
Olsen's classification of FPHL, 13
oral contraceptives
    and androgen deficiency, 104
    androgen assay interference, 185
    estrogen-progestin, 20–21
    hirsutism, 7, 8, 18, 33
oral glucose tolerance test, 30
ovarian drilling, 228–229
ovarian hyperstimulation syndrome, 33,
        220, 221
    prevention, 222
    risk factors, 221
ovarian tumors, 29

and androgen replacement therapy, 45
androgen-producing, 136–145
    hirsutism, 4, 18
    sex cord-stromal, *see* sex
            cord-stromal tumors
ovarian vein sampling, 138
ovarian wedge resection, 31, 33
ovariectomy, effects of, 99, 100, 101
ovary, androgen production by, 67–69
ovulation induction, 226–232
    clomiphene citrate, *see* clomiphene citrate
    gonadotropins, 229–230
    insulin sensitizers, 229
    letrozole, 228
    ovarian drilling, 228–229
ovulatory dysfunction in PCOS, 31–33

p23 protein, 57
P450 oxidoreductase, 111
painful intercourse, *see* dyspareunia
PCOS, *see* polycystic ovary syndrome
pelvic ultrasound, 30
penicillamine, 172
perioral dermatitis, 83
Pharmacy Compounding Accreditation
        Board, 44
phenothiazine, 172
phenytoin, 19, 172
pheochromocytoma, 119
    MRI, 197
photoepilation, 7, 22
pilosebaceous unit, 1
pioglitazone, 21
pituitary irradiation, 132
pituitary tumors, 119
placental aromatase deficiency, 6
plasminogen activator inhibitor-1, 92
polycystic ovary syndrome (PCOS), 2, 4, 6,
        8, 13, 25–37, 104, 161–179, 208–215
    cardiovascular risk, 89–96, 168
    clomiphene-resistant, 32
    depression in, 169
    diagnosis, 26, 226
    diagnostic criteria, 74–78, 89, 162
    differential diagnosis, 171–174
    etiology, 25–26, 161
        androgen excess, 209
        psychological stress, 211
        uterine environment, 209
    evaluation, 26–28, 173
    laboratory investigations, 28–31
    lifestyle interventions, 211–213
        diet, 212
        exercise, 212–213
    neoplasms, 171
        breast cancer, 171
        endometrial cancer, 31, 33
    and obesity, 91, 92, 166–167, 208–209
        disease presentation, 210–211
        pathophysiology, 209–210

polycystic ovary syndrome (PCOS) (cont.)
ovulation induction, 226–232
clomiphene citrate, 227–228
gonadotropins, 229–230
insulin sensitizers, 229
letrozole, 228
ovarian drilling, 228–229
phenotypes, 75, 162
physical examination, 174
pregnancy complications, 170
gestational diabetes, 170
hypertension, 171
neonatal, 171
preterm birth, 171
recurrent miscarriage, 170
prevalence, 76–77, 227
screening, 93
symptoms
acanthosis nigricans and skin tags, 164–165
acne, 163–164
alopecia, 164
cutaneous, 161–162
hirsutism, 18, 19, 85, 162–163, 210
hypertension, 91, 167
infertility, 170, 208, 218
insulin resistance, 167
metabolic syndrome, 165–166
nonalcoholic fatty liver disease, 91, 168
obstructive sleep apnea, 91, 168
ovulatory dysfunction, 165
psychiatric, 169–170
treatment, 31–35, 216–225, 227
aromatase inhibitors, 216–217
clomiphene citrate, 216
gonadotropins, 219–222
hirsutism and acne, 33–34
metformin, 217–219
ovulatory dysfunction, 31–33
see also metabolic syndrome
polymenorrhea, 165
postmenopausal women
androgen effects, 64–74, 105, 106, 200–207
bone, 201
breast, 203–205
endometrium, 202–203
muscle and lean body mass, 201–202
quality of life, 202
circulating DHEA levels, 66–69, 70, 106
PPNAD, see primary pigmented nodular adrenal disease
pregnancy
androgen-producing ovarian tumors, 143
and CAH, 116
luteomas of, 143
PCOS, 170
gestational diabetes, 170
hypertension, 171
neonatal complications, 171

preterm birth, 171
recurrent miscarriage, 170
Pregnancy in Polycystic Ovary Syndrome I (PPCOS I) trial, 218
premenopausal women
androgen deficiency, 104
idiopathic, 105
oral contraceptives, 104
androgen replacement therapy, 42
prenatal treatment of CAH, 116
preterm birth in PCOS, 171
Prevalence of Female Sexual Problems Associated with Distress and Determinants of Treatment Seeking (PRESIDE), 38
primary pigmented nodular adrenal disease (PPNAD), 120
progesterone, 6
progesterone receptor, 55, 56
progestins, 61
withdrawal bleeding, 28
prolactin, 6
high-dose hook effect, 29
measurement, 29
pro-opiomelanocortin (POMC), 118
Propionibacterium acnes, 82, 163
protein catabolism, 122
protein kinase A, 120
pseudo-Cushing's syndrome, 30
pseudohermaphrodism in male fetuses, 8
psoralens, 172
Psychosexual Daily Questionnaire (PDQ), 155

5α-reductase, 80, 81
inhibitors, 83, 85
see also individual drugs
reporter gene assays, 183
rheumatoid factor, 187
rosacea, 83
rosiglitazone, 21, 219

SARMs, 57, 61
sclerosing stromal tumor, 141
sebaceous glands
androgen effects, 81
androgen metabolism, 79
sebaceous hyperplasia, 83
selective androgen receptor modulators. See SARMs
selective estrogen receptor modulators. See SERMs
selective venous catheterization and hormonal sampling (SVCHS), 138
SERMs, 32
Sertoli cells, 139
tumors of, 142
Sertoli–Leydig cell tumors, 141–142
Sertoli-stromal cell tumors, 140, 141–142
sex cord-stromal tumors, 139–142

adult granulosa cell tumor, 139–141
juvenile granulosa cell tumor, 141
sclerosing stromal tumor, 141
Sertoli-stromal cell tumors, 141–142
steroid cell tumors, 142
thecoma, 141
sex determination, 97
sex hormone-binding globulin, see SHBG
sexual abuse, 41
sexual arousal, 101–103
sexual arousal disorder, 38
sexual development, 97–99
external genitalia, 99
sexual differentiation, 97–98
sexual dysfunction, 38–53, 204
anorgasmia, 40
comorbidities, 39, 41
definition, 39
dyspareunia, 40–41
examination, 41
history, 39
HSDD, 39–40
treatment, see testosterone replacement therapy
sexual trauma, 41
sexually dimorphic cognitive behaviors, 146
SHBG, 1, 6, 14, 29, 33, 83, 181, 201
bound testosterone, 184
insulin effects, 209
pregnancy, 143
Sheehan's postpartum necrosis, 104
short-tau inversion recovery (STIR), 193
sialic acid, 102
sialo-glycoproteins, 102
skin
androgen effects on, 79–82
hair follicles, 81
sebaceous glands, 81
sweat glands, 81–82
androgen metabolism in, 80–81
skin disorders, androgen-related, 82–87
acne, 82–84, 163–164
androgenetic alopecia, 12, 84–85
hidradenitis suppurativa, 86–87
hirsutism. See hirsutism
in PCOS, 161–162
skin tags, 28
in PCOS, 164–165
skin-related androgens, 79
sleep apnea, 31
sodium valproate, see valproate
spinal and bulbar muscular atrophy, see Kennedy disease
spin-spin relaxation, 193
spironolactone, 8, 15–16, 20, 21, 33, 61
acne vulgaris, 84
hirsutism, 86
SRC proteins, 58
Stein–Leventhal syndrome, see polycystic ovary syndrome

steroid cell tumors, 140, 142
steroidogenic acute regulatory (StAR) protein, 111
sterol-regulatory element-binding proteins (SREBPs), 82
streptomycin, 172
stress, and PCOS, 211
stromal luteoma, 140, 142
Study of Women's Health Across the Nation (SWAN) study, 202
sweat glands
    androgen effects, 81–82
    androgen metabolism, 79

telogen effluvium, 14, 84
terminal hair, 1, 11
test tubes for androgen assays, 186
testicular feminization, 98
testicular feminization syndrome, *see* androgen insensitivity syndrome
Testim 1% transdermal gel, 44
testosterone, 1, 6, 79, 180
    abuse, 157
    androgen-producing ovarian tumors, 137
    assays, *see* androgen assays
    and FPHL, 14
    free, 184
    and hair growth, 12
    and hirsutism, 18, 19
    measurement, 19, 29
        mass spectrometry, 64
    normal range, 105
    prenatal exposure, 98
    protein binding, 181
    total, 183–184
    vaginal muscular function, 103

testosterone cream, 44–45
testosterone deficiency, 154
    hypogonadism, 153–155
    and metabolic syndrome, 154
    treatment, *see* testosterone replacement therapy
testosterone patch, 43, 44
testosterone replacement therapy, 38, 42–49, 105, 153
    contraindications, 49
    health risks, 153
    lack of US FDA approval, 44
    men, 151–160
    patient monitoring, 157–158
    postmenopause, 42–43, 203
    pre- and perimenopause, 42
    pre-treatment evaluation, 156
    route of delivery, 43
    safety of, 106–107
    side effects, 45–48
        cancer, 45
        cardiovascular, 45–49
    treatment options, 43–44, 156–157
        cream, 44–45
        transdermal patch, 43, 44
        see also hormone replacement therapy
testosterone undecanoate, 43, 61
tetracosactide, 172
thecoma, 140, 141
thermolysis, 7
thrifty phenotype theory, 25
thyroid disease, differential diagnosis, 172
thyrotropin, 6
tibolone, 201
total testosterone, 183–184
transdermal testosterone patch, 43, 44, 204

transforming growth factor beta (TGFβ), 84
trans-sphenoidal microadenectomy, 130–131
tumor necrosis factor-alpha, 92
tumors, *see specific types*
type 2 diabetes mellitus, *see* diabetes mellitus, type 2

ultrasound, pelvic, 30
urethritis, 40
urinary free cortisol, 124–126

vagina
    androgen effects, 99–100
    atrophy, 41
    blood flow, 101
Vaginal Maturation Index, 28
vaginismus, 38
valproate, 172
    hirsutism, 6
vascular endothelial growth factor, 92
vasointestinal polypeptide, 103
vellus hair, 1, 11
virilization, 6, 19, 28, 137
virilizing tumors, *see* androgen-producing ovarian tumors
visfatin, 92
visuo-spatial processing, 147

wigs and hairpieces, 17
Women's Health Initiative, 152

zidovudine, 172
zinc deficiency, and hair loss, 14